New Generation Vaccines
The Role of Basic Immunology

NATO ASI Series

Advanced Science Institutes Series

A series presenting the results of activities sponsored by the NATO Science Committee, which aims at the dissemination of advanced scientific and technological knowledge, with a view to strengthening links between scientific communities.

The series is published by an international board of publishers in conjunction with the NATO Scientific Affairs Division

A	**Life Sciences**	Plenum Publishing Corporation
B	**Physics**	New York and London
C	**Mathematical and Physical Sciences**	Kluwer Academic Publishers
D	**Behavioral and Social Sciences**	Dordrecht, Boston, and London
E	**Applied Sciences**	
F	**Computer and Systems Sciences**	Springer-Verlag
G	**Ecological Sciences**	Berlin, Heidelberg, New York, London,
H	**Cell Biology**	Paris, Tokyo, Hong Kong, and Barcelona
I	**Global Environmental Change**	

Recent Volumes in this Series

Volume 255 —Quantitative Assessment in Epilepsy Care
edited by Harry Meinardi, Joyce A. Cramer, Gus A. Baker, and Antonio Martins da Silva

Volume 256 —Advances in the Biomechanics of the Hand and Wrist
edited by F. Schuind, K. N. An, W. P. Cooney III, and M. Garcia-Elias

Volume 257 —Vascular Endothelium: Physiological Basis of Clinical Problems II
edited by John D. Catravas, Allan D. Callow, and Una S. Ryan

Volume 258 —A Multidisciplinary Approach to Myelin Diseases II
edited by S. Salvati

Volume 259 —Experimental and Theoretical Advances in Biological Pattern Formation
edited by Hans G. Othmer, Philip K. Maini, and James D. Murray

Volume 260 —Lyme Borreliosis
edited by John S. Axford and David H. E. Rees

Volume 261 —New Generation Vaccines: The Role of Basic Immunology
edited by Gregory Gregoriadis, Brenda McCormack, Anthony C. Allison, and George Poste

Series A: Life Sciences

New Generation Vaccines
The Role of Basic Immunology

Edited by

Gregory Gregoriadis and Brenda McCormack

School of Pharmacy
University of London
London, United Kingdom

Anthony C. Allison

Syntex Research
Palo Alto, California

and

George Poste

SmithKline Beecham Pharmaceuticals
Surrey, United Kingdom

Springer Science+Business Media, LLC

Proceedings of a NATO Advanced Study Institute on
New Generation Vaccines: The Role of Basic Immunology,
held June 24–July 5, 1992,
at Cape Sounion Beach, Greece

NATO-PCO-DATA BASE

The electronic index to the NATO ASI Series provides full bibliographical references (with
keywords and/or abstracts) to more than 30,000 contributions from international scientists
published in all sections of the NATO ASI Series. Access to the NATO-PCO-DATA BASE is
possible in two ways:

—via online FILE 128 (NATO-PCO-DATA BASE) hosted by ESRIN, Via Galileo Galilei,
I-00044 Frascati, Italy

—via CD-ROM "NATO Science and Technology Disk" with user-friendly retrieval software in
English, French, and German (©WTV GmbH and DATAWARE Technologies, Inc. 1989). The
CD-ROM also contains the AGARD Aerospace Database.

The CD-ROM can be ordered through any member of the Board of Publishers or through
NATO-PCO, Overijse, Belgium.

Library of Congress Cataloging-in-Publication Data

New generation vaccines : the role of basic immunology / edited by
  Gregory Gregoriadis ... [et al.].
     p.   cm. -- (NATO ASI series. Series A. Life sciences ; v.
  261)
     "Published in cooperation with NATO Scientific Affairs Division."
     "Proceedings of a NATO Advanced Study Institute on New Generation
  Vaccines: the Role of Basic Immunology, held June 24-July 5, 1992,
  at Cape Sounion Beach, Greece"--T.p. verso.
     Includes bibliographical references and index.
     ISBN 978-1-4613-6281-4   ISBN 978-1-4615-2948-4 (eBook)
     DOI 10.1007/978-1-4615-2948-4
     1. Viral vaccines--Congresses. 2. Synthetic vaccine--Congresses.
  3. Immunological adjuvants--Congresses.   I. Gregoriadis, Gregory.
  II. North Atlantic Treaty Organization. Scientific Affairs
  Division. III. NATO Advanced Study Institute on New Generation
  Vaccines--the Role of Basic Immunology (1992 : Ákra Sóunion, Greece)
  IV. Series.
     [DNLM: 1. Vaccines, Synthetic--immunology--congresses.
  2. Adjuvants, Immunologic--congresses. 3. Vaccination--congresses.
  QW 805 N53162 1993 /]
  QR189.5.V5N47   1993
  615'.37--dc20
  DNLM/DLC
  for Library of Congress   93-48801
                                                          CIP

Additional material to this book can be downloaded from http://extra.springer.com.

ISBN 978-1-4613-6281-4

©1993 Springer Science+Business Media New York
Originally published by Plenum Press, New York in 1993
Softcover reprint of the hardcover 1st edition 1993

PREFACE

It is widely accepted that vaccination still remains the best answer to most infectious diseases. Recently, vaccine development has been greatly facilitated by advances in molecular and cell biology which have laid the foundations of a new generation of vaccines. These are exemplified by subunit vaccines produced through gene cloning and synthetic peptides mimicking small regions of proteins on the outer coat of viruses and capable of eliciting virus neutralizing antibodies. However, subunit and peptide vaccines are only weakly or non-immunogenic in the absence of immunological adjuvants. The latter are a diverse array of agents that augment specific cell-mediated immune responses to the antigens and the formation of protective antibodies.

This book contains the proceedings of the 3rd NATO Advanced Studies Institute (ASI) "New-Generation Vaccines: The Role of Basic Immunology" held at Cape Sounion Beach, Greece, during 24 June-5 July, 1992. It deals with recent developments in the understanding of immunity at the molecular and cellular levels and the application of such knowledge in the search for novel immunological adjuvants and the formulation of new-generation vaccines for experimental and clinical use. We express our appreciation to Professor K. Dalsgaard and H. Snippe for their cooperation in planning the ASI and to Mrs. Concha Perring for her excellent production of the manuscripts. The ASI was held under the sponsorship of NATO Scientific Affairs Division and generously co-sponsored by SmithKline Beecham Pharmaceuticals (Philadelphia). Financial assistance was also provided by Pasteur Merieux (Marcy L'Etoile), British Biotechnology Ltd. (Oxford), Ingenasa (Madrid), Syntex Research (Palo Alto), Wyeth-Ayerst (Philadelphia) and Merck and Co. Inc. (Rahway).

Gregory Gregoriadis
Brenda McCormack
Anthony C. Allison
George Poste

August 1993

v

CONTENTS

Dendritic Cells: Nature's Adjuvants
    J.M. Austyn 1

The Multiple Accessory Cell Concept: Its Relevance to the
    Development of Adjuvants and Vaccines
    N. Van Rooijen 11

Synthetic Peptides and the Role of T-Helper Cell
    Determinants
    M.J. Francis 23

Carriers for Peptides: Theories and Technology
    M.J. Francis 33

Co-entrapment of T-Cell and B-Cell Peptides in Liposomes
    Overcomes Genetic Restriction in Mice and Induces
    Immunological Memory
    G. Gregoriadis, Z. Wang and M.J. Francis 43

Preparation and Characterization of Stable Liposomal
    Hepatitis B Vaccine
    D. Diminsky, Z. Even-Chen and Y. Barenholz 51

Initiation of Immune Response with ISCOM
    B. Morein, M. Villacres-Eriksson, L. Åkerblom
    and K. Lövgren 61

Nanoparticles as Potent Adjuvants for Vaccines
    J. Kreuter 73

Optimization of Carriers and Adjuvants: A Model Study
    Using Semliki Forest Virus Infection of Mice
    A. Snijders, I.M. Fernandez, C.A. Kraaijeveld
    and H. Snippe 83

Immunotargeting as an Adjuvant-Independent Subunit
    Vaccine Design Option
    D.L. Skea and B.H. Barber 101

BCG Vaccine: The Next Generation of Targeted Drug
    Delivery Systems
    M.J. Groves, Y. Lou and M.E. Klegerman 111

Significance of Virulence Factors and Immuno-Evasion for the
    Design of Gene-Deleted Herpesvirus Marker Vaccines
    S. Kit 117

Eradication of Sylvatic Rabies Using a Live Recombinant
    Vaccinia-Rabies Vaccine
    M.P. Kieny, B. Brochier and P.-P. Pastoret 131

Vaccination of Wildlife: The Role of Recombinant Poxvirus
    Vaccines
    E.P.J. Gibbs 139

Recombinant Protein Antigens with "Built-in" Adjuvanticity
    P. Ghiara, L. Villa, R. Rappuoli, S. Gonfloni
    L. Castagnoli and G. Cesareni 149

Vaccination against Malaria; the Anti-Disease Concept
    J.H.L. Playfair, J. Taverne and C. Bate 155

Progress in the Development of Epstein-Barr Virus Vaccines
    A.J. Morgan 163

Requirements for Induction of Semliki Forest Virus
    Neutralizing Antibodies by a Non-Internal Image
    Monoclonal Antibody
    C.A. Kraaijeveld, T.A.M. Oosterlaken and H. Snippe 175

Polysaccharide Vaccines
    M.R. Lifely 185

Liposomal Delivery of Cytokines: A Means to Improve their
    Therapeutic Performance
    Y. Barenholz, O. Palgi, G. Golod, N. Emanuel,
    Y. Rutkowski, E. Braun and E. Kedar 201

Participants Photograph 211

Contributors 213

Index 217

DENDRITIC CELLS: NATURE'S ADJUVANTS

Jonathan M Austyn

Nuffield Department of Surgery, University of Oxford
John Radcliffe Hospital
Headington, Oxford OX3 9DU, UK

INTRODUCTION

Dendritic cells (DC) were first identified as a trace cell type in lymphoid tissues two decades ago (Steinman and Cohn 1973). Quite soon, it became clear that cells of this lineage, sometimes termed dendritic leuko- cytes (DL), play a central and probably unique role in the initiation of adaptive immune responses. For this reason, Steinman has referred to these cells as Nature's adjuvants. In the context of vaccinology, it seems likely that the development of successful vaccines and optimal immunization strategies will require targeting of antigens to these cells in particular.

The purpose of this brief article is threefold: first, to review the functions of DL and the relationship between various cells of this lineage; second, to outline the events that are likely to occur in relation to these cells following antigen administration until an immune response has been generated; and third, to present some thoughts on the optimal design of vaccines in the light of our present knowledge of DL. We focus on data obtained in the mouse, but note that similar observations have been made in other species including humans.

Because of space limitations and to establish general principles this article is nesessarily didactic, and is presented in note form. It is not possible to provide extensive literature citations, but detailed reports on the areas covered in the following sections and related topics are in Research in Immunology Volume 140; International Reviews in Immunology Volume 6; Seminars in Immunology Volume 4; Austyn, 1992; Steinman, 1991; Metlay et al, 1989; and Austyn, 1989.

DENDRITIC LEUKOCYTES

Distribution and Functions of DL

DL originate in the bone marrow and are distributed throughout most non-lymphoid and all lymphoid tissues, and migratory forms are also present in body fluids. Examples are as follows:

a) Non-lymphoid tissues: Langerhans cells (LC) in epidermis of skin and similar cells in other topologically external sites (eg. epithelia lining

New Generation Vaccines, Edited by G. Gregoriadis
et al., Plenum Press, New York, 1993

the lung and airways, gut and urogenital tract); interstitial dendritic cells (DC) in the interstitial spaces of all solid organs except for most of the central nervous system;
b) Body fluids: veiled cells of afferent lymph; blood dendritic cells in peripheral blood;
c) Lymphoid tissues: interdigitating cells (IDC) in T areas of spleen (central white pulp) and lymph nodes (interfollicular areas), and in medulla and corticomedullary junction of thymus; marginal zone DC (MZDC) in the marginal zone of spleen.

All these cells express CD45 and high levels of MHC class II molecules but can have distinct phenotypes in relation to other markers (Austyn 1992). In general they do not express a variety of lineage-restricted markers of T cells (eg. T cell receptors, CD3), B cells (eg. membrane-bound immunoglobulin, CD19-21), mononuclear and polymorphonuclear phagocytes (eg. CD13-CD16 and CD64), or NK cells (CD56/57).

It is most important to stress that follicular dendritic cells (FDC), which are localized in B areas of secondary lymphoid tissues, are almost certainly unrelated to DL: they probably do not have a bone marrow origin, and play a completely different role in the immune system. Functions of DL can be summarized, and compared to those of FDC and macrophages (Mph), as follows:

a) DL internalize and process nonself antigens for presentation as foreign peptide-MHC complexes to primary (resting) T cells, and initiate T cell and T-dependent immune responses;
b) the exception to a) is thymic DL which most probably present self peptide-MHC complexes to developing thymocytes and induce tolerance (Fairchild and Austyn, 1990).
c) In contrast, FDC present native antigens to B cells, and are likely to be involved in affinity maturation and the generation of B cell memory (Szakal et al, 1989); and
d) Mph have a primary role in scavenging and eliminating antigens, but are unable to initiate primary responses.

In relation to a) above, different DL play specialized roles during the induction of immune responses: DL progenitors enter non-lymphoid tissues and develop into "immature" DL which can take up and process antigens; subsequently these cells migrate into secondary lymphoid tissues and the now "mature" DL present antigens to resting T cells and have acquired the capacity to deliver unique signals for T cell activation (immunostimulation) which cannot be supplied by other cell types.

Precursor DL

The bone marrow origin of DL is well established from studies with radiation-induced bone marrow chimeras. An MHC class II-negative precursor has been identified in mouse bone marrow and peripheral blood that proliferates and gives rise to MHC class II-positive mature, immunostimulatory DL in the presence of granulocyte-macrophage colony stimulating factor (GM-CSF) in culture (Inaba et al, 1992). It seems most likely that this blood precursor, which presumably seeds the tissues and gives rise to immature non-lymphoid DL for example, is distinct from the monocyte which is the progenitor for tissue Mph. A DL precursor has also been identified in mouse thymus, but this cell requires interleukin-1 (IL1) for its development in culture.

Immature DL

It seems most probable that DL in non-lymphoid tissues are

specialized for the uptake and processing of foreign antigens, but they lack the capacity to initiate T cell responses, ie, they lack immuno-stimulatory functions; for the latter reason, DL at this stage are sometimes, for convenience, termed immature DL.

To date, the best characterized immature DL are freshly-isolated Langerhans cells of skin epidermis. These cells can endocytose and process native soluble protein antigens for presentation to activated T cells or T cell clones (but not resting T cells), phagocytose particulates such as certain yeasts and bacteria (Reis e Sousa et al, 1993) and internalize Leishmania major and subsequently present Leishmania antigens to sensitized T cells (Will et al, 1992 and personal communication).

Freshly-isolated LC have little or no ability to initiate primary T cell responses such as the allogeneic MLR or oxidative mitogenesis. However, this capacity is acquired during culture in the presence of GM-CSF which promotes their phenotypic and functional maturation (below). At the same time, their capacity to phagocytose particles and to process native antigens is markedly decreased. For some of these functions, similar ob-servations have been made for DL isolated from lung and airways and from hearts and kidneys suggesting that, in situ, non-lymphoid DL in general are immature.

The maturation of LC in culture is accompanied by important changes in the cell surface, cytoplasmic compartments, and biochemistry of the cells. These include the following:

a) Down-regulation of certain membrane receptors, such as Fcγ RII (CDw32) and mannose receptors, and molecules of unknown function such as the mouse F4/80 antigen. These events correlate for example with the loss of phago-cytic capacity of the cells.
b) Loss of acidic compartments, particularly early endosomes and Birbeck granules which are relatively abundant in freshly isolated LC but which es-sentially disappear during culture in GM-CSF to leave a perinuclear vesicle of unknown function resembling a late endosome. These changes correlate with the loss of antigen processing capacity by the cells.
c) A rapid reduction in the biosynthetic rate of MHC class II, and invariant chains, which is very high in fresh LC but virtually undetectable after culture in GM-CSF for 24h. This correlates with the loss of the capacity to process and present native antigens.
d) An overall increase in the levels of MHC class II molecules expressed at the cell surface. After uptake and processing of antigens by fresh LC, foreign antigens can be expressed in immunogenic form at the cell surface, presumably as peptide-MHC complexes, for several days.
e) In addition, expression of B7/BB1 (the ligand for CD28 and CTLA-4 on helper and cytotoxic T cells) is induced during maturation of LC (Larsen et al, 1992) correlating with the acquisition of immunostimulatory functions.

Various cytokines act on DL, and the three studied most closely to date are GM-CSF, tumour necrosis factor (TNF) and IL1. In the case of LC:

a) GM-CSF promotes phenotypic and functional maturation (above) and the viability of LC in culture;
b) TNF sustains LC viability in culture but does not promote maturation (LC cultured in TNF have elevated cytoplasmic levels of the invariant chain, the consequences of which are under investigation; C. Schuler, personal communication), and may be involved in LC migration in vivo.
c) IL1 synergises with GM-CSF and/or TNF in sustaining the viability of LC and may also be involved in LC migration in vivo (this cytokine can also potentiate the immunostimulatory activity of splenic DC in vitro).
d) In addition, expression of the IL2R p55 chain is induced or upregulated

during maturation in culture, but there are no reported effects of interleukin-2 (IL2) on LC, or other DL populations, to date.

Migratory DL

Migratory DL, often termed veiled cells, are present in afferent but not efferent lymph and there is a considerable flux of these cells draining sites of inflammation and/or infection, and from transplants (Austyn and Larsen 1990). This migratory route for DL from non-lymphoid tissues via lymph into lymph nodes (a) was first directly demonstrated in studies tracing the fate of radiolabelled DL that were injected into rodent foot-pads and which were subsequently localized in paracortical regions (T areas) of popliteal nodes, apparently as IDC; (b) was inferred from studies in which fluoresceinated contact sensitizing agents were administered to skin and labelled DL were isolated from draining lymph nodes; and (c) was suggested from studies of the behaviour of LC in skin transplants and explants, in which epidermal LC were observed to migrate into dermal lymphatics and out of the tissue concomitant with their phenotypic and functional maturation.

An additional migratory route for DL via blood into spleen was defined in studies tracing the migration of radiolabelled or fluorochrome labelled DL that were injected intravenously and observed to migrate specifically to the spleen where they homed to central white pulp (T areas), apparently to become IDC. Subsequently, DL were found to migrate from fully-vascularized cardiac allografts into recipient spleens where they associated with CD4+ T cells in peripheral white pulp probably as MZDC. However, in none of these or related studies were DL detectable in non-lymphoid tissues (or thymus), so it seems likely that two forms of DL are present in peripheral blood: progenitors originating directly from bone marrow to seed the tissues including the thymus, and relatively mature DL that have resided in non-lymphoid organs and which are trafficking to the spleen.

It is most probable that antigens delivered to non-lymphoid tissues during immunization are acquired by immature DL and that, under the influence of locally-produced inflammatory cytokines, these cells subsequently begin to mature and migrate into secondary lymphoid tissues to initiate antigen-specific immune responses.

Mature DL

Most DL isolated from lymphoid tissues, termed lymphoid DC, can be considered "mature", because these cells have the capacity to initiate primary T cell and T-dependent responses, but the majority at least are non-phagocytic and have relatively little ability to process native antigens. It seems likely that many lymphoid DC originated from non-lymphoid tissues and while it is also possible that some may be derived directly from blood precursors, we shall consider only this simpler hypothesis.

Within mouse spleen, two phenotypically distinct subsets of DL have been identified using lymphoid DC-restricted markers. Interdigitating cells (IDC) are localized in central white pulp (T areas) with the pheno-type N418+ NLDC145+ 33D1- J11d- (Austyn 1992). Within marginal zones, bordering T areas and close to their intersection with B areas, "nests" of DL have been identified with the phenotype N418+ NLDC145- 33D1+ J11d+ (Metlay et al, 1990). For convenience the latter cells can be termed marginal zone dendritic cells, and this subset predominates in conventionally isolated populations of splenic DC. It seems likely that

4

similar subsets are present in lymph nodes, although these have not yet been examined in detail.

There is now overwhelming evidence from a wide variety of systems in different species including rodents and humans that mature DL (eg lymphoid DC and LC cultured in GM-CSF) can initiate primary immune responses in vitro and in vivo. Other populations of leukocytes, including M and small B cells, lack this capacity, with the possible exception of B blasts. The immune responses that have been examined in this context in vitro include:

a) alloreactive T cell responses (reviewed by Austyn and Steinman, 1988) such as proliferation in the allogeneic mixed leukocyte reaction (MLR) and the generation of alloreactive CTL (the latter even in the absence of CD4+ T cells suggesting that mature DL can stimulate both CD4+ and at least a subset of CD8+ T cells directly);
b) primary T-dependent antibody formation to hapten-carrier complexes and heterologous erythrocytes, during which T cells and histocompatible B cells cluster with DL, and the antigen-specific T cells are activated and help development of antigen-specific B cells into lymphoblasts and plasma cells through cognate (direct) interactions (Inaba and Steinman 1985);
c) anti-idiotypic responses (Francotte and Urbain 1985); and
d) polyclonal proliferative responses, particularly oxidative mitogenesis (Austyn 1989) in which proliferation of T cells treated with sodium periodate or with neuraminidase-galactose oxidase is exquisitely dependent on the presence of mature DC (other polyclonal responses, eg concanavalin A responses, may not have this requirement).

Importantly, DL can also induce antigen-specific T cell sensitization after administration in vivo (ie after adoptive transfer to naive recipients). For these experiments, DC have been pulsed with proteins in vitro (Inaba et al, 1990), or isolated from lymph nodes draining skin to which contact sensitizers were applied (Macatonia et al, 1986), from pseudoafferent lymph draining regions of gut into which antigens were injected (Liu and MacPherson 1991), or from lung after administration of antigens in aerosol form (Holt et al, 1987). It seems likely these antigens were acquired by immature DL, which then matured into immuno-stimulatory cells in vivo or in culture. In addition, administration of DL can, in certain cases, overcome Ir gene defects (Boog et al, 1985), possibly by activation of antigen-specific T cells normally present at very low precursor frequencies.

Therefore, the activation of resting T cells (and probably also memory T cells;) absolutely requires mature DL, but there is good evidence that activated T cells can subsequently respond to other types of APC (eg Mph and B cells) that express the appropriate peptide-MHC complexes (Inaba and Steinman 1984).

DENDRITIC LEUKOCYTES AND IMMUNIZATION

Based on the information outlined in the previous section, it is possible to put forward a scheme for the likely events that occur during immunization, from the time of antigen administration until an immune response is induced. There are many unknowns in this process, and some that are relevant to the role of DL in vaccination will be noted.

Antigens Administered into Non-Lymphoid Tissues are Internalized and Processed by Immature DL at These Sites

For example, antigens injected
a) intramuscularly may be directed to interstitial DL;
b) subcutaneously or intradermally, to LC within or migrating from the

epidermis, or to DL developing from precursors within or recruited into
the dermis from the blood;

c) into the lung and airways (eg as aerosols), to cells resembling LC in
   the lining epithelium; or

d) directly into the gut (or administered orally), to DL in the lamina
   propria or Peyer's patches (in these cases, the potential for induction
   of oral tolerance rather than immunity should also be considered).

An important question is to what extent antigen uptake by Mph may
modulate the subsequent response, either directly (e.g. by scavenging
antigens and reducing the effective load delivered to DL) or indirectly
(e.g. by producing cytokines that alter the function of DL). Mph are
present in the majority of these sites but their precise localization
compared to that of DL differs in different tissues. For example, Mph are
not present in the skin epidermis, but they are juxtaposed to DL within the
epithelia lining the lung and airways, and in other tissues they can be
widely separated (for example in the liver, DL are localized to portal
triads whereas Kupffer cells are situated in sinusoids). At least in lung
and airways, however, there is evidence that Mph can regulate the function
of DL (Holt et al, 1987). When these cells are isolated, Mph suppress the
development of immunostimulatory function (i.e. maturation) of DL from the
same tissues in culture, apparently through the cytokines they secrete.
Conceivably, in situ, these Mph may play a role in maintaining DL of the
lung and airways in an immature form (i.e. with the capacity to internalize
and process antigens), and maturation of these DL may commence only after
they migrate from the epithelia and are released from the effect of
locally-produced, Mph-derived cytokines.

Inflamatory Cytokines Promote Migration of Non-Lymphoid DL into Secondary Lymphoid Tissues

While there is increasing information about the particular cytokines
that can induce DL maturation in vitro, those required for migration of
these cells are not well defined. It seems likely that DL from the skin
epidermis (ie LC) and other topologically external sites (e.g. lung and
airways epithelia) migrate exclusively or primarily via afferent lymph into
lymph nodes, whereas DL from solid organs (including muscle tissue) migrate
both via this route and via the blood into the spleen.

The contributions of the two migratory routes of DL to the quantity
and quality of immunity generated against different types of antigen are
unclear. The relationship between subsets of lymphoid DC that have been
defined in spleen (and thymus), and which probably also exist in lymph
nodes, is also unclear. However, one may speculate that DL migrating from
non-lymphoid organs via blood into spleen could be represented by DL
identified close to the marginal zone and near the intersection of T and B
areas (Metlay et al, 1990): these MZDC would apear to be ideally placed to
interact with antigen-specific T cells, which could then help the appropriate B cells initiate primary antibody responses. This is consistent
with observations that, in the spleen, most antibody-forming cells develop
in the outer periarteriolar sheath (white pulp). Splenic IDC, deep within
T areas, would seem to be less likely to interact with B cells.

DL Present Foreign Peptide-MHC Complexes to Resting T Cells and Deliver Signals for T Cell Activation

Antigens acquired by DL in non-lymphoid tissues are processed to
produce peptide-MHC complexes which can be stably expressed at the cell
surface for several days, presumably for the lifetime of the MHC molecule
and almost certainly for the time required for DL to migrate into secondary
lymphoid tissues and to mature into competent immunostimulatory cells.

It seems likely that antigens administered to non-lymphoid tissues are directed into the exogenous antigen - class II pathway of DL, for presentation primarily to helper T cells. After activation by DL, these T cells can initiate the effector phase of the immune response, such as (T-dependent) antibody formation by B cells. It is unclear how one may be able to route antigens to the endogenous antigen - class I pathway of DL, which may also be required in order to optimize CTL generation.

Antigen-Specific T Lymphoblasts Can Respond to Other Types of Antigen-Presenting Cells (APC)

For example, within secondary lymphoid tissues, activated T cells can secrete cytokines required for activation of B cells expressing the relevant peptide-MHC complexes and their development into B lymphoblasts and/or plasma cells. It may also be necessary for activated helper T cells and cytotoxic T cells to cluster with the same DL in these tissues because, at least in the mouse, it is difficult to envisage how a cytotoxic T cell that does not express MHC class II molecules could otherwise receive the necessary cytokines from class II-restricted helper T cells for development into a CTL.

Activated helper T lymphoblasts also migrate to peripheral sites (non-lymphoid tissues), including the original site of antigen deposition. Here they could potentially respond to any APC, such as epithelial and endothelial cells that have internalized and processed the antigen, and on which MHC class II molecules have been induced in the course of the response. However, the importance of antigen presentation in non-lymphoid compared to secondary lymphoid tissues for optimal generation of memory T cells is not clear, nor is it known to what extent this (or the interaction between T lymphoblasts and different types of APC) controls the generation of Th1 versus Th2 responses. A detailed understanding of the former may be required for the optimal generation of secondary responses, ie during antigen boosting, and of the latter in order to control eg the preferential induction of DTH and cytotoxic responses compared to antibody responses, and the production of particular antibody isotypes during immunization.

The Generation of Secondary T Cell Responses May Require a Specific Interaction Between Memory T Cells and DL

In other words, memory T cells may resemble resting T cells in their requirement for activation by DL, whereas activated T cells can respond to these and other types of APC.

Presumably, antigens administered to non-lymphoid tissues can be transported in processed form (as foreign peptide-MHC complexes) by DL to secondary lymphoid tissues and be presented to memory T cells. Conceivably, however, the increased recirculation of memory T cells and their greater ease and rate of activation, compared to resting T cells, may permit them to interact with DL that have begun to mature within, but which have not yet migrated from, non-lymphoid tissues.

Native Antigens are Retained on the Surface of Follicular Dendritic Cells which may be Required for Affinity Maturation of B Cells and/or Generation of Memory B Lymphocytes, and Consequently for Optimal Secondary B Cell Responses

FDC are localized in B areas of secondary lymphoid tissues where germinal centres develop (Szakal et al, 1989). However, FDC are unlikely to be involved in the initiation of primary antibody responses since germinal centres are instead the sites of memory B cell proliferation and differentiation. Antigens in native (non-processed) form can only associate with FDC after formation of antigen-antibody-complement complexes, and hence only after primary T cell and antibody responses have been initiated, at least in the case of T-dependent antigens. This implies that, for optimal secondary antibody responses, native antigens must persist for a sufficient time to allow the primary antibody response to be initiated (by DL) and their subsequent transport in complexed form to FDC. Once associated with FDC, antigens are retained for considerable lengths of time (weeks to months) and the cells are very long lived, whereas DL in spleen for example have a half-life of only a few days.

Antigens Administered Directly Into Secondary Lymphoid Tissues may not Elicit Optimal Immune Responses

It must be stressed that this is a speculative statement. However it seems possible that, for example, the bulk of an antigen load that is administered intravenously may be cleared by splenic Mph, and that injection via this route may circumvent uptake by DL at an immature stage, first because the majority of lymphoid DC appear to be mature, and secondly because any less mature DC may be sequestered in sites that are relatively inaccessible to free antigens. Similarly, antigens that gain direct access to lymph nodes (via afferent lymph, or conceivably after intraperitoneal injection unless DL are recruited to this site) may be cleared by Mph in the subcapsular region.

For optimal immunization via the intravenous route (and perhaps after intraperitoneal administration) it may be necessary first to incubate the antigen with purified DL in vitro, and then to administer these cells instead of free antigen. In fact, there is some evidence that antigens administered intravenously on cells other than DL can induce a state of antigen unresponsiveness rather than immunity (Mueller et al, 1989).

TARGETING DENDRITIC LEUKOCYTES FOR VACCINATION

In order to vaccinate successfully, the following factors should be optimized: the form of antigen (intact, attenuated or genetically-engineered defective microorganisms, subunit vaccines, recombinant molecules, etc); delivery systems (vehicles such as liposomes, ISCOMS, nanoparticles, etc.); and adjuvants (lipopolysaccharide or muramyl dipeptide analogues, cyto-kines, etc). Based on the preceding sections, it would seem desirable to target vaccines specifically to DL.

Antigens

One potential advantage of targeting antigens to DL is that it should be possible to immunize against a variety of naturally processed antigens rather than first having to define T cell epitopes etc. However, depending on the structure of the antigen, relevant peptides that could potentially bind to MHC molecules and be recognized by T cells in the repertoire, may not be generated (eg because the recognition sequences acting as targets for lysosomal hydrolases and/or cytosolic proteasomes may not be present). In addition, the generation of an immunodominant peptide from one component, for example in a multisubunit vaccine, may prevent ef-fective presentation of another (potentially protective) epitope. In order to elicit optimal antibody responses it may also be necessary to ensure that sufficient antigen can persist in native form for an adequate time to allow its transport to FDC; conceivably, antigens adsorbed to alum or delivered in ISCOMS for example may provide such a depot.

Vehicles

It is tempting to speculate that the most successful vehicles would have the following properties in relation to DL:

a) their stability in tissues and/or fluids (blood, lymph) may optimize their delivery to, and uptake by, DL;

b) their particulate nature may facilitate uptake (phagocytosis) by DL and the delivery of antigens to lysosomal components for processing; and moreover,

c) if they could be targeted to receptor(s) that are expressed specifically by DL and which mediate phagocytosis, but which are not expressed by Mph, it may be possible to optimize delivery of antigens for initiation of immune responses (by DL) rather than their scavenging and clearance (by Mph).

Adjuvants

We have demonstrated that systemically-administered LPS induces migration of DL from mouse heart and kidneys, and recruits (MHC class II-negative) DL precursors into these tissues; these effects can be mimicked by TNF and/or IL1 (J Roake et al, manuscripts in preparation). Conceivably, the most potent adjuvants would have the following properties:

a)   the induction of inflammatory cytokines that act on immature DL at the site of injection and arrest maturation to permit optimal antigen uptake and processing (eg TNF);

b)   perhaps more importantly, in relation to our observations on the effects of LPS, the induction of cytokines that act on non-lymphoid DL that have internalized antigens, and which promote their subsequent maturation (eg GM-CSF) and migration into secondary lymphoid tissues (eg TNF, IL1?), and/or which recruit DL precursors that can handle more of the antigen load; and it follows that

c)   it may be possible to engineer vaccines containing the immunogenic peptides of interest linked to cytokines or bioactive cytokine peptides that alter the maturation, migration and/or functions of DL and potentiate initiation of the immune response.

REFERENCES

Austyn, J.M., 1989, "Antigen-presenting cells", In Focus Series, Male D, ed., IRL Press, Oxford.

Austyn, J.M., 1992, Antigen uptake and presentation by dendritic leukocytes, Seminars in Immunology, 4: 227.

Austyn, J.M. and Larsen, C.P., 1990, Migration patterns of dendritic leukocytes. Implications for transplantation, Transplantation, 49: 1.

Austyn, J.M. and Steinman, R.M., 1988, The passenger leukocyte: a fresh look, Transplant. Revs., 2: 139.

Boog, C.J.P., Kast, W.M., Timmers, H.T.M., Boes, J., De Waal, L.P. and Melief, C.J.M., 1985, Abolition of specific immune response defect by immunization with dendritic cells, Nature, 318: 59.

Fairchild, P.J. and Austyn, J.M., 1990, Thymic dendritic cells: phenotype and function, Intern.Rev,Immunol., 6: 187.

Francotte, M. and Urbain, J., 1985, Enhancement of antibody responses by mouse dendritic cells pulsed with tobacco mosaic virus or with antiidiotypic antibodies raised against a private rabbit idiotype, Proc.Natl.Acad.Sci.USA., 82: 8149.

Holt, P.G., Schon-Hegrad, M.A. and Oliver, J., 1987, MHC class II antigen-bearing dendritic cells in pulmonary tissues of the rat. Regulation of antigen presentation activity by endogenous macrophage population, J.Exp.Med., 167: 262.

Inaba, K., Metlay, J.P., Crowley, M.T. and Steinman, R.M., 1990, Dendritic cells pulsed with protein antigens in vitro can prime antigen-specific, MHC-restricted T cells in situ, J.Exp.Med., 172: 631.

Inaba, K. and Steinman, R.M., 1984, Resting and sensitized T lymphocytes exhibit distinct stimulatory (antigen presenting cell) requirements for growth and lymphokine release, J.Exp.Med., 160: 1717.

Inaba, K. and Steinman, R.M., 1985, Protein-specific helper T-lymphocyte formation initiated by dendritic cells, Science, 229: 475.

Inaba, K., Inaba, M., Romani, N., Aya, H., Deguchi, M., Ikehara, S., Muramatsu, S. and Steinman, R.M., 1992, Generation of large numbers of dendritic cells from mouse bone marrow cultures supplemented with granulocyte/macrophage colony-stimulating factor, J.Exp.Med., 176: 1693.

Larsen, C.P., Ritchie, S.C., Pearson, T.C., Linsley, P.S. and Lowry, R.P., 1992, Functional expression of the costimulatory molecule, B7/BB1, on murine dendritic cell populations, J.Exp.Med., 176: 1215.

Liu, L. and MacPherson, G.G., 1991, Lymph-borne (veiled) dendritic cells can aquire and present intestinally administered antigens, Immunology, 73: 281.

Macatonia, S.E., Edwards, A.J. and Knight, S.C., 1986, Dendritic cells and the initiation of contact sensitivity to fluorescein isothiocyanate, Immunology, 59: 509.

Metlay, J.P., Pure, E. an Steinman, R.M., 1989, Control of the immune response at the level of antigen-presenting cells: a comparison of the function of dendritic cells and B lymphocytes, Adv.Immunol., 47: 45.

Metlay, J.P., Witmer-Pack, M.D., Agger, R., Crowley, M.T., Lawless, D. and Steinman, R.M., 1990, The distinct leukocyte integrins of mouse spleen dendritic cells as identified with new hamster monoclonal antibodies, J.Exp.Med., 171: 1753.

Mueller, D.L., Jenkins, M.K. and Schwartz, R.H., 1989, Clonal expansion versus functional clonal inactivation: a costimulatory signalling pathway determines the outcome of T cell antigen receptor occupancy, Annu.Rev.Immunol., 7: 445.

Reis e Sousa, C., Stahl, P.D. and Austyn, J.M., 1993, Phagocytosis of antigens by Langerhans cells in vitro, J.Exp.Med., 178: in press.

Steinman, R.M. and Cohn, Z.A., 1973, Identification of a novel cell type in peripheral lymphoid organs of mice. (I) Morphology, quantitation, tissue distribution, J.Exp.Med., 137: 1142.

Steinman, R.M., 1991, The dendritic cell system and its role in immuno-genicity, Annu.Rev.Immunol., 9: 271.

Szakal, A.K., Kosko, M.H. and Tew, J.G., 1989, Microanatomy of lymphoid tissue during humoral immune responses: structure function relationships, Annu.Rev,Immunol., 7: 91.

Will, A., Blank, C., Rollinghof, M. and Moll, H., 1992, Murine epidermal Langerhans cells are potent stimulators of an antigen-specific T cell response to Leishmania major, the cause of cutaneous leishmaniasis, Eur.J.Immunol., 22: 1341.

THE MULTIPLE ACCESSORY CELL CONCEPT: ITS RELEVANCE TO THE DEVELOPMENT OF

ADJUVANTS AND VACCINES

Nico Van Rooijen

Dept. of Cell Biology, Faculty of Medicine
Vrije Universiteit, van der Boechorstraat 7
1081 BT Amsterdam, The Netherlands

INTRODUCTION

When studying the mechanism of immunoadjuvant activity, it is important to realize that various non-lymphoid accessory cells are required for the induction of the antibody response and for the generation of immunological memory. These cells have a crucial function in the processing and/or presentation of antigens to lymphoid cells. For that reason, they have to be considered one of the most obvious targets in the development of adjuvants and vaccines.

During evolution of host defense mechanisms, the lymphoid system of vertebrates has developed to an extremely efficient immune apparatus, in which T- and B-lymphocytes are the well known effector cells. Several non-lymphoid cells have acquired an important role in the processing and/or presentation of antigens to these lymphoid cells. It seems that during evolution, these cells have specialized, and different non-lymphoid cells have acquired different functions in the initial steps of immune responses, or in the establishment of immunological memory (Claassen, 1991). Among these non-lymphoid cells which are involved in the induction of immunity are e.g. Langerhans cells in the skin, the so called M-cells of the mucosal immune system, dendritic cells in the T-cell areas of lymphoid organs, follicular dendritic cells in the lymphoid follicles of lymphoid organs and macrophages, which occur in all compartments of lymphoid organs (Tew, 1992). The interactions between the latter three cell types and the T cells and B cells in the spleen as well as the localization of these non-lymphoid cells in various splenic compartments are shown in Figure 1.

MACROPHAGES

Macrophages form an important group of cells in the host defense system. Their primary task is based on their activity as scavengers. A large part of e.g. bacteria that have entered into the body, aged erythrocytes in the circulation and antibody complexed antigens produced during an immune response, are cleared by macrophages. These cells are particularly well equipped to degrade all of these ingested particles or macromolecular complexes with the aid of their large panel of lysosomal enzymes. In spite of the fact that macrophages are responsible for the non-immune defense system of the body, there is a lot of evidence that they

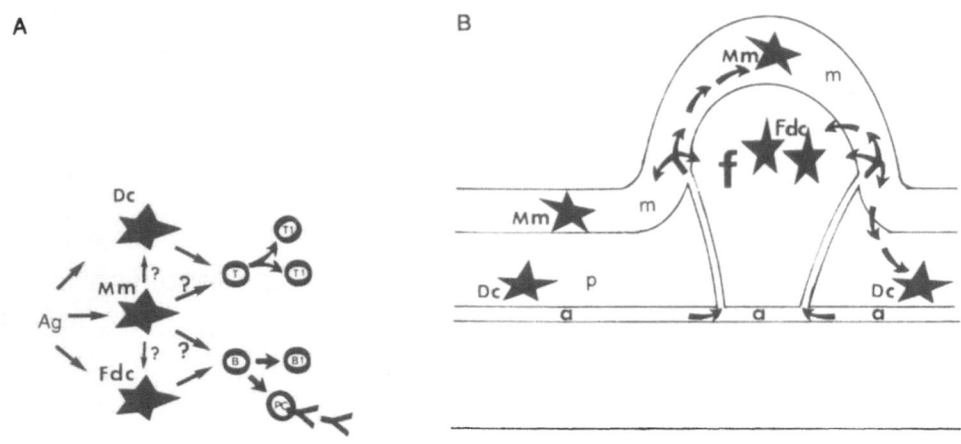

Fig. 1. Schematic representation of cellular interactions (A) and splenic
localization (B) of different accessory cells. A. Thymus-dependent
antigens (Ag) include soluble antigens and particulate antigens.
Dendritic cells (Dc) can process soluble antigens and present the
processed antigens to T cells which are proliferating (Tl) in
response (see e.g. Steinman, 1991 and Austyn, this volume).
Antibody-complexed antigens will be trapped on the surfaces of cell
processes protruding from the follicular dendritic cells (Fdc).
Free antigenic determinants will be exposed to B cells, leading to
the generation of memory B cells (Bl) in the follicles (see e.g.
Klaus et al, 1980 and Van Rooijen, 1990c). Antigens and antibody-
complexed antigens can also be ingested by macrophages. The
macrophages will digest the bulk of the antigens and may be
involved in the suppression of the immune response against soluble
antigens. However, they have an obligatory role in the immune
response against particulate antigens. There is evidence that
macrophages are able to present their processed antigens to T cells
(Unanue, 1984). There is also some evidence that soluble antigens,
which are released from macrophages after ingestion and processing
of particulate antigens and are consecutively complexed with antibodies, are trapped on the surfaces of Fdc (Enriquez-Rincon and
Klaus, 1984). Furthermore it is hypothesized that macrophages can
transfer their (pre)processed antigens to B cells and to Dc (Van
Rooijen, 1990b). B cells may (re)process these (pre)processed
antigens and present them to T cells in order to receive T cell
help for their differentiation into antibody forming plasma cells
(PC). B. Antigens arriving in the spleen via the arterioles (a),
are discharged from the terminals of the white pulp capillary
network between the cells in the marginal zone (m). They can then
easily reach the marginal zone macrophages (Mm). To reach the
dendritic cells (Dc) in the periarteriolar lymphocyte sheath
(PALS=p) and the follicular dendritic cells (Fdc) in the follicles
(f), they must enter these splenic compartments from the marginal
zone.

have as well acquired a role in the immune defense system. A role for
macrophages in the processing of antigens as an initial step in immunity
has been the subject of numerous studies since the early 1960s (Weissmann
and Dukor, 1970). The in vitro ability of macrophages to process antigens
and to present these processed antigens to T-cells has been confirmed by

Unanue and colleagues (1984, 1987). Most of the results pointing to a role of macrophages in the processing and presentation of TD antigens to T-lymphocytes were obtained during in vitro studies, i.e. in the absence of the normal cell populations and their microenvironment. Under these conditions, many macrophage populations obtained from various organs of the body are able to initiate an antigen specific stimulation of T cells (Unanue, 1984). These in vitro studies form an indispensable contribution to our knowledge of the mechanism of antigen processing and presentation. However, the finding that in vitro a particular population of macrophages can present TD antigens to T cells does not warrant that this is one of its in vivo functions. Which of the cell populations having this capability in vitro is really involved in antigen presentation in vivo mainly depends on microenvironmental factors as e.g. their localization with respect to the T-cell migration pathways, the conditions under which they meet these T-cells, and the possible in situ presence of other cell types which can present the antigens more efficiently.

Macrophage Functions "in vivo"

To get more insight into the role of macrophages in the onset of the immune response, in vivo studies thus seem unavoidable. According to the approach, that is frequently chosen, macrophages are depleted in order to investigate whether or not they are required in vivo. We have recently developed a novel approach, based on the in vivo elimination of macrophages using the liposome encapsulated drug dichloromethylene diphosphonate (Cl2MDP). Macrophages ingest the liposomes, followed by the phospholipase mediated disruption of the liposomal bilayers and release of the Cl2MDP in the cell (Van Rooijen and Claassen, 1988; van Rooijen, 1989; 1992). Cl2MDP, released within the cell is believed to cause elimination of the macrophages due to its activity as a chelator of Ca^{2+} and/or metal ions (Van Rooijen, 1991). Liposomal encapsulation of the drug both warrants its selective targeting to macrophages and minimalizes its effect on non-phagocytic cells. Cl2MDP released in the circulation from dying cells is rapidly removed in the kidneys and the half life of the free drug is in the order of minutes (Fleish, 1988). Extensive studies have shown that intra-venous administration of the liposome encapsulated drug eliminates macro-phages in spleen and liver, subcutaneous footpad administration eliminates macrophages in draining lymph nodes and intratracheal administration eliminates alveolar macrophages in the lung. Macrophage populations in several other organs may be eliminated by administration of liposome encap-sulated Cl2DMP via the appropriate routes (Van Rooijen, 1992).

Macrophages in "in vivo" Immune Responses

We have studied immune responses against several trinitrophenyl (TNP)-conjugated antigens in mice and rats. Serum antibody titers were measured and specific anti-TNP antibody forming cells were detected in tissue sections of lymphoid organs in order to correlate between the presence of macrophages and specific antibody forming cells in situ. Both intravenous injection and subcutaneous footpad administration of the thymus dependent (TD) antigen TNP-keyhole limpet hemocyanin (KLH) and the thymus independent (TI) antigens TNP-lipopolysaccharide (LPS,TI-type 1) and TNP-Ficoll (TI-type 2) elicited strong anti-TNP responses in normal mice. High anti-TNP serum antibody titers were found and anti-TNP antibody forming cells were detected in spleen (Claassen et al, 1986) and popliteal lymph nodes (Delemarre et al, 1990) respectively. When the mice had been injected with liposome encapsulated Cl2MDP one day before and along the same route as used for antigen administration, anti-TNP responses were not reduced and their in situ localization was not altered. On the contrary, after primary as well as secondary immunization with the TD antigen TNP-KLH, serum levels of anti-TNP antibodies as well as numbers of anti-TNP

13

antibody forming cells were increased and the in situ response in the lymph nodes of liposome-treated mice was prolonged (Delemarre et al, 1990).

Elimination of alveolar macrophages by intratracheal administration of liposome encapsulated C12MDP had a dramatic effect on the pulmonary immune response against TNP-KLH. An increase in the numbers of anti-TNP antibody forming cells in lung-associated lymph nodes and a prolongation of the response were found, as well as the induction of antibody forming cells in the lung tissue. It was concluded from these results that alveolar macrophages are likely to play a role in controlling the pulmonary immune response in a suppressive way (Thepen et al, 1990).

Depletion of peritoneal macrophages as well as macrophages in the draining parathymic lymph nodes by intraperitoneal inoculation of liposome encapsulated C12MDP resulted in an enhanced immune reaction in the para-thymic lymph nodes to intraperitoneally administered TNP-KLH (Soesatyo et al, 1991). The above reported experiments have demonstrated that macro-phages are not required for the induction of immune reactions against non-particulate antigens. On the contrary macrophages were shown to have a regulatory role in these responses by suppression. Possible modes of action of macrophages in the suppression of immune responses have been re-viewed recently (Vassilev and Paskov, 1991).

As did immunization with the non-particulate antigen TNP-KLH, intra-venous injection with the particulate antigen TNP-sheep red blood cells (SRBC) elicited a strong anti-TNP response in the spleen of mice. When the antigen was administered in mice that had been intravenously treated with liposome encapsulated C12MDP one day before antigen, splenic macrophages had disappeared and the anti-TNP response was severely suppressed (Delemarre et al, 1991a, b). Specifically the numbers of anti-TNP antibody forming cells of the IgG isotype were strongly decreased. Further studies were performed to investigate which of the subpopulations of splenic macro-phages were involved in the humoral immune response against TNP-SRBC. These results clearly point to the marginal zone macrophages (MZM) as the cells required for an antigen processing step in the induction of the splenic immune response against the particulate antigen TNP-SRBC (Delemarre, 1990). MZM have a strategic position in the marginal zone, close to the site where antigens are discharged from the white pulp capillaries (Van Rooijen et al, 1989).

Liposomes Mediated Targeting of Antigens into Macrophages as a Mechanism of Immunopotentiation

Protein antigens, that are associated with, or encapsulated within liposomes, are converted from soluble antigens to particulate antigens. Hence, association of small soluble antigens with liposomes is an approach to target these antigens into macrophages. Several authors have shown that liposomes are responsible for a substantial enhancement of the immune response elicited against such liposome associated antigens (Gregoriadis, 1990; Van Rooijen, 1990a; Alving, 1991). Since liposomes are biodegradable structures, that may be composed of non-toxic and immunologically inert phospholipids, they have been suggested as promising carriers for antigens. The toxicity of some antigens may be reduced by their incorporation in liposomes while at the same time their immunogenicity is increased by this procedure. Immunopotentiation has been established both for antigens exposed on liposomal outer surfaces and for antigens encapsulated within the liposomes. Since liposome encapsulated antigens are masked and thus prevented from recognition by surface receptors on lymphoid cells, and liposomes are avidly phagocytosed by macrophages, phagocytosis of liposomes followed by unmasking of the encapsulated antigens seems to be a logical first step in the induction of an immune response. Indeed, in vitro antigen presentation experiments have shown that macrophages are necessary and sufficient for antigen presentation of liposome encapsulated antigens to T-cells, whereas B-cells are incapable of presenting liposome encapsulated antigens (Dal Monte and Szoka, 1989; Harding et al, 1991; Szoka, 1992). Just as B-cells, dendritic cells have only a limited activity of phagocytosis (Steinman, 1991) and are not expected to take up any appreciable amount of liposomes. Also, injection into mice of macrophages that had been fed liposome encapsulated antigen in culture, enhanced the immune response (Beatty et al, 1984), confirming the early experiments that injection of macrophages with ingested antigen induced strongly enhanced immune responses as compared to administration of similar amounts of free antigen (Weissman and Dukor, 1970).

We have confirmed the obligatory role of macrophages in the in vivo immune response against liposome associated antigens using the liposome mediated macrophage "suicide" technique to deplete splenic macrophages and liposome associated albumin antigen for immunization (Su and Van Rooijen, 1989; Van Rooijen and Su, 1989). Liposome encapsulated drugs and liposome encapsulated antigens will be targeted to the same macrophages in the same compartments of the same organs, provided that the liposomal composition and the administration routes are the same. So, problems related to a possible differential distribution of cytotoxic drugs and antigens are minimized. Depletion of splenic macrophages by intravenous treatment with Cl2MDP liposomes caused a substantial reduction of the immune response against consecutively administered liposome associated antigens. These results confirmed earlier results in which the role of macrophages in the in vivo immune response against liposome associated antigens was studied after functional elimination of macrophages with carrageenan (Shek and Lukovich, 1982). The intracellular ways of processing of liposome encapsulated antigens and the possibilities to target the liposomal contents to specific subcellular compartments, using pH-sensitive liposomes, have been discussed recently (Harding et al, 1991; 1992).

Co-operation between Macrophages and other Specialized Accessory Cells

Based on the results reported above, we have proposed a model for co-operation among B-cells, T-cells, dendritic cells and macrophages in the humoral immune response against particulate TD antigens in the spleen (Van Rooijen, 1990b). In this model, the marginal zone macrophage (MZM), having a strategic position at the end of the white pulp capillaries, is responsible for (pre)processing of particulate antigens and soluble antigens which are converted to particulate antigens e.g. by association with liposomes. The processed antigenic fragments are then transferred to B-cells in the marginal zone followed by migration of these antigen-activated B-cells into the outer part of the periarteriolar lymphocyte sheaths (PALS). Here, the B-cells present the processed antigens to T-cells of appropriate specificity and receive T-cell help. This in turn induces their differentiation into antibody forming plasma cells. In the inner part of the PALS, dendritic cell-T-cell clusters are activated by antigen, resulting in the proliferation of antigen-specific T-cells. However, just as B-cells, dendritic cells are not expected to be able to handle the liposome encapsulated and other particulate antigens without (pre)processing by macrophages. So it may be that either the dendritic cells in the T-cell areas of the spleen have obtained their (pre)processed antigens from macrophages in the white pulp, or they may have migrated into the T-cell compartment after obtaining the antigens elsewhere, e.g. in the marginal zone (Austyn et al, 1988; Kraal, 1992). Although it is well known that macrophages, B-cells and dendritic cells can all act as antigen-presenting cells in vitro (Kulkarni et al, 1991), it is not obvious that in vivo these cells, each with their characteristic morphology, subcellular equipment, surface receptors and position within one of the different

lymphoid compartments, all have a similar function, without any special-
ization. In conclusion, we believe that macrophages are involved (and
specialized) in preprocessing (fragmentation) of particulate antigens only.
These antigenic fragments are then transferred to either B-cells or
dendritic cells. The latter cells are responsible for (and specialized in)
their presentation to T-cells. Dendritic cells mediate the proliferation
of T-cells and B-cells required to present the antigens to T-cells in order
to get help in their own differentiation into antibody forming plasma
cells.

Macrophages as Accessory Cell for Class I MHC-Restricted Immunity

There is increasing evidence that a class I MHC-restricted cytotoxic
T cell response to soluble antigen can be induced, if the antigen is
artificially introduced into the cell cytoplasm, using liposomes as a
carrier (Huang et al, 1992; Lopes and Chain, 1992). A role of macrophages
as accessory cells in class I MHC-restricted immune responses has been
confirmed by several authors (Debrick et al, 1991; Harding et al, 1991,
1992; Huang et al, 1992; Zhou et al, 1992). Apart from the possibility
that the macrophage is the main accessory cell in the liposome mediated
induction of class I MHC-restricted immunity, a co-operation between two
different accessory cells, as proposed above for the class II MHC-
restricted immune response against liposome encapsulated antigens has also
been considered (Huang et al, 1993; Zhou et al, 1992). In the latter case,
liposomal antigen would be transferred to dendritic cells after a first
processing step by macrophages. In turn, the dendritic cell would present
the antigen to the T cells.

Follicular Dendritic Cells in the Generation of Memory B-Cells

Antigens entering into the body induce a humoral immune response by
the lymphoid system, ultimately leading to the production of specific
antibodies against the antigen. Most of the antigen-antibody complexes,
formed as soon as the antibodies appear in the circulation are rapidly
ingested by cells of the mononuclear phagocyte system. A small part,
however, is carried to and trapped by the cell processes of follicular
dendritic cells (FDC) in the follicles of lymph nodes and spleen (Nossal et
al, 1968, 1971). This is the only mechanism by which undegraded antigen
may survive for long periods of time in the body. It is now generally
assumed that this antibody complexed, and FDC immobilized antigen is
playing a crucial role in the generation of memory B cells (Klaus et al,
1980; Van Rooijen et al, 1980) and in the phenomenon of "affinity
maturation" (Apel and Berek, 1990; Jacob et al, 1991; MacLennan, 1991).
IgG antibodies can enhance or inhibit the humoral immune response against
its specific antigen (Heyman, 1990). Enhancement by IgG is obtained for
soluble protein antigens, whereas IgG in combination with protein antigens
in adjuvants or with heterologous erythrocytes induces suppression
(Wiersma, 1991). As early as in 1959, an enhancing influence of pre-
existing antibodies on the magnitude of the antibody response to an
antigenic stimulus has been noted by Terres and Wolins (1959). In the
1960s, Terres and colleagues performed a series of experiments to
investigate the immunopotentiating effect of specific antibodies on the
immune response against albumin antigens (Terres and Wolins, 1961; Terres
and Morrison, 1967). Their results strongly suggested that immune
complexes were able to establish some kind of a "primed state" within a few
days, identical to the primed state observed weeks after a single antigen
injection (Terres et al, 1972; Terres et al, 1974). Further evidence for a
role of immune complexes, in the establishment of immunological memory came
from a series of experiments performed by Klaus and colleagues (Klaus,
1978; Klaus et al, 1980; Kunkl and Klaus, 1981). They demonstrated that,
under certain conditions, antigen-antibody complexes were far more ef-

fective than antigen alone in the generation of memory B cells. Optimal
priming by complexes required the integrity of the Fc portion of the
antibody, since F(ab)2 antibody fragments were less effective. Moreover
the capacity of complexes to induce the generation of memory B cells was
abrogated by depriving the animals of C3. Obviously, the same conditions
exist for immune complexes to localize in follicle centres and to generate
memory B cells (Klaus et al, 1980; Klaus and Humphrey, 1986). Both, the
appearance of a so-called 'early primed state' in mice following the
concomittant injections of antigen and specific antiserum in the early
studies of Terres and colleagues as well as the more recent results of
Klaus and colleagues showing that immunization with antigen-antibody
complexes accelerates the development of memory B cells, the maturation of
antibody affinity and the formation of germinal centres have focussed on
the lymphoid follicle as the place where antibodies modulate the response
to antigens. Recent studies showing that one and the same monoclonal anti-
TNP-specific IgG-antibody can enhance the anti-TNP-Keyhole limpet haemocyanin (KLH = carrier) response up to 38-fold whereas it suppresses the
anti-TNP-SRBC (SRBC = carrier) response by more than 10-fold (Wiersma et
al, 1989) suggest a different and antigen dependent processing of these
antibody complexed antigens. Selective enhancement of the IgG anti-carrier
(KLH) responses in mice (up to 1000-fold) by IgG2 anti-hapten antibodies
(Coulie and Van Snick, 1985) may well be based on the efficient targeting
to and retention by FDC of such IgG2 anti-TNP-TNP-KLH complexes. FDC
generally retain complexes containing IgG2 better than those formed with
IgG1 or IgG3, whereas IgM is rarely retained (Heinen et al, 1986). Im-
mobilization of IgG2-anti-TNP-TNP-KLH complexes on the cell processes of
FDC occurs in a matter of hours and leads to the intrafollicular exposition
of numerous free antigenic determinants of the carrier (KLH). The accelerated forming of anti-KLH memory-B cells and their consecutive stimulation by soluble TNP-KLH molecules still in the circulation may be
responsible for the enhanced IgG response (Van Rooijen, 1990c). It has
been shown that antigen complexed with antibody can induce germinal centre
formation and B-cell memory without evoking a detectable antibody response
(Kraft et al, 1989). Isolated FDC in vitro appeared to stimulate
proliferation of B cells (Cormann et al, 1986; Petrasch et al, 1991) and to
suppress their Ig production (Cormann et al, 1986). The latter findings
indicate that the role of antibodies in the generation of memory B cells
and affinity maturation on the one hand and its role in the enhancement of
the antibody response on the other hand may well be regulated by completely
different mechanisms. For two reasons, we do not believe that the IgM-
mediated enhancement of the antibody response, as demonstrated for erythrocyte antigens (Heyman et al, 1982; 1988) depends on lymphoid follicles.
First, it has been shown that IgM-antigen complexes do not localize in the
follicles (Heinen et al, 1986; Klaus, 1979). Moreover, there is some
evidence that antibody complexed erythrocyte antigens, that are trapped and
retained in the follicles, contain soluble erythrocyte antigens released
from macrophages, rather than intact erythrocytes (Enriquez-Rincon and
Klaus, 1984).

Antibody Mediated Targeting of Antigens to Follicular Dendritic Cells and Macrophages: Two Different Mechanisms of Immunopotentiation?

Arguments have been given that small immune complexes composed of
only one antibody molecule and one antigen molecule are neither phago-
cytosed (Van Oss et al, 1974), nor trapped by FDC (Laman et al, 1992).
Such immune complexes may be formed early in the immune response, when the
range of antibody specificities (to the different epitopes on the same
antigen) is still small and their average affinity is still low. In a
later phase of the response, antibodies against different epitopes will
appear and have higher affinities, so that complexes are formed in which
more antibody molecules participate. These larger complexes will be

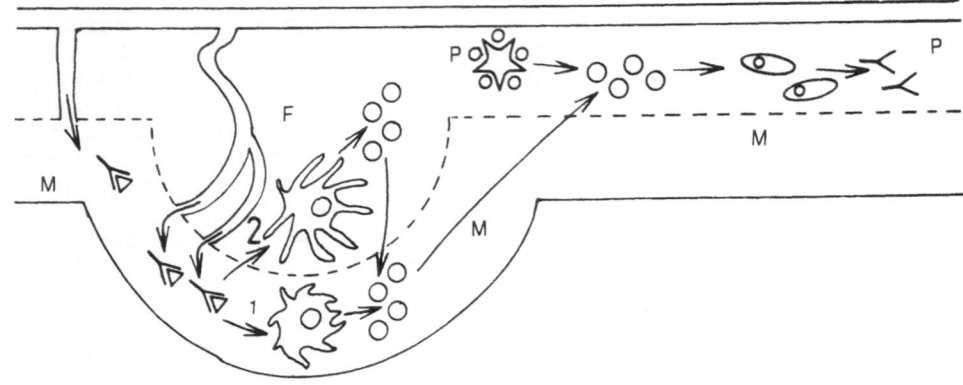

Fig. 2. Dual handling of antigen-antibody complexes by marginal zone
     macrophages and follicular dendritic cells in the spleen. Left
     side of figure: Antigen-antibody complexes are carried into the
     spleen by arterial vessels and discharged by the terminal vessels
     of the white pulp capillary network between the cells in the
     marginal zone (M). These immune complexes may induce two different
     immune activities which are mediated by two different types of
     accessory cells: 1. A large part of the complexes is ingested by
     marginal zone macrophages. After an obligatory (pre)processing
     step, the antigens are transferred to B cells (circles) in the
     marginal zone. The activated B cells migrate into the peri-
     arteriolar lymphocyte sheaths (PALS=P). In the PALS, they meet
     the appropriate T cells (small circles), which have been stimulated
     in the dendritic cell (star)-T cell clusters. With their help they
     develop into plasma cells (ellipsoids) which are responsible for
     the production of specific antibodies (see Van Rooijen 1990c for
     details of this hypothesis). 2. A small part of the complexes is
     carried from the marginal zone into the follicles and trapped in
     the cell processes of follicular dendritic cells (Fdc). The
     antibody complexes and Fdc exposed antigen is involved in the
     generation of memory B cells. Contrary to immunization with
     antigen alone, when generation of memory B cells must await the
     immunization with preformed complexes, immunization with preformed
     complexes may induce an early generation of memory B cells. This
     may explain the "early primed state" observed by Terres and
     colleagues after immunization with immune complexes. When mobile
     antigen-antibody complexes are still in the circulation, the newly
     formed memory B cells may continue to differentiate into plasma
     cells in the follicles (see Van Rooijen 1990c for details). If
     mobile antigen-antibody complexes are no longer present, memory B
     cells may be activated by antigen (pre)processing marginal zone
     macrophages (see 1).

ingested by macrophages (Van Oss et al, 1974) and trapped by FDC (Laman et
al, 1992). Both processes are mediated by the Fc portions of the
participating antibodies and as far as is known, the larger the complexes,
the more they are preferred by both macrophages and FDC. Many of the early
experiments on immunization with immune complexes or liposome associated
antigens have been performed with albumin antigens. Albumins are small and
soluble weak protein antigens. Obviously none of the possible cellular
candidates for their processing and presentation is able to handle these
small antigens efficiently, if the antigen is given in its free form (Su
and Van Rooijen, 1989; Van Rooijen and Su, 1989). However complexing of
the antigen with specific antibody or its association with liposomes

greatly enhances its palatability to the immune system. That free albumins are not processed efficiently, may be caused by a limited ability of the cellular candidates to internalize this small molecule. Obviously this bottleneck in the induction of humoral immunity is circumvented by any of these pretreatments. Whereas antibody complexing will target the antigen to both FDC and macrophages, encapsulation in liposomes will target the antigen into macrophages exclusively. Since both pretreatments of the antigen have proven to accelerate the antibody response and to increase its magnitude, it seems probable that at least macrophages are able to mediate these immunopotentiating effects.

A difference between liposome encapsulated albumins and antibody complexed albumins in the induction of humoral immunity is that the latter pretreatment of the antigen may induce a macrophage mediated acceleration of the induction of the antibody response and, simultaneously, a FDC mediated acceleration of the generation of memory B cells (early primed state). So, both accessory cells may act synergistically, since on the one hand, the generation of memory B-cells is accelerated and on the other hand, the available memory B-cells may be quickly activated by the antigenic fragments which are transferred to them by the macrophages (see Figure 2, for a schematic representation of the proposed involvement of marginal zone macrophages and FDC in the handling of immune complexes in the spleen).

REFERENCES

Alving, C. R., 1991, Liposomes as carriers of antigens and adjuvants, J.Immunol. Meth., 14:1.
Apel, M., Berek, C., 1990, Somatic mutations in antibodies expressed by germinal centre B cells early after primary immunization, Int. Immunol., 2:813.
Austyn, J. M., Kupiec-Weglinsky, J. W., Hankins, D. F., Morris, P. J., 1988, Migration patterns of dendritic cells in the mouse. Homing to T cell dependent areas of spleen and binding within marginal zone, J. Exp. Med., 167:646.
Beatty, J. D., Beatty, B. G., Paraskevas, F., Froese, E., 1984, Liposomes as immune adjuvants; T cell dependence, Surgery, 96:345.
Claassen, E., 1991, 38th Forum in Immunology: Histological organization of the spleen: Implications for immune functions in different species, Res. Immunol., 142:315.
Claassen, E., Kors, N., Van Rooijen, N., 1986, Influence of carriers on the development and localization of anti-trinitriphenyl antibody-forming cells in the murine spleen, Eur. J. Immunol., 16:271.
Cormann, N., Lesage, F., Heinen, E., Schaaf-Lafontaine, N., Kinet-Denoel, C., Simar, L. J., 1986, Isolation of follicular dendritic cells from human tonsils and adenoids. V. Effect on lymphocyte proliferation and differentiation, Immunol. Lett., 14:29.
Coulie, P. G., Van Snick, J., 1985, Enhancement of IgG anti-carrier responses by IgG2 anti-hapten antibodies in mice, Eur. J. Immunol., 15:793.
Dal Monte, P., Szoka, Jr. F. C., 1989, Effect of liposome encapsulation on antigen presentation in vitro; comparison of presentation by peritoneal macrophages and B cell tumors, J. Immunol., 142:1437.
Debrick, J. E., Campbell, P. A., Staerz, U. D., 1991, Macrophages as accessory cells for class-II MHC restricted immune responses, J. Immunol., 147:2846.
Delemarre, F. G. A., 1990, The role of macrophages in the humoral immune response in lymph nodes and spleen of mice, Academic Thesis, Vrije Universiteit, Amsterdam, The Netherlands.
Delemarre, F. G. A., Kors, N., Van Rooijen, N., 1990, The in situ

immune response in popliteal lymph nodes of mice after macrophage depletion. Differential effects of macrophages on thymus-dependent and thymus-independent immune responses, Immunobiology, 180:395.

Delemarre, F. G. A., Kors, N., Van Rooijen, N., 1991a, Elimination of spleen and of lymph node macrophages and its difference in the effect on the immune response to particulate antigens, Immunobiology, 180:395.

Delemarre, F. G. A., Kors, N., Van Rooijen, N., 1991b, The role of the spleen and of lymph node macrophages in the immune response to particulate antigens, in "Lymphatic Tissues and In Vivo Immune Responses," Imhof et al., eds., Marcel Dekker Inc., New York, pp 843-847.

Enriquez-Rincon, F., Klaus, G. G. B., 1984, Follicular trapping of hapten-erythrocyte-antibody complexes in mouse spleen, Immunology, 52:107.

Fleisch, H., 1988, Biphosphonates: a new class of drugs in diseases of bone and calcium metabolism, Handbook of Experimental Pharmacol., 83:441.

Gregoriadis, G., 1990, Immunological adjuvants: a role for liposomes, Immunology Today, 11:89.

Harding, C. V., Collins, D. S., Slot, J. W., Geuze, H. J., Unanue, E. R., 1991, Liposome encapsulated antigens are processed in lysosomes, recycled and presented to T cells, Cell, 64:393-401.

Harding, C. V., Collins, D. S., Kanagawa, O., Unanue, E. R., 1991, Liposome encapsulated antigens engender lysosomal processing for class-II MHC presentation and cytosolic processing for class-I presentation, J. Immunol., 147:2860.

Harding, C. V., Collins, D. S., Unanue, E. R., 1992, Processing of liposome-encapsulated antigens targeted to specific subcellular compartments, Res. Immunol., 143:188.

Heinen, E., Coulie, P., Van Snick, J. Braun, M., Cormann, N., Moeremans, M., Kinet-Denoel, C., Simar, L. J., 1986, Retention of immune complexes by murine lymph node or spleen follicular dendritic cells: role of antibody isotype, Scand. J. Immunol., 24:327.

Heyman, B., 1990, The immune complex: possible ways of regulating the antibody response, Immunol Today, 11:310.

Heyman, B., Andrighetto, G., Wigzell, H., 1982, Antigen-dependent IgM-mediated enhancement of the sheep erythrocyte response in mice. Evidence for induction of B cells with specificities other than that of the injected antibodies, J. Exp. Med., 155:994.

Heyman, B., Pilstrom, L., Schulman, M. J., 1988, Complement activation is required for IgM-mediated enhancement of the antibody response, J. Exp. Med., 167:1999.

Huang, L., Reddy, R., Nair, S. K., Zhou, F. Rouse, B. T., 1992, Liposomal delivery of soluble protein antigens for Class I MHC mediated antigen presentation, Res. Immunol., 143:192.

Jacob, J., Kelsoe, G. Rajewsky, K., Weiss, U., 1991, Intraclonal generation of antibody mutants in germinal centres, Nature, 353:389.

Klaus, G. G. B., 1978, The generation of memory cells. II. Generation of B memory cells with preformed antigen-antibody complexes, Immunology, 34:643.

Klaus, G. G. B, Humphrey, J. H., 1986, A re-evaluation of the role of C3 in B-cell activation, Immunol. Today, 7:163.

Klaus, G. G. B, Humphrey, J. H., Kunkl, A., Dongworth, D. W., 1980, The follicular dendritic cell: its role in antigen presentation in the generation of immunological memory, Immunol. Rev., 53:3.

Kraal, G., 1992, Cells in the marginal zone of the spleen, Int. Rev. Cytol., 132:31.

Kraft, R., Buerki, H., Schweizer, T., Hess, M. W., Cottier, H., Stoner, R. D., 1989, Tetanus toxoid complexed with heterologous antibody can induce germinal centre formation and B cell memory in mice without evoking a detectable anti-toxin response, <u>Clin. Exp. Immunol.</u>, 76:138.

Kulkarni, A. B., Mullbacher, A., Blanden, R. V., 1991, Functional analysis of macrophages, B cells and splenic dendritic cells as antigen-presenting cells in West Nile virus-specific murine T lymphocyte proliferation, <u>Immunol. Cell Biol.</u>, 69:71.

Kunkl, A., Klaus, G. G. B., 1981, The generation of memory cells. IV. Immunization with antigen-antibody complexes accelerates the development of B memory cells, the formation of germinal centres and the maturation of antibody affinity in the secondary response, <u>Immunology</u>, 43:371.

Laman, J. D., Ter Hart, H., Boorsman, D. M., Claassen, E., Van Rooijen, N., 1992, Production of a monomeric antigen-enzyme conjugate to study requirements for follicular immune complex trapping, <u>Histochemistry</u>, 97:189.

Lopes, L. M., Chain, B. M., 1992, Liposome-mediated delivery stimulates a class I-restricted cytotoxic T cell response to soluble antigen, <u>Eur. J. Immunol.</u>, 22:287

MacLennan, I., 1991, The centre of hypermutation, <u>Nature</u>, 354:352.

Nossal, G. J. V., Abbot, A., Mitchell, J., Lummus, Z., 1968, Antigens in immunity. XV. Ultrastructrual features of antigen capture in primary and secondary follicles, <u>J. Exp. Med.</u>, 127:277.

Nossal, G. J. V., Ada, G. L., 1971, "Antigens, Lymphoid Cells and the Immune Response, F. J. Dixon and A. Kunkel, eds., Academic Press, New York.

Petrasch, S. G., Kosco, M. H., Perez-Alvarez, C. J., Schmitz, J., Brittinger, G., 1991, Proliferation of germinal center B lymphocytes <u>in vitro</u> by direct membrane contact with follicular dendritic cells, <u>Immunobiol.</u>, 183:451.

Shek, P. N., Lukovich, S., 1982, The role of macrophages in promoting the antibody response mediated by liposome-associated protein antigens, <u>Immunol. Lett.</u>, 5:305.

Soesatyo, M., Biewenga, J., Van Rooijen, N., Kors, N., Sminia, T., 1991, The <u>in situ</u> immune response of the rat after intraperitoneal depletion of macrophages by liposome-encapsulated dichloromethylene diphosphonate, <u>Res. Immunol.</u>, 143:533.

Steinman, R. M., 1991, The dendritic cell system and its role in immunogenicity, <u>Ann. Rev. Immunol.</u>, 9:271.

Su, D., Van Rooijen, N., 1989, The role of macrophages in the immunoadjuvant action of liposomes: effects of elimination of splenic macrophages on the immune response against intravenously injected liposome associated albumin antigen, <u>Immunology</u>, 66:466.

Szoka, F. C. Jr., 1992, The macrophage as the principal antigen presenting cell for liposome encapsulated antigens, <u>Res. Immunol.</u>, 143:186.

Terres, G., Wolins, W., 1959, Enhanced sensitization in mice by simultaneous injection of antigen and specific rabbit antiserum, <u>Proc. Soc. Exp. Biol. Med.</u>, 102:632.

Terres, G., Wolins, W., 1961, Enhanced immunological sensitization of mice by the simultaneous injection of antigen and specific antiserum. I. Effect of varying the amount of antigen used relative to the antiserum, <u>J. Immunol.</u>, 86:361.

Terres, G., Morrison, S. L., 1967, Enhanced immunologic sensitization of mice by the simultaneous injection of antigen and specific antiserum. III. The role of antigen in controlling the immune response elicited with immune complexes, <u>J. Immunol.</u>, 98:584-592.

Terres, G., Morrison, S. L., Habicht, G. S., Stoner, R. D., 1972, Appearance of an early "primed state" in mice following the concomitant injections of antigen and specific antiserum <u>J. Immunol.</u>, 108:1473.

Terres, G., Habicht, G. S., Stoner, R. D., 1974, Carrier-specific enhancement of the immune response using antigen-antibody complexes, J. Immunol., 112:804.

Tew, J. G., 1992, Antigen trapping and presentation in vivo, Semin Immunol., 4:4.

Thepen, T., Van Rooijen, N., Kraal, G., 1990, Alveolar macrophage elimination in vivo is associated with an increase in pulmonary immune responses in mice, J. Exp. Med., 170:499.

Unanue, E. R., 1984, Antigen-presenting function of the macrophage, Ann. Rev.Immunol., 2:395.

Unanue, E. R., Allen, P. M., 1987, The basis for the immunoregulatory role of macrophages and other accessory cells, Science, 236:551.

Van Oss, C. J., Gillman, C. F., Neumann, A. W., 1974, Phagocytosis as a surface phenomenon. IV. The minimum size and composition of antigen-antibody complexes that can become phagocytosized, Immunol. Commun., 3:77.

Van Rooijen, N., 1980, Immune complex trapping in lymphoid follicles: a discussion on possible functional implications, in: "Phylogeny of Immunological Memory," M. J. Manning, ed., Elsevier, Amsterdam, pp 281-290.

Van Rooijen, N., 1989, The liposome mediated macrophage 'suicide' technique, J.Immunol. Meth., 124:1.

Van Rooijen, N., 1990a, Liposomes as carrier and immunoadjuvant of vaccine antigens, Adv. Biotechnol. Processes, 13:255.

Van Rooijen, N., 1990b, Antigen processing and presentation in vivo: the microenvironment as a crucial factor, Immunology Today, 11:436.

Van Rooijen, N., 1990c, Direct intrafollicular differentiation of memory B cells into plasma cells, Immunol. Today, 11:154.

Van Rooijen, N., 1991, High and low cytosolic Ca2+ induced macrophage death? Cell Calcium, 12:381.

Van Rooijen, N., 1992, Liposome mediated elimination of macrophages, Res. Immunol., 143:215.

Van Rooijen, N., Claassen, E., 1988, In vivo elimination of macrophages in spleen and liver, using liposome encapsulated drugs: methods and applications, in: "Liposomes as Drug Carriers: Recent Trends and Progress," G. Gregoriadis, ed., John Wiley and Sons, Chichester, UK, pp 131-143

Van Rooijen, N., Claassen, E., Kraal, G., Dijkstra, C. D., 1989, Cytological basis of immune functions in the spleen: immunocytochemical characterization of lymphoid and non-lymphoid cells involved in the in situ immune repsonse, Progr. Histochem. Cytochem., 19 (No. 3):1.

Van Rooijen, N., Su D., 1989, Immunoadjuvant action of liposomes: mechanisms, in: "Immunological Adjuvants and Vaccines," NATO ASI Series G. Gregoriadis, ed., Plenum Publ. Comp. USA, pp 95-106.

Vassilev, T. L., Paskov, A. D., 1991, Modes of action of immune response adjuvants, Int. J. Immunopath. Pharm., 4:191.

Weissmann, G., Dukor, P., 1970, The role of lysosomes in immune responses, Adv. Immunol., 12:283.

Wiersma, E. J., 1991, Effects of IgG antibodies and complement on the immune response, Academic Thesis, Uppsala University, Uppsala, Sweden.

Wiersma, E. J., Coulie, P. G., Heyman, B., 1989, Dual immunoregulatory effects of monoclonal IgG-antibodies: suppression and enhancement of the antibody response, Scand. J. Immunol., 29:439.

Zhou, F., Rouse, B. T., Huang, L., 1992, Induction of cytotoxic T lymphocytes in vivo with protein antigen entrapped in membranous vehicles, submitted.

SYNTHETIC PEPTIDES AND THE ROLE OF T-HELPER CELL DETERMINANTS

M.J. Francis

Department of Virology and Process Development
Pitman-Moore Ltd, Breakspear Road South, Harefield
Uxbridge, Middlesex, UB9 6LS, UK

INTRODUCTION

Relatively short chains of amino acids (peptides) have been widely used in recent years to mimic antigenic sites of a wide variety of viruses and thus act as surrogate immunogens for vaccine purposes (Francis, 1990). However, in the past it was generally assumed that because of their relatively small molecular size many synthetic peptides would behave like haptens and would require coupling to a large "foreign" protein carrier in order to enhance their immunogenicity. Immunisation with such conjugates often resulted in the production of anti-peptide antibodies that totally failed to recognise the native protein or infectious agent due to the method of peptide-carrier linkage. Other problems that could be encounter-ed, of particular relevance to vaccination, were hyspersensitivity to the "foreign" carrier, carrier induced suppression and poor batch to batch reproducibility of the conjugates.

It was therefore the goal of many immunologists to dispense with such undefined carrier proteins and to produce a totally synthetic immunogen. It is now clear that synthetic peptides can be highly immunogenic in their free form provided they contain appropriate antibody recognition sites (B-cell epitopes) as well as sites capable of eliciting help for antibody production (Th-cell epitopes) (Mitchison 1971) (Fig. 1). These Th-cell epitopes must be capable of binding class II major histocompatability complex (MHC) molecules on the surface of host antigen-presenting cells (APC) and B-cells, and subsequently interacting with the T-cell receptor in the form of a trimolecular complex (Rosenthal 1978, Babbit et al, 1985) in order to induce B-cells to differentiate and proliferate. Indeed there are good examples of peptides that contain B- and Th-cell epitopes which un-doubtedly account to a large extent for their success as immunogens (Francis et al, 1985; Francis et al, 1987[a]). If, however, a free peptide is a poor immunogen or produces an immune response that is genetically restricted appropriate Th-cell epitopes may be added (Good et al, 1987; Francis et al, 1987 [b]). The now widespread use of such combinations of B-cell and T-cell peptides will be presented and discussed.

POLYMERISATION OF B AND T CELL PEPTIDES

The simplest method of combining B- and T-cell epitopes is to produce

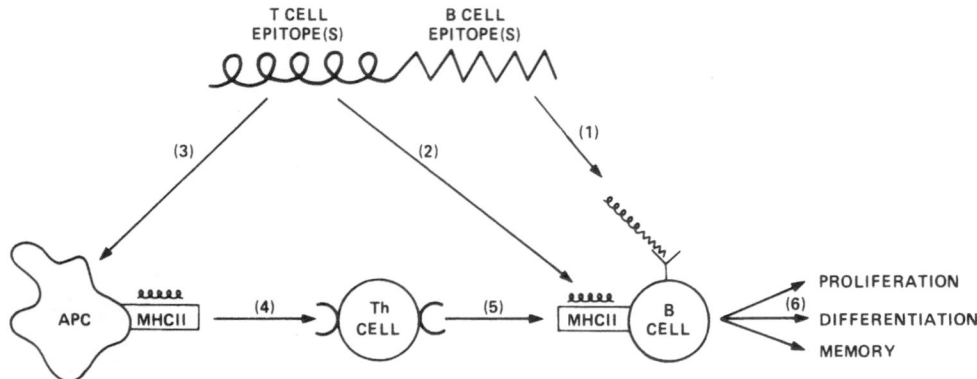

PEPTIDE

Fig. 1. T-cell help for B-cell antibody production to uncoupled peptides. (1) B-cell epitopes are recognised by immunoglobulin receptors on B-cells; (2) the T-cell epitopes within the same peptide are presented on the surface of the B-cells in association with class II major histocompatability complex molecules (MHC II); (3) the same T-cell epitopes are presented on the surface of antigen-presenting cells (APC) in association with MHC II; (4) Th-cells recognise the peptide-MHCII complex; and (5) the B-cells proliferate and differentiate into antibody-secreting plasma cells. (From Francis and Clarke 1989, Copyright 1989, Academic Press Inc).

co-polymers of a number of individual peptides by chemical means. A good example of this approach is provided by the work of Chedid and his colleagues who used glutaraldehyde to polymerise four peptides from two bacterial antigens (streptococcus pyrogens M protein and diptheria toxoid), one viral antigen (hepatitis B surface antigen) and one parasitic antigen (circumsporozoite protein of plasmodium knowlensii) (Chedid et al, 1984). In this study it was shown that the association of several peptides enhanced their respective immunogenicities as compared with those of their homopolymers. The importance of Th-cell epitopes in these polymers was highlighted in a later study (Jolivet et al, 1990). In a refinement of this technique Leclerc et al, (1987) co-polymerised a streptococcal protein peptide (S34), which contained B- and T-cell epitopes, with a hepatitis B virus surface antigen (HBsAg) peptide, containing only a B-cell epitope, and demonstrated an antibody response to the HBsAg peptide in BALC/c ($H-2^d$) mice. It also appeared from their studies that the co-polymer may have acquired a new Th-cell determinant in addition to that within the S34 sequence. The main disadvantages of this approach are the uncontrolled nature of the polymerisation reaction and the risk of affecting the antigenic properties of the peptides.

CHEMICAL LINKAGE

A more controlled method of chemical linkage between B- and T-cell epitopes is provided by using a heterobifunctional cross-linking reagent, such as M-maleimidobenzoyl-N-hydroxysuccinimide ester (MBS). This has an amino-reactive NHS-ester as one functional group and a sulphydryl reactive group as the other. Amino groups on one peptide (e.g. B-cell epitope), are acylated with the NHS-ester via the hyroxysuccinimide group and then a second peptide (Th-cell epitope) is introduced that possesses a free sulfhydryl group that can react with the malemide group of the coupling reagent. This method has been used to link the malaria-encoded sequence (NANP)n from the circumsporozoite (CS) protein, which can elicit an antibody response and also stimulate Th-cells from mice carrying the I-A^b gene, to another peptide from the CS protein which is an amphipathic helical segment covering residues 326-343 (Good et al, 1987). The resultant conjugate raised anti-(NANP)n antibodies in B10.BR and B10.A(4R) mice which both express the I-A^k gene, and were therefore non-responders to the (NANP)n sequence alone. However, this technique may require the synthesis of a specific peptide with a non-natural cysteine residue added to its carboxy terminus. Furthermore, the presence of essential natural cysteine or lysine residues within either peptide is likely to affect the nature and final antigenicity of the conjugate produced.

COLINEAR SYNTHESIS

The problems associated with the polymerisation and chemical coupling methods described above may be largely overcome by the direct colinear synthesis of a peptide containing B- and Th-cell epitopes. This technique allows for a peptide to be constructed with known immunological properties. It also offers the flexibility to alter the position of one epitope in relation to the other and to synthesise peptides containing a number of B- and/or T-cell epitopes. This approach has been used to synthesize two peptides (103-115 and 133-147) from the major coat protein VP6 of bovine rotavirus, which contained non-immunogenic B-cell epitopes, in combination with a peptide from influenza virus haemagglutinin residues 111-120, which contained an H-2^d restricted T-cell epitope (Borras-Cuesta et al, 1987). Both the peptides induced anti-rotavirus responses in BALB/c (H-2^d) mice which were greater than those elicited by the same rotavirus sequences conjugated to bovine serum albumin. This group also went on to demonstrate that linear homopolymers prepared from side-chain protected peptide monomers of both haptenic and co-linear peptides had enhanced immunogenicity (Borras-Cuesta et al, 1988). They speculated that the polymerisation might have generated a new Th-cell determinant.

In a further study the 141-160 peptide from VP1 of FMDV, which contained an important neutralisation site (B-cell epitope) as well as an H-2^k T-cell epitope, was co-synthesised with three different sequences containing H-2^d T-cell epitopes, one from ovalbumin and two from sperm whale myoglobin [Francis et al, 1987(b)]. The resultant three peptides all produced anti-FMDV peptide antibodies in both H-2^k (B10.BR) and H-2^d (B10.D2 and BALB/c) haplotype mice while a control peptide containing 141-160 plus additional FMDV residues was only immunogenic in the B10.BR mice (Fig. 2). Interestingly, however, it was shown that only two of the co-linear peptides elicited significant levels of virus neutralizing anti-peptide responses in the H-2^d mice. This suggested that the specificity of the antibodies produced may have been influenced by the T-cell epitope used or its location in relation to the B-cell epitope it was regulating. This model has also been used to study Th-cell epitopes from other viruses (influenza and hepatitis B) (Francis 1990) and mycobacteria (Cox et al, 1988). In addition preliminary findings suggested that the use of predicted T-cell epitopes from other structural proteins (VP3 and VP4) of FMDV could provide a means of overcoming non-responsiveness in cattle, the major target species for any vaccine. This observation is supported by the work of Collen et al, (1991) who used a defined cattle Th-cell epitope from residues 21-40 of FMDV VP1 in combination with a B-cell epitope to elicit neutralising antibodies, T-cell responses and protection in cattle.

Similar data have also been generated with peptides from another picoronavirus, human rhinovirus (HRV). In this study five predicted ten amino acid T-helper cell epitopes (T1 to T5) from HRV were used to improve

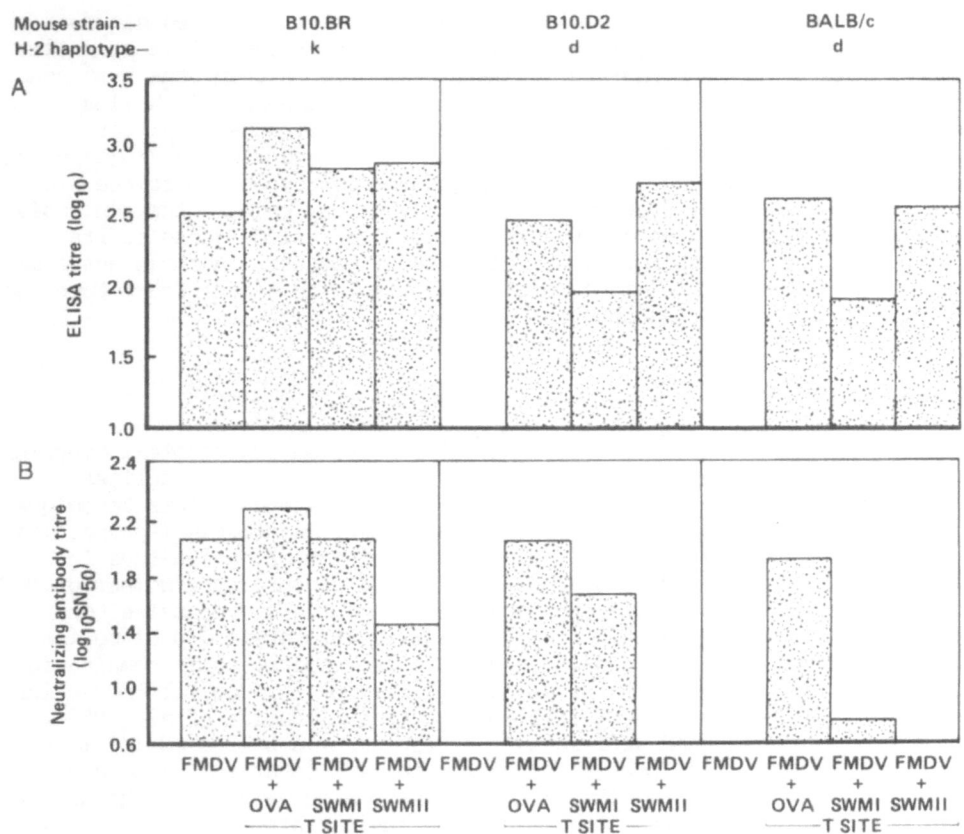

Fig. 2. Immune response of inbred $H-2^k$ and $H-2^d$ mice 42 days after inoculation of FMDV peptide with or without an added foreign Th-cell epitope. (a) Anti-peptide 141-160 response; (b) Neutralizing antibody response (from Francis et al, 1988, Copyright 1988, Cold Spring Harbor Laboratory Press).

picoronavirus, human rhinovirus (HRV). In this study five predicted ten amino acid T-helper cell epitopes (T1 to T5) from HRV were used to improve the performance of a non-immunogenic B-cell epitope peptide (B3) from the same virus (Francis et al, 1989). Studies in both outbred (MFI) and inbred (BALB/c) mice confirmed that B3 alone and B3 plus 10 extra amino acid residues towards the carboxyl terminus were indeed non-immunogenic (Fig. 3). These results indicate that the simple extension of the B3 peptide by ten amino acids did not overcome non-responsiveness to B3 in mice. However, peptides consisting of B3 plus T1, T2, T3, T4 and T5 all elicited anti-peptide responses to B3 in MFI mice (Fig. 3a). In contrast peptides B3/T1 and B3/T5 were non-immunogenic in BALB/c mice while peptides B3/T2, B3/T3 and B3/T4 were all immunogenic. Furthermore, this in vivo antibody response data in BALB/c mice was confirmed by in vitro T-cell proliferation assay results. Thus, the response to the B/T peptides was restricted in inbred mice. Similar B-cell epitope/T-cell epitope co-linear peptides have now been successfully used in a number of systems, including hepatitis B virus (HBV) (Milich et al, 1988), human immunodeficiency virus type 1 (HIV-1) (Palker et al, 1989), respiratory syncitial virus (RSV) (Levely et al, 1990), Rift valley fever virus (de La Cruz et al, 1991), measles virus

(Partidos et al, 1991a) and Murrey Valley encephalitis virus (Roehrig et al, 1992).

LOCATION OF T-CELL EPITOPE

The importance of location of the T-cell epitope in relation to the B-cell epitope it is regulating was studied using the FMDV system by moving the ovalbumin Th-cell peptide 323-339, which successfully elicited neutralizing antibodies in H-2^d mice when it was positioned at the carboxy terminus of the FMDV 141-160 peptide, to the amino terminus or leaving it at the carboxy terminus with five glycine residues between itself and the FMDV 141-160 peptide (Francis 1990). While the amino terminal location had no effect on immunogenicity in H-2^d mice, the five amino acid spacer at the carboxy terminus abolished helper activity for a neutralizing antibody response in H-2^d mice. Therefore it appears that the location of the Th-cell epitope on the co-linear peptide and the immunogen configuration are critical. Similar findings have been observed using FMDV/mycobacterial chimeric peptides (Cox et al, 1988), HIV-1 gp120/influenza or sperm whale myoglobin chimeric peptides (Golvano et al, 1990), RSV peptides (Levely et al, 1990) and measles virus peptides (Partidos et al, 1991b). It may be possible in the future to overcome problems of combining B-cell epitopes and T-cell epitopes in the same peptide by presenting the unlinked peptides to the immune system in the same delivery vehicle. Recent studies using peptides from HBV incorporated into liposomes have suggested that cross help can occur between unlinked peptides using this system (Gregoriadis et al, 1990, Gregoriadis, This volume).

BROADLY REACTIVE T-CELL EPITOPES

The final success or failure of the "free" peptide approach will depend to a large extent on the selection of appropriate Th-cell epitopes with broad activity in outbred populations. In this respect there are now a number of examples of "promiscuous" (broadly reactive) Th-cell sites which would appear to offer the possibility of effective immunization using

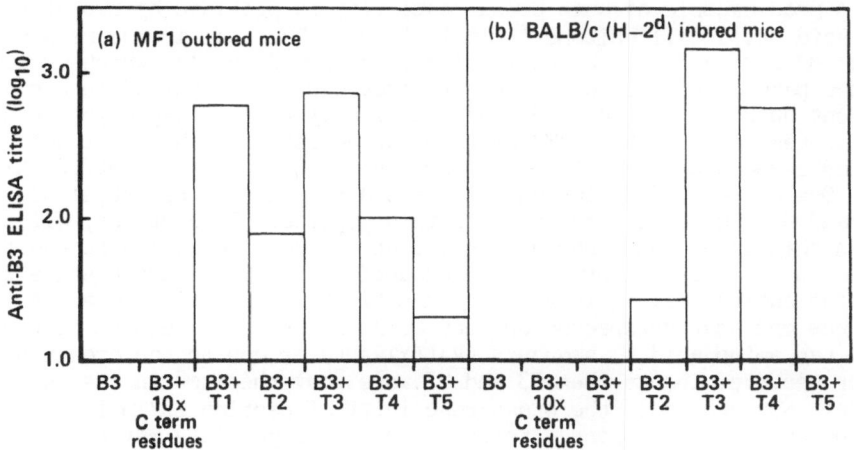

Fig. 3. Response of inbred and outbred mice to human rhinovirus peptide B3 with and without added T-cell epitopes (28 days after inoculation) (from Francis et al 1989, Copyright 1989, Cold Spring Harbor Laboratory Press).

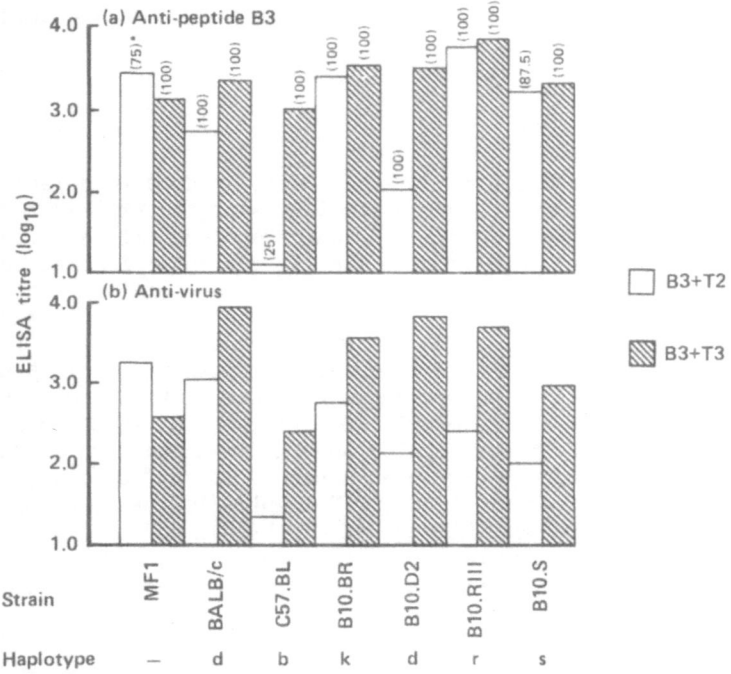

* Percent responder animals given in parenthesis

Fig. 4. Response of various mouse strains to human rhinovirus peptides with added T-cell epitopes (84 days
after inoculation) (from Francis et al 1989,
Copyright 1989, Cold Spring Harbor Laboratory Press).

synthetic peptides without the need for carrier coupling.

A good example of such broad reactivity is provided by two of the ten
amino acid HRV T-cell epitope (Francis 1990). Thus the immune response to
peptide B3 linked to T2, which is highly conserved, or T3, which varies in
sequence between viruses of different serotypes, was examined in a range of
different outbred, inbred, and congenic mice of various haplotypes (Fig.
4). The response to the B3/T2 peptide was variable, with 75% of outbred
MF1 mice responding and only a low response in 25% of the C57BL/10 (H-2^b)
mice. One of the B10.S mice was also a non-responder. Nevertheless, H-2^d,
H-2^k, and H-2^r mice all responded to this peptide. In contrast, all mice
used in this experiment, whether inbred or outbred, responded to the B3/T3
peptide in a fairly uniform manner, producing high-titer antipeptide B3 and
antivirus antibodies. This very broad activity across five mouse haplotypes was somewhat unexpected and may be an interesting feature of the
chosen ten amino acid T3 sequence (RALEYTRAHR) or due to the creation of
further neodeterminants when B3 and T3 were cosynthesized in the same
peptide. Nevertheless, the broad reactivity of peptide B3/T3 indicates
that the restricted responses to other uncoupled peptides may be
circumvented by the addition of appropriate sequences. T-cell peptides
with similar broad reactivity have now been described for a number of
antigens including malaria (Singaglia et al, 1988), tetanus toxin (Panina-
Bordignon et al, 1989, Ho et al, 1990), HIV (Hale et al, 1989), measles
(Partidos and Steward 1990), streptococcus mutans (Childerstone et al,
1991) and hepatitis B (Greenstein et al, 1992).

IDENTIFICATION OF SUITABLE Th-CELL EPITOPES

A number of Th-cell epitopes have been defined from a wide range of proteins and infectious agents (Berzofsky et al, 1987, Rothband and Taylor 1988). Such epitopes may be used to improve the immunogenicity of a peptide in order to raise antipeptide antibodies for experimental purposes despite the fact that they will come from "foreign" proteins. A good example would be the use of an H-2^d restricted Th-cell epitope to facilitate the production of monoclonal antibodies against an important peptide sequence in BALB/c mice. Th-cell epitopes from "foreign" proteins may also be used for vaccines in situations where protective levels of antibody need to be maintained by repeated inoculation of the population at risk; for example, FMDV prophylaxis requires regular revaccination at intervals of 6-12 months.

In situations where memory responses are required for immunity natural Th-cell epitopes from the infectious agent should be used. A number of algorithms now exist which may improve the chances of selecting appropriate peptide sequences with T-cell stimulating activity from the primary sequence of a protein (DeLisi and Berzofsky, 1985; Rothband and Taylor, 1988, Sette et al, 1989). However, in identifying these sites there are few shortcuts, and generally a detailed analysis, using in vitro T-cell stimulation techniques, of component proteins, protein fragments, and peptides, will be required to identify appropriate sequences.

In carrying out such an analysis it should be remembered that a number of isolated peptides may function as Th-cell epitopes in facilitating antibody production without actually being able to elicit or provoke recall memory to the native antigen i.e. they are not "natural" Th-cell epitopes. In order to identify such epitopes it is necessary to immunise with the native antigen and test resultant lymphocytes with a range of peptides, and then to use those identified peptides in vivo to see if they prime for a response to the native antigen. Not all identified peptides will have this ability.

CONCLUSIONS

In conclusion, during the development of synthetic vaccines we must recognise the essential role played by Th-cell epitopes and the difficulties in selecting and incorporating suitable peptide sequences. Nevertheless, there are clearly a number of advantages associated with replacing large ill-defined protein carriers which make the concept of a fully synthetic immunogen worth pursuing. By selecting the correct Th-cell epitopes it should be possible in the future to both overcome genetically restricted non-responsiveness and to elicit recall memory to the native antigen. There are also clear indications that the helper effect may be directed to important B-cell epitopes and that the choice of appropriate Th-cell epitopes may influence class switching. The overall result would be more defined, controllable and effective peptide vaccines with enhanced immunogenicity.

REFERENCES

Babbit, B.P., Allen, P.M., Matsueda, G., Haber, E. and Unanue, E.R., 1985, Binding of immunogenic peptides to Ia histocompatibility molecules, Nature, 317:359.
Berzofoky, J.A., Cease, K.B., Cornette, J.L., Spouge, J.L., Margolit, H., Berkower, I.J., Good, M.F., Miller, L.H. and Delisi, 1987, Protein antigenic structures recognised by T cells: Potential applications

to vaccine design, Immunol.Rev., 98:9.

Borras-Cuesta, F., Petit-Camurdan, A. and Fedon, Y., 1987, Engineering of immunogenic peptides by co-linear synthesis of determinants recognised by B and T cells, Eur.J.Immunol., 17:1213.

Borras-Cuesta, F., Petit-Camurdan, A. and Fedon, Y., 1988, Enhancement of peptide immunogenicity by linear polymerization, Eur.J.Immunol., 18:199.

Chedid, L., Audibert, F., Jolivet, M., Corelli, C., Uzan, M., Gras-Masse, H. and Tutar, A., 1984, Epitope specific immunity elicited by totally synthetic monovalent or polyvalent vaccines, Ann Sclavo, 2:77.

Childerstone, A., Altmann, D., Haron, J.A., Wilkinson, D., Trowsdale, J. and Lehner, T., 1991, An analysis of synthetic peptide restriction by HLA-DR alleles in T-cells from human subjects, naturally sensitised by streptococcus mutans, J.Immunol., 146:1463.

Collen, T., Di Marchi, R. and Doel, T.R., 1991, A T-cell epitope in VP1 of foot-and-mouth disease virus is immunodominant for vaccinated cattle, J.Immunol., 146:749.

Cox, J.H., Ivanyi, J., Young, D.B., Lamb., J.R., Syred, A.D. and Francis, M.J., 1988, Orientation of epitopes influences the immunogenicity of synthetic peptide dimers, Eur.J.Immunol., 18:2015.

de la Cruz, V.F., Benzon, L.A., Woods, R.M. and Collett, M.S., 1991, Synthetic peptides as subunit immunogens: Parameters for combinating T-helper sites with B-cell antigenic determinants, in: "Vaccines 91 Modern approaches to new vaccines including prevention of AIDS" R.M. Chanok, H.S. Ginsberg, F. Brown and R.A. Lerner, eds., Cold Spring Harbor Laboratory, New York, 345.

Delisi, C., Berzofsky, J.A., 1985, T-cell antigenic sites tend to be amphipathic structures, Proc.Natl.Acad.Sci.USA, 82:7048.

Francis, M.J., 1990, Peptide vaccines for viral diseases, Sci Progress, 74:115.

Francis, M.J. and Clarke, B.E., 1989, Peptide vaccines based on enhanced immunogenicity of peptide epitopes presented with T-cell determinants or hepatitis B core protein, Meth.Enzymol., 178:659.

Francis, M.J., Fry, C.M., Rowlands, D.J., Brown, F., Bittle, J.L., Houghten, R.A. and Lerner, R.A., 1985, Immunological priming with synthetic peptides of foot-and-mouth disease virus, J.Gen.Virol., 66:2347.

Francis, M.J., Fry, C.M., Rowlands, D.J., Bittle, J.L., Houghten, R.A. Lerner, R.A. and Brown, F., 1987a, Immune response to uncoupled peptides of foot-and-mouth disease virus, Immunology, 61:1.

Francis, M.J., Hastings, G.Z., Syred, A.D., McGinn, B., Brown, F. and Rowlands, D.J., 1987b, Non responsiveness to a foot-and-mouth disease virus synthetic peptide overcome by addition of foreign helper T-cell determinants, Nature (Lond)., 330:168.

Francis, M.J., Hastings, G.Z., Syred, A.D., McGinn, B., Brown, F. and Rowlands, D.J., 1988, Peptides with added T-cell epitopes can overcome genetic restriction of the immune response, in: "Vaccines 88: New Chemical and genetic approaches to vaccination", H. Ginsberg, F. Brown, R.A. Lerner and R.M. Chanok, eds., Cold Spring Harbor Lab, N.Y. 1.

Francis, M.J., Hastings, G.Z., Campbell, R.O., Rowlands, D.J., Brown, F. and Peat, N., 1989, T cell help for B cell antibody production to rhinovirus peptides, in: "Vaccines 89", Cold Spring Harbor Laboratory, New York, 437.

Golvano, J., Lasarte, J.J., Sarobe, P., Gullon, A., Prieto, J. and Borras-Cuesta, F., 1990, Polarity of immunogens: implications for vaccine design, Eur.J.Immunol., 20:2363.

Good, M.F., Maloy, W.F., Lunde, M.N., Margalit, H., Cornette, J.L., Smith, G.L., Moss, B., Miller, L.H. and Berzofsky, J.A., 1987, Construction of synthetic immunogen: use of new T-helper epitope on malaria circumsporozoite protein, Science, 235:1059.

Greenstein, J.L., Schad, V.C., Goodwin, W.H., Brauer, A.B., Bollinger, B.K., Chin, R.D. and Mei-Chang Kuo, 1992, A universal T-cell epitope – containing peptide from hepatitis B surface antigen can enhance antibody specific for HIV gp120, J.Immunol., 148:3970.

Gregoriadis, G., Wang, Z. and Francis, M., 1990, Liposomes as imunological adjuvants for proteins and peptides, in: Proceedings NATO ASI "Vaccines Recent Trends and Progress" meeting, Cape Sounion Beach, Greece, 25.

Hale, P.M., Cease, K.B., Houghten, R.A., Ouyang, C., Putney, S., Javaherian, K., Margalit, H., Cornette, J.L., Spouge, J.L., Delisi, C. and Berzofsky, J.A., 1989, T-cell multideterminant regions in human immunodeficiency virus envelope: towards overcoming the problem of major histocompatibility complex restriction, Int.Immun., 1:409.

Ho, P.C., Mutch, D.A., Winkel, K.D., Saul, A.J., Jones, G.L., Doran, J.J. and Rzepczyk, C.M., 1990, Identification of two promiscuous T-cell epitopes from tetanus toxin, Eur.J.Immunol., 20:477.

Jolivet, M., Lise, L., Gras-Masse, H., Tarter, A., Audibert, F. and Chedid, L., 1990, Polyvalent synthetic vaccines: relationship between T epitopes and immunogenicity, Vaccine, 8:35.

Leclerc, C., Przewlocki, G., Schutze, M. and Chedid, L., 1987, A synthetic vaccine constructed by copolymerization of B and T cell determinants, Eur.J.Immunol., 17:269.

Levely, M.E., Mitchell, M.A. and Nicholas, J.A., 1990, Synthetic immunogens constructed from T-cell and B-cell stimulating peptides (T.B. Chimeras): Preferential stimulation of unique T- and B-cell specificities is influenced by immunogen configuration, Cell Immunol., 125:65.

Milich, D.R., Hughes, J.L., McLachlan, A., Thornton, G.B. and Moriarty, A., 1988, Hepatitis B synthetic immunogen comprised of nucleocapsid T-cell sites and an envelope B-cell epitope, Proc.Natl.Acad.Sci,USA, 85:1610.

Mitchison, N.A., 1971, The carrier effect in the secondary response to hapten-protein conjugates. II Cellular cooperation. Eur.J.Immunol., 1:18.

Palker, T.J., Matthews, T.J., Langlois, A., Tanner, M.E., Martin, M.E., Scearce, R.M., Kim, J.E., Berzofsky, J.A., Bolognesi, D.P. and Haynes, B.F., 1989, Polyvalent human immunodeficiency virus synthetic immunogen composed of envelope gp120 T helper cell sites and B cell neutralisation epitopes, J.Immunol., 142:3612.

Panina-Bordignon, P., Tan, A., Termijtelen, A., Demotz, S., Corradin, G. and Lanzavecchia, A., 1989, Universally immunogenic T-cell epitopes: promiscuous binding to human MHC class II and promiscuous recognition by T-cells, Eur.J.Immunol., 19:2237.

Partidos, C.D. and Steward, M.W., 1990, Prediction and identification of a T-cell epitope in the fusion protein of measles virus immunodominant in mice and humans, J.Gen.Virol., 71:2099.

Partidos, C.D., Stanley, C.M. and Steward, M.W., 1991a, Immune responses in mice following immunisation with chimeric synthetic peptides representing B and T cell epitopes of measles virus proteins, J.Gen,Virol., 72:1293.

Partidos, C. Stanley, C. and Steward, M., 1992, The effect of orientation of epitopes on the immunogenicity of chimeric synthetic peptides representing measles virus protein sequences, Mol.Immunol., 29:651.

Rosenthal, A.S., 1978, Determinant selection and macrophage function in genetic control of the immune response, Immunol.Rev., 40:136.

Rothband, J.B. and Taylor, W.B., 1988, A sequence pattern common to T cell epitopes, EMBO.J., 7:93.

Sette, A., Buus, S., Appella, E., Smith, J.A., Chesnut, R., Miles, C., Colon, S.M. and Grey, H.M., 1989, Prediction of major histocompatibility complex binding regions of protein antigens by sequence pattern analysis, <u>Proc,Natl.Acad,Sci.USA</u>, 86:3296.

Sinigaglia, F., Guttinger, M, Kilngus, J., Doran, D.M., Matile, H., Etlinger, H., Trzeciak, A., Gillessen, D. and Pink, R.J.L., 1988, A malaria T cell epitope recognised in association with most mouse and human MHC class II molecules, <u>Nature,</u> 336:778.

CARRIERS FOR PEPTIDES: THEORIES AND TECHNOLOGY

M.J. Francis

Dept of Virology and Process Development
Pitman-Moore Ltd, Breakspear Road South, Harefield
Uxbridge, Middlesex UB9 6LS, UK

INTRODUCTION

Many studies in recent years have been directed towards producing fully synthetic peptide immunogens (Francis and Clarke 1989; Francis 1993, this volume) which contain the necessary B- and T-cell epitopes for immunogenicity in vivo. However, there is little doubt that carrier proteins still offer the most straightforward and convenient way of presenting poorly immunogenic peptides to the immune system. This may be achieved either by chemical linkage of synthetic peptides to "foreign" protein molecules or by direct linkage to proteins for expression as recombinant fusion proteins. However it should be recognised that the method used to link the peptide to the carrier can greatly influence both the quantitative and qualitative nature of the immune response observed.

ROLE OF THE CARRIER

The classical role of a carrier protein in immunology is to provide T-cell help for B-cell antibody production to a poor/non-immunogenic antigen or hapten (Ovary and Benacerraf 1967; Mitchison 1971). For peptide delivery they are also generally used to recruit helper T-cells for poorly immunogenic B-cell epitopes. However, in some cases the peptides may be immunogenic in their own right, due to the presence of B- and T-cell epitopes within the same sequence, and the carrier molecule is simply acting as a polymeric delivery system (Francis et al, 1985). In addition, peptide carriers may act by increasing the molecular mass of the peptide and thus improve the uptake by antigen presenting cells. Small peptides also generally have a very rapid half-life within the body which may be prolonged by linkage to a carrier protein.

So peptide carriers may actually function in a number of different ways but their overall role is to increase the immunogenicity of the peptide and thus assist in producing an antibody response against the peptide in vivo. Two forms of carriers will be discussed in this article, first natural or synthetic molecules linked to natural or synthetic peptides and secondly recombinant proteins linked to recombinant peptides as fusion proteins.

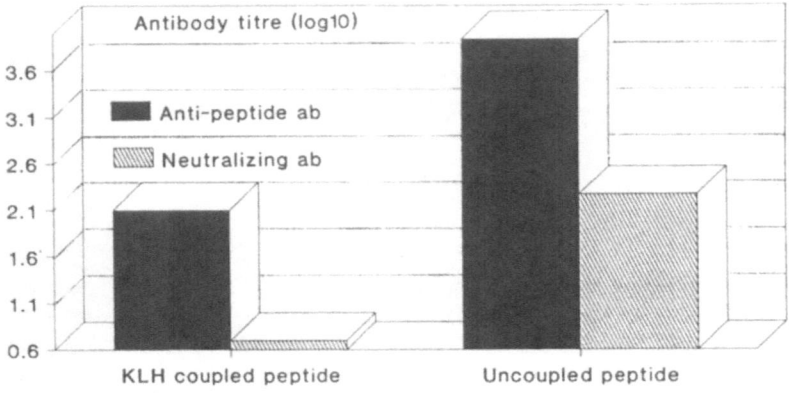

Fig. 1. Effect of carrier linkage on antibody production to
an FMDV SAT2 peptide (Francis et al, 1990a)

CARRIERS FOR SYNTHETIC OR ISOLATED NATURAL PEPTIDES

General considerations – when selecting a suitable carrier there are
a number of important points which one should consider. These are as
follows:

1. Purpose of coupling – What is the reason for coupling? Is it simply to
elicit high titre antipeptide antibodies or to immunise an animal against
infection? Does the peptide require a carrier? This last point is well
illustrated by a study carried out with an FMD serotype SAT2 peptide
(Francis et al, 1990a). In this study coupling of the peptide to a carrier
via a carboxy-terminal cysteine residue reducedimmunogenicity and abolished
its ability to elicit neutralising antibodies against the virus (Figure 1).

2. Nature of the peptide – Study the primary sequence of the peptide. Is
it hydrophilic or hydrophobic? Are certain residues likely to be important
for its antigenicity? Will these residues be affected or masked by carrier
coupling? Are cysteine residues present that might form disulphide bridges?

3. Choice of carrier – Will you use a natural or artificial carrier? Should
the carrier come from the same organism as the peptide? Should it be a
common protein or an unusual protein? Is there a risk of eliciting auto
immunity or hypersensitivity? Will carrier antibodies interfere with
subsequent in vitro analysis? Is there a possibility that pre-priming
against the carrier has occured?

4. Method of conjugation – Some of the most commonly used methods of
conjugation will have a significant affect on the antigenicity of many
peptides (see below). It is therefore important to consider what effect
the conjugation procedure will have and how you wish to present the peptide
to the immune system. The peptide to carrier ratio can also be very
important.

5. Method of immunisation – A number of factors can affect the qualitative
and quantitative response following peptide-carrier immunisation. These
include route, dose, choice of adjuvant, frequency, interval between
primary and subsequent booster inoculations, age and sex.

6. Species – The choice of species can have a marked effect on the nature

of the antipeptide antibodies produced. The response in laboratory animals may be very different to the response in target species for a vaccine. Differences may also be observed between different species of laboratory animal and even different strains within a species.

Commonly used carriers - Traditional carriers include proteins such as keyhole limpet haemocyanin (KLH) and sperm whale myoglobin (SWM). These are generally chosen because they are likely to be highly foreign to the species being immunised and also unlikely to elicit cross-reactive or interfering antibodies. They are also well established model immunogens and in the case of SWM they have been extensively characterised. Interestingly KLH has also been used in a number of studies in man without any obvious side effects (Devey et al, 1990; Ward et al, 1990). Other common carriers include albumins such as bovine serum albumin and ovalbumin. These appear to work well and are freely available. However, it should be noted that they are also commonly used as blocking proteins in immunoassays. Such assays would therefore have to be modified for screening antipeptide antisera generated with these carriers. Furthermore, since antipeptide antibodies are commonly used for screening mixtures of proteins, for example by Western blotting of PAGE gels, the anticarrier antibodies may cross react. This will be particularly important if bovine serum is used to culture an organism and BSA is used as the carrier. Further possible carriers which have been used successfully in the past include bacterial toxoids such as tetanus toxoid and diptheria toxoid and other bacterial proteins used in vaccines such as PPD from BCG (Davis et al, 1990). One of the reasons for this is that a pre-primed human population will exist which may produce an enhanced response to the peptide or hapten linked to the same carrier (Katz et al, 1970; Milich et al, 1987a). However, it should be noted that other reports suggest that no pre-priming occurs (Gupta et al, 1986) or even active suppression may be the result (Hertzenberg et al, 1980; Schutze et al, 1985). Clearly this needs to be studied on a case by case basis.

Coupling methods - A wide range of methods now exist for coupling peptides to protein carriers some of the more common ones are given in Table 1. Perhaps the most popular method, largely due to its simplicity, is to use gluteraldehyde. However this is relatively uncontrolled and may under certain circumstances involve primary reactions with alpha-amino groups, epsilon-amino groups on lysine residues and sulfhydryl groups on cysteine residues, as well as secondary reactions with phenolic hydroxyl groups on tyrosine residues and imidazole groups on histidine residues. Therefore, if Lys, Cys, Tyr or His residues are involved in antigenic sites on the peptide these may be significantly altered by this coupling method. The use of heterobifunctional cross-linkers can largely overcome this problem by facilitating specific linkages. For example the cross-linker MBS has an amino-reactive NHS-ester as one functional group and a sulfhydryl reactive group as the other. Amino groups on the carrier are acylated with the NHS-ester via the hydroxysuccinimide group, and then a peptide is introduced that possesses a free sulfhydryl group which can react with the malemide group of the coupling reagent. This may require synthesis of a specific peptide with a non-natural cysteine residue added to its carboxy- or amino-terminus, or indeed anywhere else within the sequence. The effect of peptide orientation on the immune response is well illustrated by the work of Dyrberg and Oldstone (1987). In this study peptide linked via its amino-terminus elicited antibodies that would only recognise amino-linked peptide _in vitro_ and vice versa for the carboxy-linked peptide (Figure 2). Thus linkage can have a marked influence on the specificity of the antipeptide antibody response _in vivo_.

Guidelines - While it is difficult to provide any firm guidelines for peptide delivery, attempts have been made in the past to compare published

Table 1. Commonly used reagents for coupling

- Glutaraldehyde (amino gps, sulfhydryl gps, phenolic hydroxyl gps, imidazole gps).

- Carbodiimides (amino gps, carboxyl gps, phenolic hydroxyl gps, sulfhydryl gps).

- Bis-imido esters (amino gps).

- Heterobifunctional cross-linkers e.g. SPDP, SMCC, MBS

  (amino groups, sulfhydryl gps).

- Homobifunctional NHS esters (amino groups).

Brackets indicate functional groups involved in the coupling reaction.

data in order to reach some sort of consensus (Palfreyman et al, 1984). The results of this study suggest that peptides should be 10-15 residues long, although it is the experience of this author that longer peptides of 20-30 residues often make the best immunogens for vaccine purposes. They can be positively or negatively charged but should be hydrophilic. In this limited comparison BSA was better than KLH, but in practice it may be worth trying a range of carriers for any given antigen and BSA may not be the best choice (see above). MBS is better than gluteraldehyde or carbodiimide. It is also likely to be a more controlled process (see above). Finally C or N-terminal sequences of protein may be the best first choice for raising anti-protein antibodies.

Potential problems - While chemical coupling offers a quick and convenient method for presenting peptides in a more immunogenic form there are a number or potential problems that should be recognised. It is often

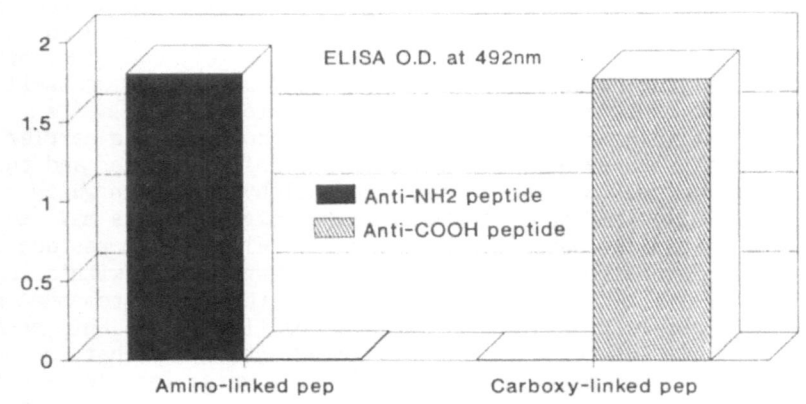

Fig. 2. The influence of orientation of carrier linkage on the specificity of the anti-peptide antibodies produced (Dryberg and Oldstone, 1987)

poorly defined and the reaction is difficult to control. As a result the reproducibility from batch to batch is likely to be poor. There is also a strong risk of actually masking important antigenic sites and of modifying the peptides during the coupling process. Problems of carrier induced suppression and hypersensitivity to the carrier may also be encountered.

RECOMBINANT PEPTIDE CARRIERS

General considerations - The recombinant fusion protein approach does offer a number of distinct advantages over chemical coupling. It is both reproducible and controlled and should therefore offer a more straightforward production process. Since peptides may be placed in chosen sites on the carrier and in different forms there is a good opportunity for selecting for improved immunogenicity and also structured presentation of an epitope. Finally this technology should be well suited to producing multiple constructs with a range of peptides on the same carrier.

Bacterial fusion proteins - The use of peptide sequences fused to bacterial proteins as immunogens has the potential advantage of a completely uniform and defined structure compared with the uncharacterised and variable nature of peptide/carrier conjugates prepared by chemical cross-linking. This approach has been used independently by two groups (Broekhuijsen et al, 1986a; Winther et al, 1986) to express FMDV peptides fused to the N-terminus of β-galactosidase in E.coli cells. β-galactosidase was chosen because it had been shown that antibodies can be produced to foreign proteins located at the N-terminus and it has been shown to contain a number of helper T-cell sites (Krzych et al, 1982; Manca et al, 1985). Preliminary experiments with β-galactosidase (Broekhuijsen et al, 1986b) and TrpLE (Kleid et al, 1985) fusion proteins had indicated that multiple copies of the inserted peptide sequence may be beneficial. Therefore, in collaboration with Dr B E Enger-Valk's group from Medical Biological Laboratory TNO, The Netherlands, we have examined the immunogenicity of one, two or four copies of FMDV VP1 peptide 137 to 162 fused to the N-terminus of β-galactosidase both in laboratory animals and target species. The protein containing one copy of the viral determinant elicited only low levels of neutralising antibody whereas protective levels were elicited by proteins containing two or four copies of the determinant (Broekhuijsen et al, 1987). Furthermore, single inoculations of the two copy and four copy proteins containing as little as 2 ug or 0.8 ug of peptide respectively were sufficient to protect all the animals against challenge infection. The equivalent of 40 ug of peptide in the four copy protein also protected pigs against challenge infection after one inoculation.

Particulate fusion proteins - More recently a further development of the fusion protein concept for multiple peptide presentation has led to the production of particulate structures with epitopes repeated over their entire surface. To date work has principally concentrated on three proteins, hepatitis B surface antigen (Delpeyroux et al, 1986), hepatitis B core antigen (HBcAg) (Newton et al, 1987; Clarke et al, 1987) and yeast Ty protein (Adams et al, 1987), which spontaneously self assemble into 22, 27 and 60 particles respectively.

The use of the core antigen from hepatitis B virus (HBcAg) to present "foreign" peptide epitopes, first described at the Cold Spring Harbor Vaccines Meeting by Newton et al, in 1986 (Newton et al 1987), offers several potential advantages. The protein subunit is a 21 KDa polypeptide which when overproduced spontaneously assembles into characteristic 27 nm particles. It can be expressed in a wide range of systems, including bacterial cell, yeast cell and mammalian cell or via vaccinia virus and

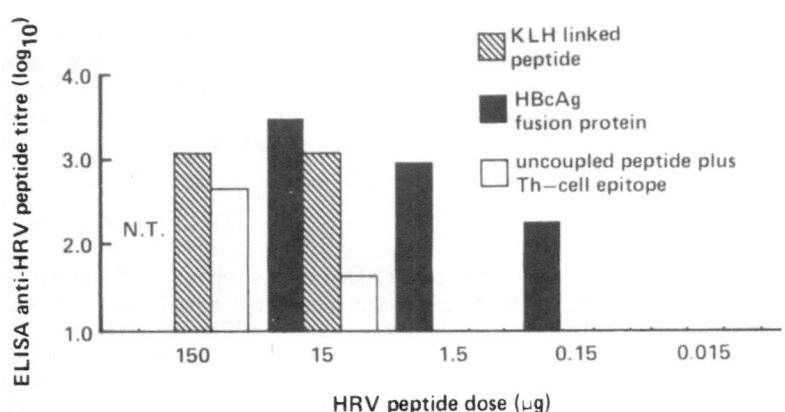

Fig. 3. Comparative immunogenicity of HRV peptide – HBcAg fusion
protein with KLH – coupled and uncoupled HRV peptide in
guinea pigs at 56 days after a single intramuscular
inoculation (from Francis et al, 1990c, Copyright 1990,
Cold Spring Harbor Laboratory Press)

baculovirus. The core particle is known to be highly immunogenic, probably
due to a combination of its polymeric nature, the presence of a number of
well defined T-cell epitopes and its ability to function as a T-cell
independent antigen. Furthermore, N-terminal (Clarke et al, 1987) and C-
terminal fusions (Stahl & Murray 1989) plus insertions into surface loop
structures (Schodel et al, 1990; Brown et al, 1991) have all been shown to
produce chimeric core fusion particles (CFP). In early studies using CFP
in which a peptide from foot-and-mouth disease virus (FMDV) had been fused
to be N-terminus and expressed via vaccinia virus in mammalian cells it was
shown that the peptide immunogenicity could approach that of inactivated
FMDV particles (Clark et al, 1987). In fact as little as 0.2 ug of FMDV
VP1 142-160 peptide corresponding to 10% of the fusion protein, presented
on the surface of a CFP gave full protection to guinea pigs at 56 days
after a single inoculation. However, the poor yields from this particular
vaccinia virus expression system provided insufficient material for further
immunological investigations. In order to overcome this antigen supply
problem, CFP have now been produced in bacterial (Francis and Clarke, 1989)
and yeast (Beasley et al, 1990) expression systems which have been shown to
produce in excess of 200 mg of antigen per litre of culture supernatant.
In a more detailed immunological study, using a human rhinovirus (HRV)
peptide, N-terminal CFP were shown to be one hundred-fold more immunogenic
than a free disulphide dimer synthetic peptide containing B- and T-cell
determinants and ten-fold more immunogenic than keyhole limpet haemocyanin
carrier linked peptide (Francis et al, 1990b) (Fig. 3). This activity
appears to be dependent both on provision of T-cell help from the core
(Milich et al, 1987b) and on particle formation (Clarke et al, 1990). The
CFP could be readily administered without adjuvant or with adjuvant
acceptable for human and/or veterinary use. They could also be given by
nasal or oral routes and could act in a T-independent manner. It is
interesting to note that core antigen has now been shown to be immunogenic
in a wide range of species including mice, guinea pigs, rabbits, cats,
swine, cattle, and monkeys (Francis et al, 1990c).

The objectives of developing a system for internal fusions were
firstly to restrict the inserted peptide in a more defined conformation and
secondly to replace an important immunogenic site on the core with a
"foreign" sequence in order to direct and enhance the immune response

Table 2. Peptide carrier technologies

- Synthetic or natural peptides -

  Chemical coupling

  Often poorly defined

  Useful primary screen

  Production of peptide antisera

 Recombinant peptides -

  Fusion proteins

  Well defined

  In depth analysis

  Peptide vaccines

---

against that sequence. In initial experiments an HRV peptide known to be capable of eliciting anti-virus neutralising antibodies was inserted into the el-loop region of core and its immunogenicity in guinea pigs was compared to that of N-terminal CFP (Brown et al, 1991). The el-insert CFP preparation was not only ten-fold more immunogenic, but it also appeared that a greater proportion of the anti-HRV peptide antibodies reacted with virus. This observation was supported by improved neutralising activity present in the antisera. Therefore, by modifying the presentation system it was possible to both enhance immunogenicity and direct the immune response against a more structured form of the inserted peptide.

In conclusion, CFP have now been produced to a number of viral antigens including FMDV, HRV, hepatitis B surface and pre-S antigens, hepatitis A, dengue virus, feline leukaemia virus, simian immunodeficiency virus, human immunodeficiency virus, respiratory syncytial virus, poliovirus and bovine leukaemia virus. They offer a highly immunogenic and versatile method for delivering peptides as vaccines. Their use is gradually becoming more widespread and their value as molecular carriers for peptide antigens is generally acknowledged.

CONCLUSIONS

Peptide carriers can broadly be classified in two main areas (Table 2). The first are those for synthetic or natural peptides which generally involve some form of chemical coupling. The principal drawback of this procedure is that it is often poorly defined. Nevertheles it is undoubtably one of the most straightforward means of carrying out an in vivo primary screen of peptides and it is still the most commonly used method for preparing anti-peptide antisera. The second method is for delivery of recombinant peptides in the form of genetically expressed fusion proteins. Such proteins are well defined and offer the means of carrying out an in depth analysis of the carrier-peptide relationship. It seems likely this type of delivery system will be the method of choice for peptide vaccine production in the future until chemical coupling processes can be further refined. In both cases it should be recognised that the carrier and type of linkage can have a profound effect on the qualitative and quantitative nature of the anti-peptide antibodies produced. The essential message when choosing a method for peptide immunisation should therefore be if at first you don't succeed try try again!

REFERENCES

Adams, S.E., Dawson, K.M., Gull, Kingsman, S. and Kingsman, A.J., 1987, The expression of hybrid HIV: Ty virus like particles in yeast, Nature (Lond), 329:68.

Beesley, K.M., Francis, M.J., Clarke, B.E., Beesley, J.E., Dopping-Hepenstal, P.J.C., Clare, J.J., Brown, F. and Romanos, M.A., 1990, Expression in yeast of amino-terminal peptide fusions to hepatitis B core antigen and their immunological properties, Biotechnol., 8:644.

Broekhuijsen, M.P., Blom, T., Kottenhagen, M., Pouwels, P.H., Meloen, R.H., Barteling, S.J. and Engel-Valk, B.E., 1986a, Synthesis of fusion proteins containing antigenic determinants of foot-and-mouth disease virus, Vaccine, 4:119.

Broekhuijsen, M.P., Blom, T., Van Rijn, J., Pouwels, P.H., Klasen, E.A., Fasbender, M.J. and Enger-Valk, B.E., 1986b, Synthesis of fusion proteins with multiple copies of antigenic determinant of foot-and-mouth disease virus, Gene, 49:189.

Broekhuijsen, M.P., Van Rijn, J.M.M., Blom, A.J.M., Pouwels, P.H., Enger-Valk, B.E., Brown, F. and Francis, M.J., 1987, Fusion proteins with multiple copies of the major antigenic determinant of foot-and-mouth disease virus protect both the natural host and laboratory animals, J Gen Virol., 68:3137.

Brown, A.L., Francis, M.J., Hastings, G.Z., Parry, P.V., Rowlands, D.J. and Clarke, B.E., 1991, Foreign epitopes in immunodominant regions of hepatitis B core particles are highly immunogenic and conformationally restricted, Vaccine, 9:595.

Clarke, B.E., Newton, S.E., Carroll, A.R., Francis, M.J., Appleyard, G., Syred, A.D., Highfield, P.E., Rowlands, D.J. and Brown, F., 1987, Improved immunogenicity of a peptide epitope after fusion to hepatitis B core protein, Nature (Lond), 330:381.

Clarke, B.E., Brown, A.L., Grace, K.K., Hastings, G.Z., Brown, F., Rowlands, D.J. and Francis, M.J., 1990, Presentation and immunogenicity of viral epitopes on the surface of hybrid hepatitis B virus core particles produced in bacteria, J Gen Virol., 71:1109.

Davies, D., Chardhri, B., Stephens, M.D., Carne, C.A., Willers, C. and Lachmann, P.J., 1990, The immunodominance of epitopes within the transmembrane protein (gp41) of human immunodeficiency virus type 1 may be determined by the host's previous exposure to similar epitopes on unrelated antigens, J Gen Virol., 71:1975.

Delpeyroux, F., Chenciner, N., Lim, A., Malpiece, Y., Blondel, B., Grainic, R., van der Wef, S. and Streeck, R.E., 1986, A polio neutralisation epitope expressed on hybrid hepatitis B surface antigen particles, Science, 233:472.

Devey, M.E., Bleasdale-Barr, K.M., Bird, P. and Amlot, P.L., 1990, Antibodies of different human IgG subclasses show distinct patterns of affinity maturation after immunisation with keyhole limpet haemocyanin, Immunology, 70:168.

Dyrberg, T. and Oldstone, M.B., 1986, Peptides as antigens, Importance of Orientation, J Exp Med., 164:1344.

Francis, M.J., 1993, Synthetic peptides and the role of T-helper cell determinants (This volume).

Francis, M.J. and Clarke, B.E., 1989, Peptide vaccines based on enhanced immunogenicity of peptide epitopes presented with T-cell determinants or hepatitis B core protein, Meth Enzymol., 178:659.

Francis, M.J., Fry, C.M., Rowlands, D.J., Brown, F., Bittle, J.L., Houghten, R.A. and Lerner, R.A., 1985, Immunological priming with

synthetic peptides of foot-and-mouth disease virus, J Gen Virol.,
66:2347.

Francis, M.J., Hastings, G.Z., Clarke, B.E., Brown, A.L., Beddell, C.R.,
Rowlands, D.J. and Brown, G., 1990a, Neutralising antibodies to all
seven serotypes of foot-and-mouth disease virus elicited by
synthetic peptides, Immunol., 69:171.

Francis, M.J., Hastings, G.Z., Brown, A.L., Grace, K.G., Rowlands, D.J.,
Brown, F. and Clarke, B.E., 1990b, Immunological properties of
hepatitis B core antigen fusion proteins, Proc Natl Acad Sci USA,
87:2545.

Francis, M.J., Clarke, B.E., Hastings, G.Z., Brown, A.L., Rowlands, D.J.
and Brown, F., 1990c, Immune response to peptide/hepatitis B core
antigen fusion proteins, in: "Vaccines 90", F. Brown, R.M. Chanock,
H.S. Ginsberg and R.A. Lerner, eds, Cold Spring Harbor Lab, N.Y.

Gupta, S.G., Hengartner, H. and Zinkernagel, R.M., 1986, Primary antibody
responses to a well-defined and unique hapten are not enhanced by
pre-immunisation with carrier: Analysis in a viral model, Proc Natl
Acad Sci USA, 83:2604.

Herzenberg, L.A., Tokuhisa, T. and Herzenberg, L.A., 1980, Carrier-priming
leads to hapten-specific suppression, Nature, 285:664.

Katz, D.H., Paul, W.E., Goidl, E.A. and Benacerraf, B., 1970, Carrier
function in anti-hapten immune responses. I. Enhancement of
primary and secondary anti-hapten antibody responses by carrier
pre-immunisation, J Exp Med., 132:261.

Kleid, D.G., Dowbenko, D.J., Bock, L.A., Hotlin, M.E., Jakson, M.L.,
Patzer, E.J., Shine, S.J., Weddell, G.N., Yansura, D.G., Morgan,
D.O., McKercher, P.D. and Moore, D.M., 1985, Production of
recombinant vaccines from micro-organisms: Vaccine for foot-and-
mouth disease, in: "Microbiology", L. Leive, ed., Washington, D.C.,
American Society for Microbiology.

Kryzch, U.A., Fowler, A.V., Miller, A. and Sercarz, E.E., 1982, Repertoires
of T cells directed against a large protein antigen, -
galactosidase, J Immunolo., 128:1529.

Manca, F., Kunki, A., Fenoglio, D., Fowler, A., Sercarz, E. and Celada, F.,
1985, Constraints in T-B cooperation related to epitope topology on
E.coli -galactosidase, Eur J Immunol., 15:345.

Milich, D.R., McLachlan, A., Thornton, G.B. and Hughes, L.L., 1987a,
Antibody production to the nucleocapsid and envelope of the
hepatitis B virus primed by a single synthetic T-cell site, Nature,
329:547.

Milich, D.R., McLachlan, A., Moriarty, A. and Thornton, G.B., 1987, Immune
response to hepatitis B virus core antigen (HBcAg): Localisation of
T-cell recognition sites within HBcAg/HBeAg, J Immunol., 139:1223.

Mitchison, N.A., 1971, The carrier effect in the secondary response to
hapten-protein conjugates. II Cellular cooperation, Eur J
Immunol., 1:18.

Newton, S.E., Clarke, B.E., Appleyard, G., Francis, M.J., Carroll, A.R.,
Rowlands, D.J., Skehel, J. and Brown, F., 1987, Novel antigen
presentation via vaccinia in: "Vaccines 87": Modern Approaches to
New Vaccines", R.M. Chanock, R.A. Lerner, F. Brown and H. Ginsberg,
eds., Cold Spring Harbor Laboratory, New York.

Ovary, Z. and Benacerraf, B., 1963, Immunological specificity of the
secondary response with dinitrophenylated proteins, Proc Soc Exp
Biol Med., 114:72.

Palfreyman, J.W., Aitcheson, T.C. and Taylor, P., 1984, Guidelines for the
production of polypeptide specific antisera using small synthetic
oligopeptides as immunogens, J Immunol Meth., 75:383.

Schodel, F., Weiner, T., Will, H. and Millich, D., 1990, Recombinant HBV
core particles carrying immunodominant B-cell epitopes of the HBV
pre-S2 region, in: "Vaccines 90: Modern Approaches to New Vaccines
Including Prevention of AIDS", F. Brown, R.M. Chanock, H.S.

Ginsberg and R.A. Lerner, eds., Cold Spring Harbor Laboratory, New York.

Schutze, M.P., Leclerc, C., Jolivet, M., Audibert, F. and Chedid, L., 1985, Carrier-induced epitopic suppression, A major issue for future synthetic vaccines, J Immunol., 135:2319.

Stahl, S. and Murray, K., 1989, Immunogenicity of peptide fusions to hepatitis B virus core antigen, Proc Natl Acad Sci USA, 68:6283.

Ward, M.M., Hall, R.P. and Pisetsky, D.S., 1990, Serum interleukin-2 receptor responses to immunisation, Clin Immunol Immunopathol., 57:120.

Winther, M.D., Allen, G., Bomford, R.H. and Brown, F., 1986, Bacterially expressed antigenic peptide from foot-and-mouth disease virus capsid elicits variable immunologic responses in animals, J Immunol., 136:1835.

CO-ENTRAPMENT OF T-CELL AND B-CELL PEPTIDES IN LIPOSOMES OVERCOMES GENETIC RESTRICTION IN MICE AND INDUCES IMMUNOLOGICAL MEMORY

Gregory Gregoriadis[1] and Zhen Wang[1] and Michael J. Francis[2]

[1]Centre for Drug Delivery Research, School of Pharmacy University of London, 29-39 Brunswick Square, London WC1N 1AX U.K.; [2]Pitman-Moore Ltd., Breakspear Road South, Harefield Middlesex, UB9 6LS, U.K.

INTRODUCTION

New-generation immunological adjuvants currently under investigation include immunostimulating complexes, block copolymers, nanoparticles and liposomes (Gregoriadis et al, 1991). The latter, used extensively since 1970 as a drug delivery system (Gregoriadis and Florence, 1993), were shown in 1974 (Allison and Gregoriadis, 1974) to promote immune responses to entrapped diphtheria toxoid. Work by numerous workers during the last two decades (Gregoriadis, 1990) has extended this observation for a growing number of bacterial, viral and protozoan antigens. In addition, as a result of the unique structural versatility of liposomes, it is now possible to manipulate their membrane fluidity (Gregoriadis et al, 1987; Davis and Gregoriadis, 1987; Therien et al, 1991), surface charge (Latiff and Bacchawat, 1987), size (Francis et al, 1985) and phospholipid to antigen mass ratio (Davis and Gregoriadis, 1987, 1989; Therien et al, 1991) so as to achieve optimal adjuvanticity for a number of antigens. Further amplification of adjuvanticity has been observed by receptor-mediated targeting to antigen presenting cells (Garcon et al, 1988) and the co-entrapment of interleukin-2 together with the antigen (Tan and Gregoriadis, 1989).

Although liposomes also induce IgG responses to small peptides containing helper T-cell (Th-cell) epitopes (Francis et al, 1985; Gregoriadis, 1990), they are unable to act similarly for peptides containing B-cell epitopes alone (Goodman-Snitboff et al, 1990). B-cell epitopes, on the other hand, can become immunogenic and induce memory when conjugated to a carrier protein containing a Th-cell epitope (Van Regenmortel et al, 1988). Unfortunately, this is often associated with immune responses to the carrier, masking of epitopes and antigenic competition (Van Regenmortel, 1988; Francis, 1993). Moreover, for the carrier protein to exert its full effect it is usually necessary to employ a potent adjuvant such as complete Freund's adjuvant (Bittle et al, 1982). Studies on peptide immunogenicity have now revealed that B-cell peptides may also become immunogenic, by co-polymerisation (Leclerc et al, 1987) or co-linear synthesis (Francis et al, 1987; Borras-Cuesta et al, 1987) to a Th-cell peptide sequence. In addition, a helper effect may be elicited by co-immunization with a mixture of B- and Th- cell peptides (eg. Partidos et al, 1992).

New Generation Vaccines, Edited by G. Gregoriadis *et al.*, Plenum Press, New York, 1993

We have recently investigated the possibility of a T-cell epitope providing help for a B-cell epitope when (a) the epitopes are co-entrapped in the same liposomes (and therefore, presented together to the immune system in the absence of covalent bonding; (b) the epitopes are co-emulsified in incomplete Freund's adjuvant (IFA). As epitope models we have chosen a peptide from the S region (S peptide) and a peptide from the pre-S_1 (pre-S_1 peptide) region of the hepatitis B surface antigen (HBsAg). S and pre-S_1 peptides are known (Neurath et al, 1987; Milich, 1987) to promote the formation of antibodies that bind to relevant sites in the HBsAg. It should be noted that the choice of the S peptide was from a region containing an H-2^S Th-cell epitope (Milich et al, 1987). (The pre-S_1 peptide was specifically designed to exclude the latter epitope.) Results from experiments in which SJL (H-2^S) mice (a strain that is unable to mount an IgG response to pre-S_1 peptide) were immunized with S and pre-S_1 peptides entrapped in liposomes or co-emulsified in incomplete Freund's adjuvant are described below.

ENTRAPMENT OF S AND PRE-S_1 PEPTIDES IN LIPOSOMES

Entrapment of peptides in liposomes was carried out by the dehydration-rehydration procedure (Kirby and Gregoriadis, 1984). This leads to the formation of multilamellar liposomes (Gregoriadis et al, 1993a) which when produced by this procedure are also known as dehydration-rehydration vesicles (DRV). In brief, small unilamellar vesicles (SUV) prepared (Kirby et al, 1980) from egg phosphatidylcholine and equimolar (32 μmoles) cholesterol were mixed with 1 mg each of synthetic S peptide (aminoacid sequence 110-137, subtype adw) and/or synthetic pre-S_1 peptide (aminoacid sequence 15-48; subtype adw) from the corresponding regions of the HBsAg, to which tracers of ^{125}I-labelled (S) and ^{131}I-labelled (pre-S_1) tracer peptides had been previously added. The mixtures were freeze-dried and rehydrated under controled conditions (Kirby and Gregoriadis, 1984). Generated peptide-containing DRV liposomes were then separated by centrifugation from non-entrapped materials for further use. Values of peptide entrapment (individually or together in the same liposomes) measured on the basis of radioactivity were 42.2-48.1% of the amounts used.

EXPERIMENTAL PROTOCOL

In a typical experiment, SJL-(H-2^S) mice in groups of 5-6 were injected intramuscularly with 20 μg of free S or pre-S_1 peptides, alone or in mixture, or with 20 μg of each of the peptides entrapped in liposomes, co-entrapped in the same vesicles or entrapped in separate vesicle populations which were mixed before immunization. Mice were boosted intramuscularly 56 days after priming with the peptides as above. In a separate experiment, animals were injected as above with 20 μg S or pre-S_1 peptide in incomplete Freund's adjuvant (IFA) or with 20 μg of each of the peptides co-emulsified in IFA. Two weeks later mice were bled and blood sera tested by Elisa for IgG$_1$ against S and pre-S1 peptide, recombinant HBsAg consisting of the S region only (226 aminoacids) and recombinant full length HBsAg consisting of the S, pre-S_1 and pre-S_2 regions (400 aminoacids) (Gregoriadis et al, 1993b).

IMMUNE RESPONSES IN MICE AGAINST LIPOSOMAL S AND PRE-S_1 PEPTIDES

Experiments were carried out in an attempt to demonstrate that T-cell help from an appropriate peptide (S) could be provided to a B-cell peptide (pre-S_1) by co-entrapping the two peptides in the same liposomes. These peptides (S and pre-S_1) contain either an H-2^S Th-cell epitope plus a B-cell

Table 1. Immune responses (IgG_1) in SJL ($H-2^S$) mice immunized with HBsAg peptides.

Peptide used for immunization	Antibody response[a]			
	Anti-S	Anti-pre-S_1	Anti-rHBsAg[b]	Anti-rHBsAg[c]
Liposomal S	153,600	N.D.	76,800	1,600
Liposomal pre-S_1	N.D.	<50[d]	N.D.	<50[d]
Liposomal (S + pre-S_1)	102,400	3,200	25,600	102,400
Liposomal S + liposomal pre-S_1	51,200	<50[d]	N.D.	6,400
IFA-S	1,200	N.D.	800	N.D.
IFA-pre-S_1	N.D.	<50[d]	N.D.	N.D.
IFA (S + pre-S_1)	300	200	800	N.D.

SJL ($H-2^S$) mice were injected with peptides on days 0 and 56 (see text). Animals were bled two weeks later and sera (fifty-fold diluted) tested by ELISA. Sera from mice immunized with free S or pre-S_1 peptides or a mixture of the two gave values of <50 when tested for anti-S, anti-pre-S_1 or anti-rHBsAg (S region only) IgG_1. For other details see Gregoriadis et al, 1993b.

a, Values are median from 5 or 6 mice in each group and denote dilutions required for readings to reach values of 0.2 or less; b, 226 amino acids (S region only); c, full length (includes S and pre-S regions); d, all animals tested gave readings of 0.2 or less.

45

epitope or a B-cell epitope alone respectively. Table 1 (and legend for results with free peptides) show that immunization of SJL-(H-2^2) mice with liposome-entrapped S peptide promotes an IgG$_1$ response. This confirms the immunoadjuvant activity of liposomes for small peptides as shown previously (Gregoriadis, 1990). Immunization with either free peptide alone or as a mixture did not elicit an immune response. Table 1 also indicates that the anti-S peptide antibodies were able to bind to rHBsAg (226 aminoacids), presumably to a region relevant to the peptide. Binding of the anti-S IgG$_1$ to the full length rHBsAg plus pre-S was, however, poor possibly as a result of partial masking of the S region epitope by the pre-S region within the 22 nm particle, with access of antibodies to the binding site thus being restricted. As anticipated, no IgG response to liposomal pre-S$_1$ was observed as the haplotype of mice used here would only recognize the pre-S$_1$ peptide as a B-cell antigen. On the other hand, when mice were immunized with both S and pre-S$_1$ peptides co-entrapped in the same liposomes, the latter peptide generated a measurable IgG$_1$ response (Table 1). These findings are in agreement with results of a study in which influenza Th-cell peptide provided help for an antibody response to a DNP-hapten incorporated in the same liposomes (Garcon and Six, 1991) and suggest that such an approach could be used to develop viral peptide vaccines. Results with co-entrapped peptides (Table 1) also support those previously reported (Goodman-Snitboff et al, 1990) using liposomes coupled to hydrophobically tailed peptides. Judging from our results, however, and in contrast to the author's suggestion, covalent linkage is not a prerequisite for immuno-genicity to occur.

Antisera to the (co-entrapped) peptides were found to bind to the full length rHBsAg + pre-S with titres being much higher (Table 1) compared to those measured for the liposome-entrapped S peptide. However, because antisera to co-entrapped peptides contain antibodies to both (see Table 1), the extent to which binding values for the full length rHBsAg plus pre-S include those for anti-S antibodies is uncertain. Nonetheless, the reduced binding values (full length rHBsAg plus pre-S) observed for sera raised against liposomal S (Table 1), strongly suggest that anti-pre-S$_1$ antibodies account for much of the binding activity elicited by the co-entrapped peptides: anti-pre-S$_1$ antibodies are expected to have full access to the pre-S$_1$ region of the rHBsAg plus pre-S (Neurath et al, 1989). Furthermore, the higher titres seen against the full length rHBsAg plus pre-S imply that the conformation adopted by the pre-S1 region of this antigen is more favourable (in terms of binding) than that adopted by the synthetic pre-S$_1$ peptide coated onto Elisa plates. It also appears that entrapment of the two peptides in the same liposomes is a prerequisite for an antibody response to occur. For instance, when the two peptides were entrapped in separate liposome populations which were mixed before use, no antibody response to pre-S$_1$ peptide was observed. Indeed, production of this helper effect by co-entrapping the peptides (as opposed to simply administering a mixture of the separately entrapped peptides), argues against the operation of a bystander help mechanism (Jensen and Knapp, 1986).

IMMUNE RESPONSE IN MICE AGAINST PEPTIDES S AND PRE-S$_1$ PEPTIDES EMULSIFIED IN IFA

IFA, being a water-in-oil emulsion can accommodate in the same water phase the two peptides. Its ability to promote an IgG response to pre-S$_1$ peptide co-emulsified with the S peptide was therefore tested. Results in Table 1 suggest that although IFA did act as anticipated, values for all three preparations tested were generally very low in comparison to those obtained with liposomes. This indicates that liposomes are more efficient than IFA in terms of adjuvant activity under the present conditions. Work with oil emulsion adjuvants incorporating other peptides (25, 26) has produced similar results although in these studies the antibody titres observed appeared to be higher than those in Table 1. This could be explained by the fact that CFA was used for the primary immunization as opposed to IFA.

PROLIFERATIVE RESPONSE OF T-CELLS FROM SJL MICE INOCULATED WITH S AND PRE-S_1 PEPTIDES

In an experiment designed to confirm that the peptides chosen behaved as expected in terms of T-cell stimulatory activity, SJL ($H-2^2$) mice were injected with S and pre-S_1 peptides entrapped together in the same liposomes or separately entrapped in different liposomal populations which were mixed before injection. Lymph nodes draining the injected sites were then tested in the presence of the peptides by the T-cell proliferation assay. Data obtained (Gregoriadis et al, 1993b) showed a significant response to liposomal S peptide. There was little or no response to the liposome-entrapped or co-entrapped pre-S_1 peptide. It can thus be concluded that IgG response to the liposomal pre-S_1 (Table 1) must have been the result of it receiving help from the (co-entrapped) S peptide, presumably taken up by the same B cells.

CONCLUSIONS

On the basis of this and other studies, it is clear that co-immunization with appropriate Th- and B-cell peptides accommodated within a vaccine delivery system such as liposomes overcomes non-responsiveness to defined B-cell epitopes and produces an IgG response. On the other hand, co-emulsification of the peptides within the aqueous phase of incomplete Freund's adjuvant, although sufficient for this purpose, appears to require the presence of reactogenic mycobacterial components to stimulate significant activity. Liposomes thus offer a safer alternative and also have the added advantage of being amenable to structural modifications for the optimal delivery of the antigen. It appears that for liposomes to be effective, neither the Th- nor B-cell epitopes need to have hydrophobic anchors (Goodman-Snitboff et al, 1990) and no additional adjuvant (Frisch et al, 1991) is required. The present findings are expected to further the concept of fully synthetic peptide vaccines.

Acknowledgements

This work was supported by a Medical Research Council project grant to Professor Gregory Gregoriadis. We thank Professor Y. Barenholz for the supply of the full length HBsAg and helpful discussion, and Dick Campbell, Wellcome Laboratories for peptide synthesis.

REFERENCES

Allison, A.C. and Gregoriadis, G., 1974, Liposomes as immunological adjuvants, Nature, 252:252.

Bittle, J.L., Houghten, R.A., Alexander, H., Schinnick, T.M., Sutcliff, J.G., Lerner, R.A., Rowlands, D.J. and Brown, F., 1982, Protection against foot-and-mouth disease by immunization with a chemically synthesised peptide predicted from the viral nucleotide sequence, Nature, 298:30.

Borras-Cuesta, F., Petit-Camurdan, A. and Fedon, Y., 1987, Engineering of immunogenic peptides by co-linear synthesis of determinants recognised by B and T cells, Eur.J.Immunol., 17:1213.

Davis, D. and Gregoriadis, G., 1987, Liposomes as adjuvants with immuno-
purified tetanus toxoid: Influence of liposomal characteristics,
Immunology, 61:229.

Davis, D. and Gregoriadis, G., 1989, Primary immune response to liposomal
tetanus toxoid in mice: The effect of mediators, Immunology, 68:277.

Francis, M.J., 1993, Carriers for peptides: Theories and technology, in:
"New Generation Vaccines: The Role of Basic Immunology",
G. Gregoriadis, B. McCormack, A.C. Allison, and G. Poste (Eds.),
Plenum Press, New York, in press.

Francis, M.J., Fry, C.M., Rowlands, D.J., Brown, F., Bittle, J.L., Houghten,
R.A. and Lerner, R.A., 1985, Immunological priming with synthetic
peptides of foot-and-mouth disease virus, J.Gen,Virol., 66:2347.

Francis, M.J., Hastings, G.Z., Syred, A.D., McGinn, B., Brown, F. and
Rowlands, D.J., 1987, Non-responsiveness to a foot-and-mouth disease
virus synthetic peptide overcome by addition of foreign helper T-cell
determinants, Nature, 330:168.

Frisch, B., Muller, S., Briand, J.P., Van Regenmortel, M.H.V. and Schuber,
F., 1991, Parameters affecting the immunogenicity of a liposome-
associated synthetic hexapeptide antigen, Eur.J.Immunol., 21:185.

Garcon, N.M.J. and Six, H.R., 1991, Universal vaccine carrier: Liposomes
that provide T-dependent help to weak antigens, J.Immunol., 146:3697.

Garcon, N., Gregoriadis, G., Taylor, M. and Summerfield, J., 1988, Targeted
immunoadjuvant action of tetanus toxoid-containing liposomes coated
with mannosylated albumin, Immunology, 64:743.

Goodman-Snitboff, G., Eisele, L.E., Heimer, E.P., Felix, A.M., Anderson,
T.T., Fiuerst, T.R. and Mannino, R.J., 1990, Defining minimal
requirements for antibody production to peptide antigen, Vaccine,
8:257.

Gregoriadis, G., 1990, Immunological adjuvants: A role for liposomes,
Immunol.Today, 11:89.

Gregoriadis, G. and Florence, A.T., 1993, Liposomal drug delivery systems:
Clinical, diagnostic and ophthalmic applications, Drugs, 45:15.

Gregoriadis, G., Davis, D. and Davies, A., 1987, Liposomes as immunological
adjuvants: Antigen incorporation studies, Vaccine, 5:143.

Gregoriadis, G., Allison, A.C. and Poste, G. (eds.), 1991, "Vaccines: Recent
Trends and Progress", Plenum, New York.

Gregoriadis, G., Garcon, N., da Silva, H. and Sternberg, B., 1993a, Coupling
of ligands to liposomes independently of solute entrapment:
Observations on the formed vesicles, Biochim.Biophys.Acta, 1147:185.

Gregoriadis, G., Wang, Z., Barenholz, Y. and Francis, M., 1993b, Liposome-
entrapped T-cell peptide provides help for a co-entrapped B-cell
peptide to overcome genetic restriction in mice and induce immunological memory, Immunology, in press.

Jensen, P.E. and Knapp, J.A., 1986, Bystander help in primary immune
reponses in vivo, J.Exp.Med, 164:841.

Kirby, C. and Gregoriadis, G., 1984, Dehydration-rehydration vesicles (DRV):
A new method for high yield drug entrapment in liposomes,
Biotechnology, 11:979.

Kirby, C., Clarke, J. and Gregoriadis, G.,1980, Effect of the cholesterol
content of small unilamellar liposomes on their stability in vivo and
in vitro, Biochem J., 186:591.

Latiff, N.A. and Bacchawat, R.K., 1987, The effect of surface-coupled
antigen of liposomes immunopotentiation, Immunol.Lett., 15:45.

Leclerc, C., Przewlocki, G., Schutze, M. and Chedid, L., 1987, A synthetic
vaccine constructed by copolymerization of B and T cell determinants,
Eur.J.Immunol., 17:269.

Milich, D.R., 1987, Genetic and molecular basis for T- and B-cell
recognition of hepatitis B viral antigen, Immunol.Reviews, 99:71.

Milich, D.R., McLachlan, A., Moriarty, A. and Thornton, G.B., 1987, A single
10-residue pre-S(1) peptide can prime T cell help for antibody

production to multiple epitopes within the pre-S(1), pre-S(2) and S regions of HBsAg, J Immunol., 138:4457.

Neurath, A.R., Jameson, B. and Huima, T., 1987, Hepatitis B virus proteins eliciting protective immunity, Microbiological Sciences, 4:45.

Neurath, A.R., Seto, B. and Strick, N., 1989, Antibodies to synthetic peptides from the pre-S1 region of the hepatitis B virus (HBV) envelope (env) protein are virus-neutralizing and protective, Vaccine, 7:234.

Partidos, C.D., Obeid, O.E. and Steward, M.W., 1992, Antibody response to non-immunogenic synthetic peptides induced by co-immunization with immunogenic peptides, Immunology, 77:262.

van Regenmortel, M.H.V., Briand, J.P., Muller, S. and Plave, S., 1988, Synthetic polypeptides as antigens, in: Laboratory Techniques in Biochemistry and Molecular Bioloby, Vol. 19, R.H. Burdon and P.H. von Knippenberg (Eds.), Elsevier, Amsterdam.

Tan, L. and Gregoriadis, G., 1989, The effect of interleukin-2 on the immunoadjuvant action of liposomes, Biochem.Soc.Trans., 17:693.

Therien, H-M., Shahum, E. and Fortin, A., 1991, Liposome adjuvanticity: Influence on dose and protein-lipid ratio on the humoural response to encapsulated and surface-linked antigens, Cell.Immunol., 136:403.

PREPARATION AND CHARACTERIZATION OF STABLE LIPOSOMAL HEPATITIS B VACCINE

D. Diminsky,[1] Z. Even-Chen,[2] and Y. Barenholz[1]

[1]Department of Biochemistry
The Hebrew University - Hadassah Medical School
P.O. Box 1172, Jerusalem 91010, Israel

[2]Biotechnology General Ltd
Rehovot 76326, Israel

INTRODUCTION

Various efficient hepatitis B vaccines are available today. Most of them are composed of 22-nm hepatitis B surface antigen (HBsAg) particles, with alum as adjuvant (Zuckerman, 1986). The HBsAg particles are prepared by recombinant DNA techniques in yeast (Butterly et al, 1989). These vaccines suffer from two major disadvantages: (i) low efficacy in non- and low-responders; (ii) low stability during freezing and at high temperature. While the problem of non- and low-responders is a general one, low stability under extreme weather conditions makes the use of the vaccine in Third World countries very problematic.

We approached these two drawbacks by using a liposomal vaccine in which the antigenic 22 nm HBsAg particles were derived from Chinese hamster ovary (CHO) cells. These particles contain 3 peptides: Small (S), Middle (M), and Long (L) peptides, and lipids (Rutgers et al, 1990; Even-Chen et al, 1990). The particles were encapsulated in liposomes which serve as adjuvant-carrier (Diminsky and Barenholz, in preparation). The three peptides are all products coded by ORF Pre-S/S genes. The S peptide corresponds to the 3' region of the ORF. The second peptide (M) includes the pre-S2 region + the S. The longest peptide, L, contains the amino acid sequences of Pre-S1, Pre-S2 and S peptides (for review see Rutgers et al, 1990; Neurath and Thanavala, 1990). It has been shown that the Pre-S2 and Pre-S1 amino acid sequences have antigenic determinants independent of the S proteins which result in enhanced immunogenicity when present in HBsAg particles (Pfaff et al, 1986). The presence of the Pre-S1 region induces a specific T cell response which can bypass non-responsiveness to the Pre-S2 and S regions of HBsAg (Milich et al, 1986; Even-Chen et al, 1990).

The liposome-based vaccines are an attractive choice because (i) liposomes are composed of natural components and are biocompatible; (ii) extensive vaccination studies which include a broad spectrum of antigens were performed with good results; (iii) liposomes may improve the uptake and processing by antigen presenting cells (APC) (Alving, 1991; Gregoriadis, 1990); (iv) liposomes can be frozen and freeze-dried without

New Generation Vaccines, Edited by G. Gregoriadis
et al., Plenum Press, New York, 1993

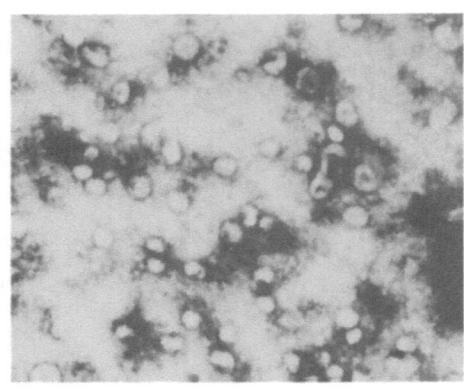

Fig. 1. Electron micrograph of purified rHBsAg particles. The rHBsAg particles produced by CHO cells were visualized by negative staining with 2% phosphotungstic acid. Average particle size is 22-nm. Magnification x 23,000

major changes in their physical properties (Crommelin et al, 1986; Gregoriadis et al, 1988; Lichtenberg and Barenholz, 1988). Therefore there is a good possibility that the liposomal hepatitis B virus (HBV) vaccine can be prepared as a stable freeze-dried formulation.

In this paper we describe the preparation of liposomal HBV vaccine which retains full activity upon freezing and freeze-drying.

SELECTION OF LIPOSOME LIPIDS

Multilamellar liposomes made of dimyristoyl phosphatidylcholine (DMPC) and dimyristoyl phosphatidylglycerol (DMPG) were chosen because these phospholipids have saturated fatty acids that make them resistant to various oxidative reactions of either the acyl chains or head group. The negative charge of DMPG results in the enlargement of the aqueous volume entrapped in the liposomes by electrostatic repulsion between the bilayer leaflets. It also prevents irreversible aggregation of the liposomes. The negative charge improves the uptake of large liposomes by APC. At body temperature (37° C) the lipid bilayer of the liposome is in a liquid crystalline state (Tm = 24° C; Silvius, 1982), which facilitates uptake and processing by APC. The DMPC and DMPG phospholipids are biodegradable by phospholipases (Lichtenberg and Barenholz, 1988; Gregoriadis, 1990; Alving, 1991).

HBsAg PARTICLES

The HBV DNA was isolated and CHO cells were transfected. Following selection and cloning, these cells secreted 22-nm HBsAg particles, which were visualized by negative staining with 2% phosphotungstic acid (Fig. 1). The particles contain 3 peptides: S (78%); M, containing Pre-S2 region (8%); and L containing Pre-S1 and Pre-S2 regions (14%). The peptide composition obtained upon sodium dodecyl sulfate polyacrylamide gel electrophoresis (SDS-PAGE) with a reducing agent reveals the glycosylation profile of the peptides (Fig. 2). The presence of Pre-S1 in the particle (L-peptide) was proven by Western blot analysis with monoclonal antibodies against Pre-S1 region (Fig. 3). It was demonstrated that in vaccines based

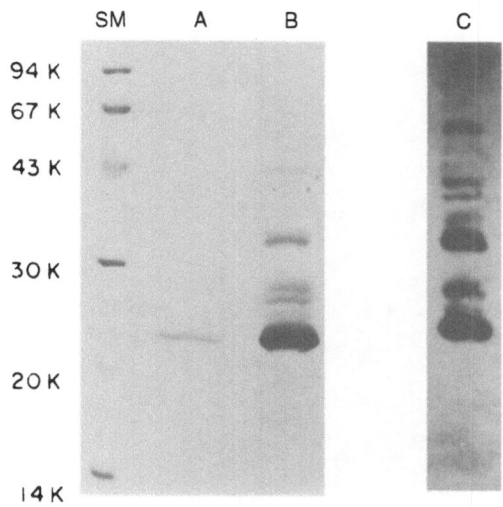

Fig. 2.   SDS-PAGE with reducing agent.   rHBsAg samples
(2-4 ug/lane) were loaded on a 15% polyacrylamide gel.

Lane SM: Molecular mass markers.
Lane A: Proteins of HBsAg stained with Coomassie blue.
Lane B: Proteins of HBsAg stained with silver stain.
Lane C: Western blot. Proteins of HBsAg challenged with rabbit
          anti-HBsAg antibodies.
24K - Nonglycosylated S (P_{25})
27K - Glycosylated S (GP_{27})
33K - Single glycosylated M (S + Pre-S_2) (GP_{33})
36K - Double glycosylated M (GP_{36})
39K - Nonglycosylated L (S + Pre-S_2 + Pre-S_1) (GP_{39})
42K - Glycosylated L (GP_{42})

(GP = glycoprotein, P = Protein)

on alum as adjuvant this peptide combination significantly enhances the
immunogenicity of HBsAg (Even-Chen et al, 1990). The recombinant CHO
derived HBsAg + alum vaccine is more effective than other commercially
available recombinant HBsAg vaccines as it elicits the production of anti-
Pre-Sl antibodies. In addition, immunization of nonresponder mice with
high vaccine dosage of recombinant HBsAg + alum resulted in a high titer of
neutralizing antibody (Even-Chen et al, 1990).

The HBsAg particles also contain lipids. For quantification and
analysis the lipids of HBsAg particles were extracted by the method
described by Bligh and Dyer (1959). Total lipid phosphorus was determined
by the method of Bartlett (1959) as modified by Barenholz and Amselem
(1993). Polar lipids were identified and quantified by using a combination
of one-dimensional thin layer chromatography (TLC) and two-dimensional TLC
(2D-TLC) under conditions described elsewhere (Yavin and Zutra, 1977;
Barenholz et al, 1981; Barenholz and Amselem, 1993).

Phospholipid analysis shows that phosphatidylcholine (PC) is the
major phospholipid. Other phospholipids detected were sphingomyelin (SPM),
phosphatidylethanolamine (PE), lysophosphatidylcholine (L-PC) and
phosphatidylserine (PS) (Fig. 4).

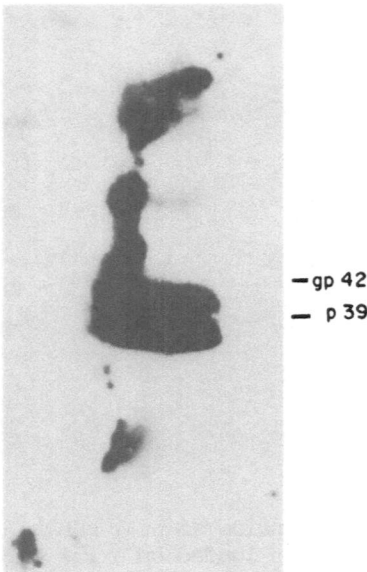

- gp 42
- p 39

Fig. 3. Western blot of rHBsAg proteins challenged with
monoclonal antibodies against Pre-S1. SDS-PAGE
was performed. The Western blot was developed
with monoclonal antibodies specific for the pre-S1
region of the HBsAg.

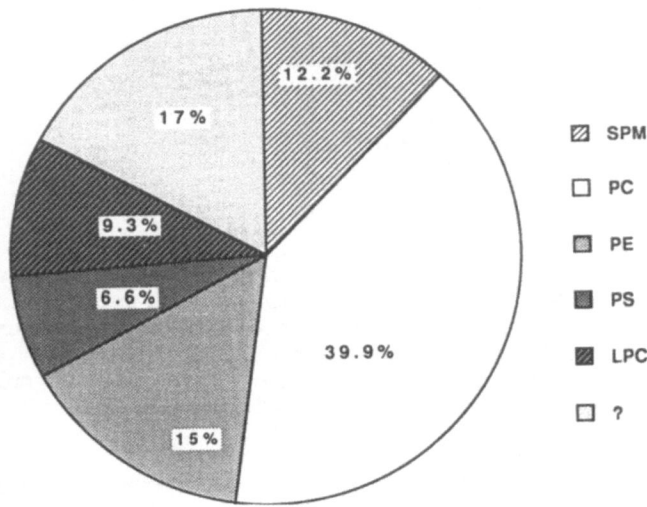

Fig. 4. Phospholipid composition of rHBsAg (Batch No: 1-38P-82):
The phospholipid composition was determined by two-
dimensional thin layer chromatography (2D-TLC). The
spots were identified using markers and detected by
primulin spray, and then analyzed for their organic
phosphorus content. Data are presented as % of total.
Abbreviations: PC = phosphatidylcholine; PE = phosphatidylethanolamine; PS = phosphatidylserine; LPC = lysophosphatidylcholine; SPM = sphingomyelin; ? = represents
unidentified phospholipid.

Table 1. Polar lipid contents of HBsAg.

Batch	2501/1	2501/3
Total phospholipids (mg/mg protein)	0.31	0.53
Free cholesterol (mg/mg protein)	0.11	0.20
Free cholesterol/phospholipid (mole/mole)	0.71	0.74

Cholesterol was identified and quantified by the HPLC method of Ansari et al, (1979) as described by Barenholz and Amselem (1993).

Table 1 shows the phospholipid to protein and free cholesterol to phospholipid ratios for two different preparations of HBsAg particles.

Neutral lipid classes were fractionated as described by Kates (1986) and using TLC of silica gel G and petroleum ether:ether:acetic acid (80:20:2 by volume) as the developing solvent. The plates were sprayed with a 0.001% primuline solution and visualized under UV light (Barenholz and Amselem, 1993), then sprayed with 10% H_2SO_4 in methanol. The spots were identified using commercial markers and by R_f of the neutral lipid classes (Kates, 1986). The main HBsAg neutral lipid constituents were cholesterol, cholesterol ester and triacylglycerols. Other lipids such as free fatty acids and diacylglycerols were also present (Diminsky and Barenholz, in preparation).

Analysis of acyl chain composition of HBsAg particles was performed by gas liquid chromatography as described by Barenholz and Amselem (1993). The major acyl chains were C18:1 > C18:0 > C16:0 > C22:5 > C14:0 > C16:1 > C20:4 = C22:6.

Our lipid analysis demonstrates that the major lipid components of HBsAg are similar to those of lipoproteins (Carolyn and Wright, 1988).

THE STRUCTURE OF THE ENVELOPE OF HBsAg

The degree of exposure of phospholipids to the external medium can be used to distinguish if the HBsAg particles are viral envelopes and behave like sealed vesicles or have a lipoprotein-like structure (reviewed in Rutgers et al, 1990). In sealed vesicles of 22 nm size the degree of phospholipid exposure expressed as exposed phospholipids/total phospholipids is about 55%. For lipoprotein-like structures in which a core of nonpolar lipid is surrounded by an envelope composed of a monolayer of polar lipids, the exposure is about 100% (Lichtenberg and Barenholz, 1988).

We determined the % exposure by two methods: (i) Exposure of ester phospholipids to Naja naja phospholipase A_2 as described by Suginara et al, (1992); (ii) Interaction of amino phospholipids (PE and PS) with 2,4,6-trinitrobenzene sulfonic acid (TNBS) (Lichtenberg and Barenholz, 1988).

We found that all phospholipids except SPM were hydrolyzed almost completely by phospholipase A_2 and that all amino lipids were exposed to TNBS. This suggests that the HBsAg particle is not a sealed vesicle. Its structure can be explained either as having a nonsealed lipid bilayer or, more probably, as a lipoprotein-like structure in which a phospholipid monolayer serves as an envelope to a core of neutral lipids and the hydrophobic part of the proteins. The exact organization of the protein has still to be determined.

LIPOSOMAL VACCINE PREPARATION

The liposomal vaccine has been prepared by four methods: (I) The phospholipids were dissolved in chloroform, dried and suspended in a salt solution containing HBsAg. (II) The liposomes were prepared as in (I) followed by ten freezing and thawing cycles between -185° C and 37° C. (III) Liposomes were prepared as in (I). t-Butanol was then added, the liposomes were dissolved and the preparation was lyophilized. (IV) The phospholipids were dissolved in t-butanol. An aqueous HBsAg solution was added and the preparation was lyophilized. This is the only method that does not involve initial preparation of liposomes. In methods III and IV the powder that is obtained has to be reconstituted before vaccination by adding the desired volume of double-distilled water.

CHARACTERIZATION OF LIPOSOME-ENTRAPPED HBsAg

It has been found that increasing the amount of lipid at constant antigen weight increased the antigen entrapment. The entrapment efficiency also increased with the concentration of antigen that was introduced, with an optimum at the HBsAg:lipid ratio (w/w) of 0.002. The best entrapment, 97%, was obtained with methods III and IV. Method II gave 80% entrapment, and I, only 73%.

The liposome size was determined with a Coulter Counter Model PCA-1. It was 4.5 um on average and independent of the method of preparation. The presence of antigen in the liposomes and their storage time had no significant effect on size. Similar size distributions were obtained with all four preparations. Upon repeated washes of the liposomes the phospholipid:antigen ratio (w/w) remained constant. This shows that there is no leakage of antigen from the liposomes. We also showed that no lipid decomposition and no size change occurs upon storage for up to 18 months.

The extent of HBsAg exposure on the liposome surface was determined by ELISA. It was found that after removing the free antigen, the extent of the exposure was less than 5%. In preparations III and IV HBsAg exposure was below the sensitivity limit (0.5%) of the method.

EFFICACY OF THE LIPOSOMAL VACCINE

The efficacy of the liposomal vaccine was examined by vaccination of BALB/c mice with different doses of HBsAg entrapped in liposomes. The antibody titer (IgG) which developed was determined 35 days after vaccination. The liposomal preparations were compared with a vaccine containing antigen (the 22-nm particles) and alum as adjuvant (alum vaccine). Liposomes prepared by method II were better than the alum vaccine in their ability to stimulate antibody formation. Liposomes prepared by method I were as efficient as alum vaccine. All the liposomal vaccines resulted in a sufficiently high antibody titer to protect against HBV infection. The liposomal vaccines have an advantage over the alum vaccine especially at low dosages (0.09 and 0.27 g of HBsAg) (Fig. 5).

Taken together, these results show that the preparation of an efficacious and stable liposomal vaccine which is stored as dry powder is feasible.

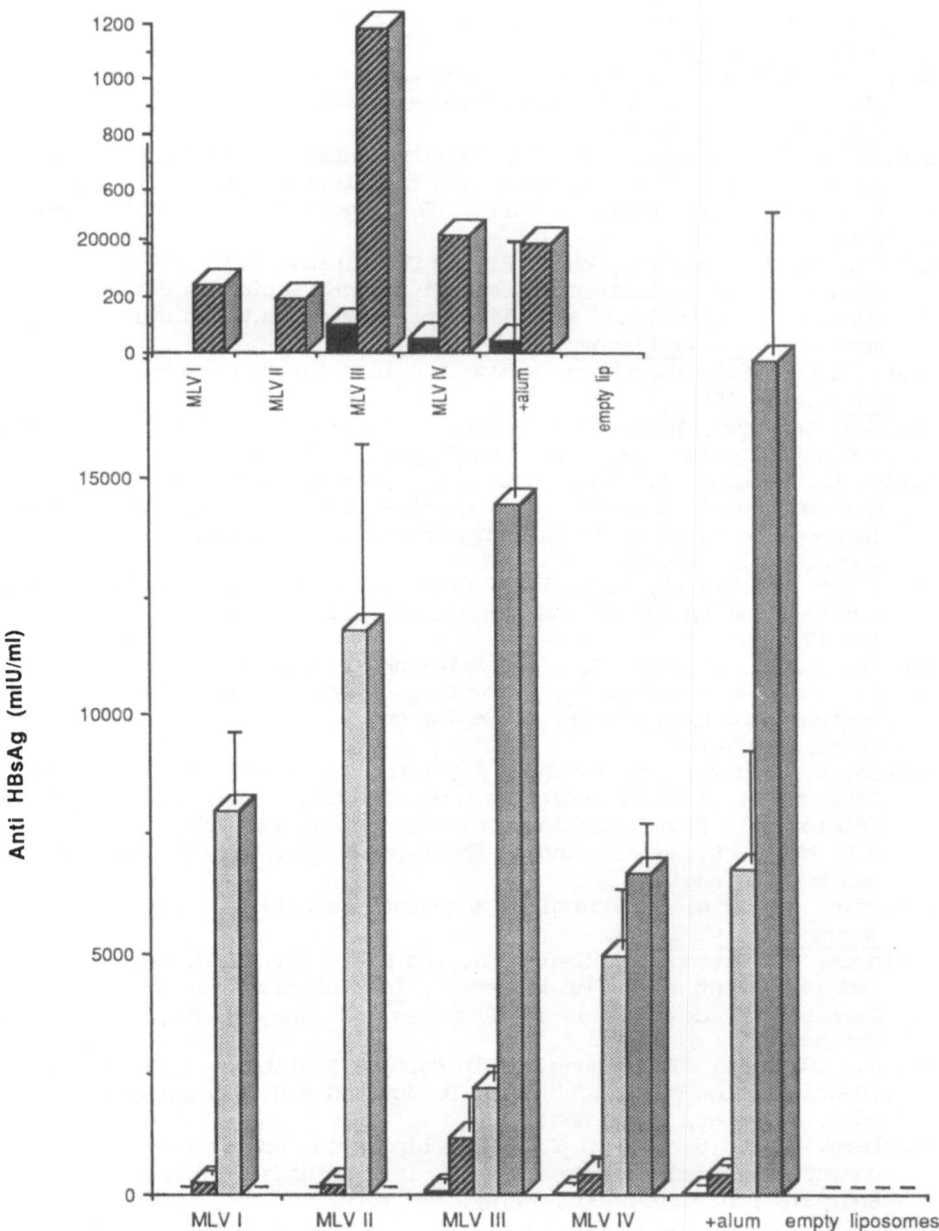

μg HBsAg

Fig. 5. Anti-HBsAg titer in mIU/ml was determined by radio-
immunoassay for the detection of antibody to HBsAg
(as measured by AUSAB[R], Abbott) tested 35 days after
intraperitoneal infection of 0.09 μg ▮ ; 0.27 μg ▨ ;
0.81 μg ▨ and 2.5 μg ▤ HBsAg protein per mouse.

REFERENCES

Alving, C.R., 1991, Liposomes as carriers of antigens and adjuvants, J.Immunol.Meths., 140:1.

Ansari, G.A., Lelands, S., Smith, L., 1979, High performance liquid chromatography of cholesterol autooxidation products, J.Chromatog., 175:307.

Barenholz, Y. and Amselem, S., 1993, Quality control assays in development and clinical use of liposome based formulations, in: "Liposome Technology", 2nd edition, Vol. 1, G. Gregoriadis, ed., CRC Press, Boca Raton, FL.

Barenholz, Y., Yechiel, E., Cohen, R. and Deckelbaum, R.J., 1981, Importance of cholesterol phospholipid interaction in determining dynamics of normal and abetalipoproteinemic red blood cell membrane, J.Cell Biophys., 3:115.

Bartlett, G.R., 1959, Phosphorus assay in column chromatography, J.Biol. Chem., 234:466.

Bligh, E.J. and Dyer, W.J., 1959, A rapid method of total lipid extraction and purification, Can.J.Biochem.Physiol., 37:911.

Butterly, L., Watkins, E. and Dienstag, J.L., 1989, Recombinant yeast derived hepatitis B vaccine in healthy adults: safety and two year immunogenicity of early investigative lots of vaccine, J.Med.Virol., 27:155.

Carolyn, E.M. and Wright, L.C., 1988, Organization of lipids in the plasma membranes of malignant and stimulated cells: a new model, TIBS, 13: 172.

Crommelin, D.J.A., Fransen, G.J. and Salemink, P.J.M., 1986, Stability of liposomes on storage, in: "Targeting of Drugs with Synthetic Systems", G. Gregoriadis, J. Senior and G. Poste, eds., Plenum Press, New York.

Even-Chen, Z., Drummer, H., Levanon, A., Panet, A. and Gorecki, M., 1990, Development of novel hepatitis B vaccine containing Pre-S1, in: "Biologicals from Recombinant Microorganisms and Animal Cells", M.D. White, S. Reuveny and A. Shafferman, eds., Balaban Publishers and VCH, Weinheim.

Gregoriadis, G., 1990, Immunological adjuvants: a role for liposomes, Immunol.Today, 11:89.

Gregoriadis, G., Garcon, N., Senior, J. and Davis, D., 1988, The immunoadjuvant action of liposomes, in: "Liposomes as Drug Carriers. Recent Trends and Progress", G. Gregoriadis, ed., Wiley, Chichester.

Kates, M., 1986, in: "Techniques of Lipidology: Isolation, Analysis and Identification of Lipids", R.H. Burdon and P.H. Van Knippenberg, eds., Elsevier, Amsterdam.

Lichtenberg, D. and Barenholz, Y., 1988, Liposomes: Preparation, characterization and preservation, in: "Methods of Biochemical Analysis", D. Glick, ed., Wiley, New York.

Milich, D.R., McLachlan, A., Chisari, F.V., Kent, B.H. and Thornton, G.B., 1986, Immune response to the Pre-S(1) region of hepatitis B surface antigen (HBsAg): A Pre-S(1) specific T cell response can bypass non-responsiveness to the Pre-S(2) and S regions of HBsAg, J.Immunol., 137:315.

Neurath, A.R. and Thanavala, Y., 1990, Hepadenoviruses, in: "Immunochemistry of Viruses II. The Basis for Serodiagnosis and Vaccines", M.H.V. Van Regenmortel and A.R. Neurath, eds., Elsevier, New York.

Pfaff, E., Klinkert, M.Q., Theilmann, L. and Schaller, H., 1986, Characterization of large surface proteins of hepatitis B virus by antibodies to preS-S encoded amino acids, Virology, 148:15.

Rutgers, T., Cabezon, T. and De Wilde, M., 1990, Production of viral antigens in prokaryotic and eukaryotic cells, in: "Immunochemistry of Viruses II. The Basis for Serodiagnosis and Vaccines", M.H.V. Van Regenmortel and A.R. Neurath, eds., Elsevier, New York.

Silvius, J.R., 1982, in: "Lipid Protein Interactions", Vol. 2, P.C. Jost and O.H. Griffith, eds., Wiley, New York.

Sugihara, T., Sugihara, K. and Hebbel, R., 1992, Phospholipid asymmetry during erythrocyte deformation: maintenance of the unit membrane, Biochim.Biophys.Acta, 1103:303.

Yavin, E. and Zutra, A., 1977, Separation and analysis of ^{32}P-labelled phospholipids by a simple and rapid thin layer chromatographic procedure and its application to cultured neuroblastoma cells, Anal.Biochem., 80:430.

Zuckerman, A.J., 1986, Novel hepatitis B vaccines, J.Infect., 13:61.

INITIATION OF IMMUNE RESPONSE WITH ISCOM

B. Morein, M. Villacres-Eriksson, L. Åkerblom and
K. Lövgren

Swedish University of Agricultural Sciences and the
National Veterinary Institute, Dept of Virology
Biomedical Centre, Box 585, S-751 23 Uppsala, Sweden

INTRODUCTION

Recent developments of vaccines focused on techniques for production of antigens, with most of the attention paid to the use of gene technology to produce replicating or non-replicating antigens. Promising results were e.g. obtained with vaccinia as a replicating virus vector and with salmonella as a bacterial vector producing cloned antigens. Generally, non-replicating purified antigens are poorly immunogenic regardless of whether they are isolated from conventional microorganisms or recovered from genetically cloned bacteria, yeast or animal cells. Common for all non-replicating antigens is the demand for forming them into a favourable immunogenic form. Consequently, the first rule for making an antigen immunogenic is to give it an immunogenic physical form, i.e. the antigen should be present in a multimeric form as small particle (Morein et al, 1978; Morein & Simons, 1985). Several adjuvants - immunomodulators - have recently been used to enhance the immunogenicity of antigens obtained from conventional microorganisms or antigens being gene technology products. However, few adjuvant systems combine the physical multimeric presentation with immunomodulation as the case is with SAF-1 or iscoms. (For references, see Allison, 1989; Morein et al, 1989).

Synthesis of oligopeptides is another recent approach for production of antigens. These antigens generally consist of few aminoacids sometimes without T-helper epitopes. In addition to a multimeric presentation, such antigens require supplementation of T-helper epitopes. For synthetic oligopeptides, gene technology products as well as for conventional antigens, iscoms have proved to be an efficient carrier and adjuvant system often enhancing the immune response a ten-fold or more (for references see Höglund et al, 1989; Morein & Åkerblom 1992).

THE ISCOM

The structure and composition of iscoms are described in detail in another chapter of this book by C. Dalsgaard. Here we just conclude that the iscom consists of a carrier structure - a matrix - into which antigens are incorporated by hydrophobic interactions. We would also like to emphasize that the building blocks of the iscom - forming the matrix - are triterpenoids, some of which are responsible for the typical 40 nm cage-like structure formed when they associate with cholesterol and a "soft" lipid, e.g. phosphatidyl choline. Other triterpenoids in the Quil A mixture have adjuvant activity, some give rise to side effects in higher doses, some cause no or negligible side effects. Structure forming triter-penoids giving rise to iscom-like particles which are virtually non-toxic for mice have been isolated.

THE INITIAL PHASE OF IMMUNIZATION

For classical adjuvants like aluminium hydroxide or oil adjuvants the depot effect at the site of injection is considered to be an important immunopotentiating factor. In contrast, antigens in iscoms are rapidly transported from the site of injection to the draining lymphatic organ. Local inflammatory reactions following injection of iscoms are, therefore,

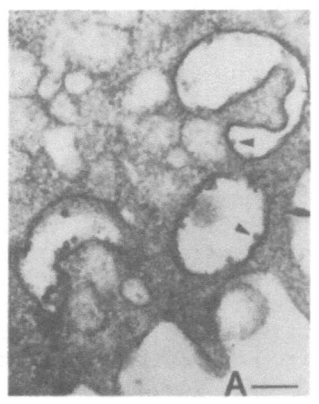

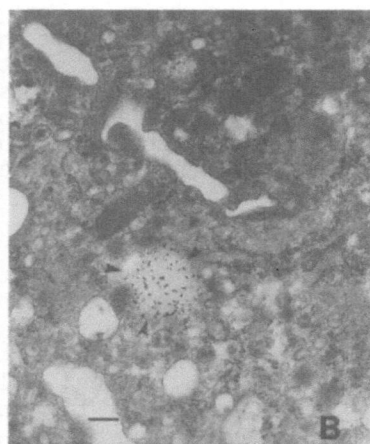
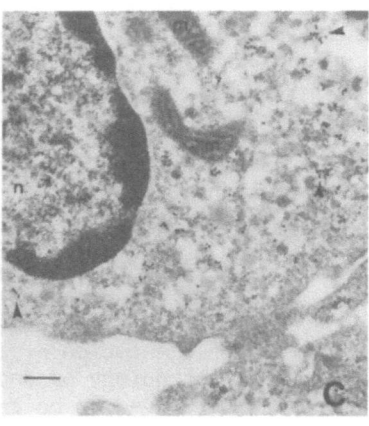

Fig. 1. Electron microphotographs of peritoneal macrophages which internalized influenza virus iscoms. In B and C iscoms were biotin-labelled and the biotinylated antigen was localized by incubation with streptavidin gold conjugate. Gold particles are 15 nm in diameter. A. Iscoms localized in endosomes (Bar 200nm). B. Gold particles localized in a vesicle probably an endosome (Bar 300 nm). C. Gold particles localized intracellularly, probably in the cytosol, i.e. not associated with intracellular membrane structures (Bar 300 nm). n= nucleus; m= mitochondrion

negligible compared to aluminium hydroxide (Speijers et al, 1988) or oil adjuvants and granuloma are not observed following injection of iscoms. A transient redness may be seen locally after subcutaneous injection of high doses of iscoms. After intraperitoneal immunization with radiolabelled antigens in iscoms, a comparatively high proportion of the antigens become cell associated and transported to the spleen where it remains for a longer period of time than the same antigen in micelle form (Watson et al, 1989).

INTERACTION OF ISCOMS WITH ANTIGEN PRESENTING CELL (APC)

The unique property of iscoms to induce immune response over both MHC class I and class II trigger speculations about the internalization of iscom borne antigens and their intracellular transport. The general dogma conceived about antigen presentation is that injected non-replicating (**exogenous**) antigens are taken up by APC. In these cells the antigens enter the endosomatic-lysosomic pathway and are processed by proteolytic enzymes at an acid pH in the lysosomes to peptides of various length before association with the MHC class II antigen (Rudensky et al, 1991). Subsequently the processed antigen is transported to the cell surface for interaction with T helper cells.

Endogenous antigens are produced in the antigen presenting cell either due to infection with e.g. virus or as a cancer antigen. These antigens are processed by cytosolic proteolytic enzymes to nonapeptides and transported by aid of the transportation molecules TAP1 and TAP2 to the endoplasmatic reticulum where they join the MHC class I molecules for further transport to the cell surface where they are available for interaction with CD8+ cytotoxic T-cells (Kelly et al, 1992, Monaco 1992, Powis et al, 1992).

By electron microscopy, iscom particles carrying influenza virus antigen could directly be localized in endosomes (Fig. 1 A, reproduced with permission from Watson et al, 1992). Recently biotinylated influenza virus iscoms have been studied by immunoelectron microscopy to follow the fate of the iscom borne antigen in purified APC populations. The biotinylated antigen was detected with streptavidin gold conjugate. In this system, peritoneal macrophages contained about 10 times more antigen than peritoneal monocytes or splenic naive B cells after 5 min. of synchronized internalization. In macrophages, a significant pro-portion of the Ag was detected inside the cells, not only bound to the plasma membrane or to other membrane structures (Fig. 1 B) as shown with whole iscom particles (Watson et al, 1992; Fig. 1 A), but also outside vesicles presumably located in the cytosol (Fig. 1 C Villacres-Eriksson et al, manuscript in preparation). A possible mechanism of the iscom to deposit antigen in the cytosol would be its integration into the plasma membrane or in the vesicle membrane. These membranes are 30 nm which the 40 nm iscom particle would span exposing antigens both on the cytosol side and in the internal endosomal space. Accessibility to proteolytic enzymes in the cytosol or lysosomes facilitates the processing for the association with MHC class I and II molecules respectively.

In comparative studies it was shown that macrophages internalize iscom borne antigen more efficiently than monocytes or naive B cells. Splenic dendritic cells were less active than macrophages but more efficient than monocytes or naive B cells in taking up iscom borne antigen. These quantitative data are at present difficult to assess since macrophages are professional phagocytic cells in contrast to the others. Furthermore, purified manipulated dendritic cells may have lost a great deal of their capacity to internalize antigens. The capacity of peritoneal macrophages, splenic naive B cells and splenic dendritic cells to present

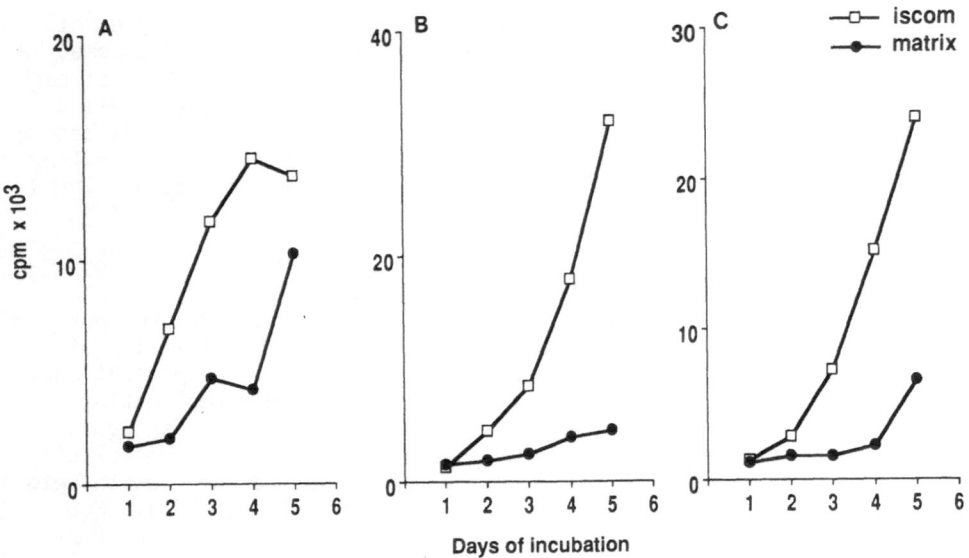

Fig. 2. Stimulation of splenocytes enriched with T-cells. Stimulation was measured by ^{3}H thymidine uptake by different antigen presenting cells (APC). A, resident peritoneal cells; B, splenic naive B-cells; C, splenic dendritic cells after uptake of influenza virus iscoms or cultured with medium only. kT cells were obtained from BALb/C mice primed once with influenza virus iscoms

antigen to primed T cells was studied <u>in vitro</u>. These purified APC populations were allowed to internalize influenza virus iscoms for 1 hr. After removal of free antigen, the APC's were able to stimulate T cells

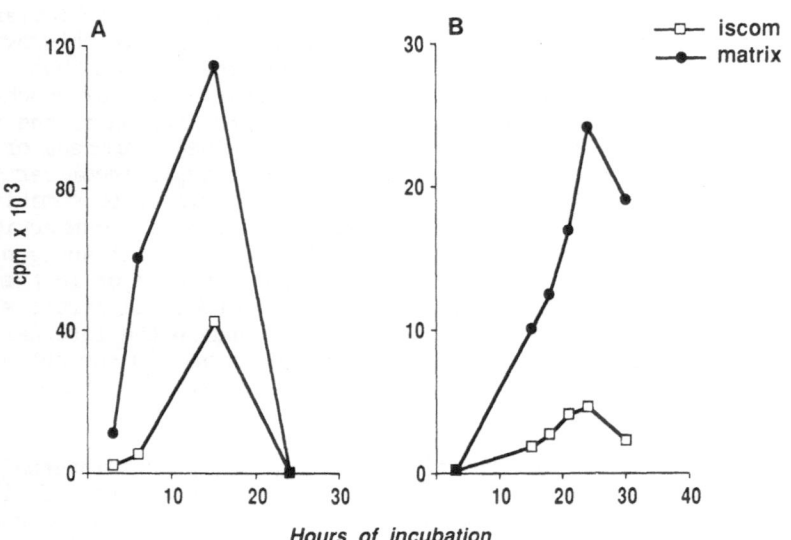

Fig. 3. Estimation of production of membrane bound (A) and soluble IL-1 (B) after incubation of splenocytes with influenza virus iscoms or the matrix of iscom (iscoms without antigens) using a bioassay measuring IL-1 dependent proliferation of D10-cells by ^{3}H thymidine uptake

from mice primed with iscoms containing homologous antigens. As measured by H^3-thymidine uptake, both naive B cells and dendritic cells proved to be more efficient than peritoneal macrophages and monocytes in the capacity to present the antigen internalized by aid of iscoms and stimulate T-memory cells (Fig. 2).

Cytokine studies showed that iscoms and the matrix of the iscoms induce macrophages to produce IL-1 (Villacres-Eriksson et al, 1992) and IL-6. Both in in vivo and in vitro experiments the production of membrane-bound IL-1 is maximal after 16 hr of stimulation. Iscoms and matrix induced, in contrast to the antigen in micelle form, also soluble IL-1, with a maximal production detected after 24 hr of culture (Fig. 3). Attempts to detect TNF-alpha in supernatants of peritoneal cells or splenocytes cultured with iscoms or matrix were unsuccessful. Nevertheless, it is possible that TNF-alpha may be present in the membrane-bound form, as preliminary assays suggest.

THE PRODUCTION OF CYTOKINES IN THE RECALL RESPONSE TO ISCOM BORNE ANTIGEN

The T cell products IL-2, IL-4, IL-5 and INF- are important factors for determination of the class and isotype distribution in the antibody response as well as for determining regulatory and effector T-cell functions including the expansion of cytotoxic lymphocyte clones. Cell culture fluids from splenocytes originating from mice immunized with influenza virus iscoms, cultured with the same iscoms contained IL-2, and INF-γ (Villacres-Eriksson et al, 1992). In comparison, micelles containing influenza virus envelope proteins induced in such spleen cell cultures low levels of IL-2 and IFN-γ. With splenocytes from micelle primed mice low or insignificant amounts of IL-2 and IFN-γ were produced in response to in vitro stimulation with iscoms or micelles. Iscoms and micelles respectively stimulate similar amounts of IL-4 in the spleen cell cultures regardless of whether or not the cells were from iscom or micelle primed mice (Villacres-Eriksson, submitted). The latter result indicates that iscoms and micelles are equally potent in priming and boosting T-cells producing IL-4, i.e. T_{H2} response. However, the iscoms are superior to micelles in inducing IL-2 and IFN-γ, i.e. evoking a T_{H1} response (Fig. 4).

The isotype profile after subcutaneous immunization with influenza virus iscoms is mainly composed of IgG1, IgG2a and IgG2b. A similar profile was detected after intranasal immunization of mice with influenza virus iscoms (Lövgren, 1988; Villacres-Eriksson et al, 1992). The pattern of lymphokine response and the profile of the antibody response observed after immunization with iscom borne antigen indicate the stimulation of both T_{H1} and T_{H2} type of T-helper lymphocytes. Possibly, the T_{H1} response is dominating. The importance of selecting the right antibody class response is exemplified in a post-exposure rabies vaccination experiment in mice. In this experiment, 78% of the mice died of anaphylactic shock after the 4th injection of the commercially available whole virus vaccine. No mice died from an iscom vaccine containing the envelope antigen of rabies virus and this vaccine awarded a 95% protection to challenge infection. No significant protection was induced by the whole virus vaccine (Fekadu et al, 1992). It is conceivable to assume that the IFN-γ production, which is shown to be induced by iscoms, would prevent the development of an IgE response.

CELL MEDIATED IMMUNE RESPONSE

A unique property of the iscom is its capacity to present antigens via both the MHC class I and class II pathways in spite of the fact that it

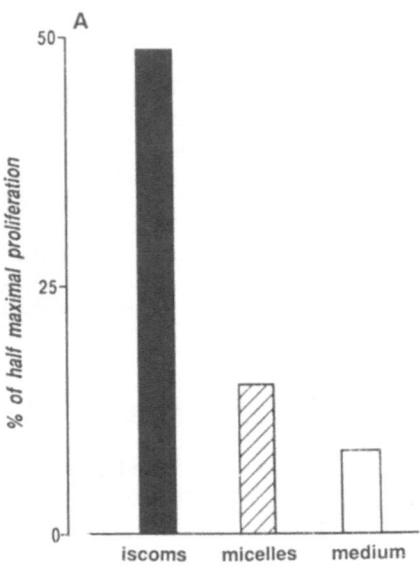

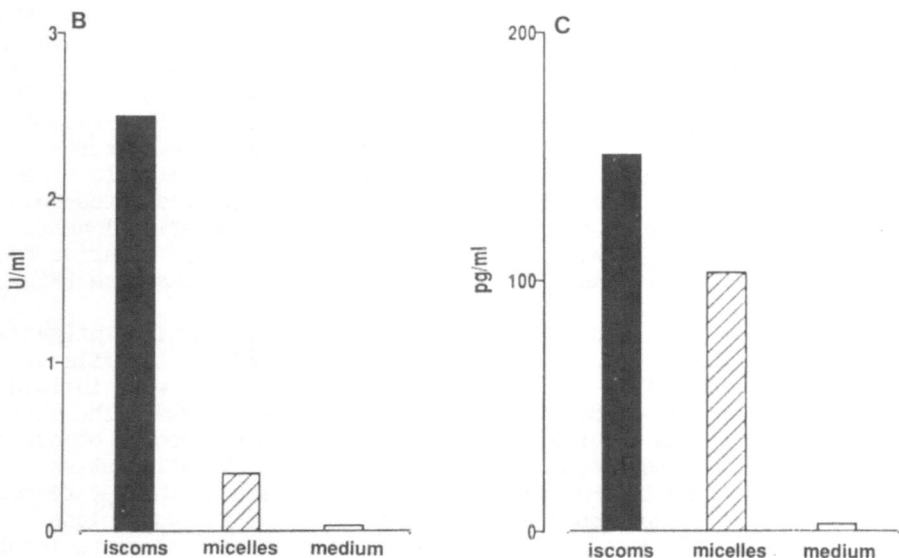

Fig. 4. Production of lymphokines as measured in culture supernatants. The cultures consisted of spleen cells from mice immunized twice 4 weeks apart, collected 4 days after the second immunization and stimulated in vitro with the homologous antigen preparation, or incubated with medium alone. A, IL-2 production measured in the Il-2 dependent CTLL cell line using the fluid of the spleen cells cultures. Results are expressed as percentage of half maximal proliferation; B, IFN-γ in the fluid of spleen cell culture, quantitated by ELISA; C, IL-4 in the fluid of spleen cell cultures, quantitated by ELISA.

does not replicate. Over the class I pathway, the iscom borne antigen efficiently induces cytotoxic T-cells (CTL), which is shown with a number of antigens injected parenterally e.g. gp160 of HIV-1, envelope proteins of influenza virus, F-protein of measles virus, heat shock protein of mycobacteria, the nucleoprotein of influenza virus and ovalbumin (Takahashi et al, 1990; van Binnendijk, 1992; Mowat et al, 1991; Heeg et al., 1991).

Also by intranasal route, influenza virus iscoms induce CTL under restriction of MHC class I, indicated by the fact that the target cell P815 used in this experiment does not carry MHC class II antigens (Jones et al, 1988). Recently, Maloy et al, (1992) demonstrated that ovalbumin in iscoms induced MHC class I restricted CTL also after oral administration provided high doses were used.

The capacity of p97 melanoma antigens carried by iscoms to cause rejection of transplanted melanoma in mice is possibly due to cell mediated immunity although CTL response to p97 is not shown so far (collaboration with I. Hellstrom and Shi Luk Ho, Bristol Myers Squibb, Seattle, USA). In another experiment performed in a monkey model (cotton top tamerin) iscom carrying the Epstein Barr virus (EBV) envelope protein gp340 induced complete protection to a 100% tumourgenic dose of EBV (Morgan et al, 1988). The protection is considered due to cell mediated immunity since EBV neutralizing antibodies are not protective. It still remains to be shown if MHC class I restricted CTL is the effector mechanism.

In a recent study, van Binnendijk (1992) showed that the fusion protein of measles virus carried by iscoms, besides inducing MHC class I restricted immune response, also induces MHC class II restricted response which was not affected by treatment with chloroquine. The chloroquine abrogates the processing in lysosomes by keeping the pH neutral, which prevents the proteolytic activity being dependent on acid pH. Most likely, therefore, the antigen in the chloroquine treated cells have been processed in the cytosol and thereafter transported to the ER for association to MHC class II molecules and eventually back to lysosomes and then out to the cell surface in the plasma membrane for the presentation to T-helper cells. These results are in accord with recent reports showing that the endogenous pathway for MHC-class II antigen is operative (Nuchtern et al, 1990; Jaraquemada et al, 1990).

The crucial point of induction of MHC class I restricted immune response is to deposit the antigen in the cytosol. Collected data from the electron microscopic studies indicate that this is the case, since it was visualized that biotinylated antigen carried by iscoms and internalized by macrophages could be located with gold streptavidin to vesicles (endosomes) as well as outside the vesicles presumably in the cytosol.

IMMUNE RESPONSE TO HIV-ISCOMS

Generally, very high antibody titres as well as strong cell mediated immune responses including MHC class I restricted CTL are induced by iscoms containing gp160 of HIV-1 integrated via the transmembrane region (Akerblom et al, 1991, Takahashi et al, 1990). Such iscoms are likely to contain several copies of gp160, which is a prerequisite for a highly immunogenic iscom. Problems arise when the external units (SU) of the envelope proteins of HIV-1, HIV-2 or SIV are to be incorporated into iscom because they lack a hydrophobic transmembrane region. In the case of gp120 of HIV-1 the proportion of gp120 conjugated to the matrix with the SPDP coupling procedure, (Eriksson et al, manuscript in preparation) was low (about 16%) resulting in few antigens per iscom particle and a suboptimal immune response. This was also the case with gp120 of HIV-1 incorporated into iscoms (Pyle et al, 1989; Eriksson 1992) by exposing hydrophobic domains using acid conditions. However, after three immunizations the same level of antibody response was obtained as compared with two immunizations of gp160 iscoms (recovery of more than 80% of gp160) (Åkerblom et al, 1991). Also Gp120 of SIV has been integrated into iscoms either by acid pH or by chemical conjugation with similar integration rates i.e. about 16%. An experimental SIV vaccine containing both of these gp120 iscoms induced protection in two out of four monkeys to challenge infection with SIV-infected leukocytes from an infected monkey (Osterhaus et al, 1992). A similar result was obtained with an MDP adjuvanted experimental vaccine although the antibody response to the MDP formulation was lower. In another experiment three out of four monkeys were protected from challenge infection with HIV-2 grown in monkey cells. In this study the monkeys were first immunized with HIV-2 iscoms and then boosted with peptides representing a V-3 region of gp120 of HIV-2 (Putkonen et al, 1992). An experimental HIV-2 vaccine adjuvanted with RIBI failed to induce protection in this study.

THE USE OF PEPTIDES FOR IMMUNIZATION

What place do synthetic peptides have in prospective vaccines? Can they be used to emphasize certain antigenic determinants immunologically not well exposed in a native antigen? Mice immunized with gp160 iscom respond with high antibody response to gp160, but only 20% of the mice elicit antibody to the V3-loop of gp120 (Åkerblom et al, 1991) harbouring the principal neutralizing determinant (PND)(La Rosa et al, 1990). The failure of individuals to respond to the V3 region seems not to be genetically controlled since this is the case with both inbred (Balb/C) and outbred (NMRI) mice. In contrast, in another species - i.e. the guinea pig - all individuals responded to the PND and with high virus neutralizing antibody titres following immunization with gp160 iscoms. In the mouse, the V3 loop is possibly cleaved at the serine cleavage sequence GPGRA during the process of presentation in the antigen presenting cells giving few individuals time to respond. In order to analyse if a peptide can induce antibody response to the PND, a peptide covering that region was conjugated to the envelope proteins of influenza virus in iscoms and was used to immunize mice. All the mice responded with antibodies to both the peptide and to the gp160, i.e. all mice can respond to the PND if this region is accentuated by a peptide. In a subsequent experiment, it was shown in guinea-pigs that a peptide covering the V3-region conjugated to the matrix of the iscom can boost conformational epitopes primed by gp160 iscoms inducing high titres of virus neutralizing antibodies which also neutralize a heterologous isolate (Åkerblom et al, submitted).

HIV-1 PEPTIDES CARRIED BY ISCOMS INDUCE EFFICIENTLY ANTIBODY RESPONSE BY MUCOSAL INTRANASAL ROUTE

It is previously shown that influenza virus iscoms containing envelope proteins haemagglutinin and neuraminidase induce via the intranasal route a strong immune response and protection to challenge infection (Lövgren 1988; Lövgren et al, 1990). Consequently, these iscoms penetrate mucus to reach APC in the mucosal membrane and subsequently induce a powerful immune response.

In a recent study by Åkerblom et al, an influenza virus iscom was tested for its capacity as a carrier for mucosal administration of a peptide. An HIV-1 peptide covering the V3-region including the PND conjugated to influenza virus iscoms induced clear-cut antibody response to the peptide already after one intranasal immunization and a high serum

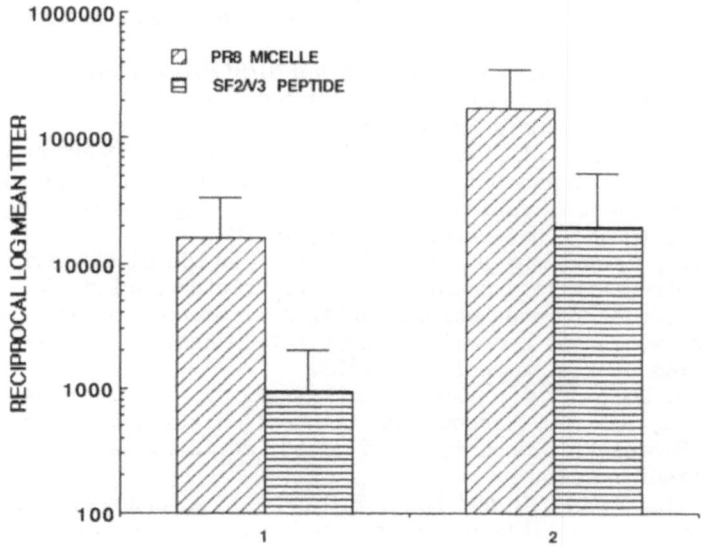

Fig. 5. HIV-peptide conjugated to influenza virus iscoms
efficiently induces antibody response by intranasal immunization of mice (n=10). Mean serum antibody responses
were measured by ELISA for peptide SF2/V3 (amino acid
sequence 296-335) and the carrier protein (the envelope
proteins of influenza virus serotype A, H1N1, PR8 isolate)
following intranasal immunizations six weeks apart

antibody response in most mice after the second administration (Fig. 5).
The serum antibody titres showed greater variation between the mice after
intranasal immunization than after parenteral administration, a problem
which most likely will be overcome with improved intranasal administration
techniques.

PROSPECTIVES

The iscom is an immunological carrier system useful for defined
antigens no matter if they are produced in conventional microorganisms, by
gene technology techniques or chemical synthesis. Iscoms induce response
both over MHC-class I and class II molecules and can be formulated for
parenteral and mucosal administration.

The prospectives for future vaccine development seem interesting
(also for human use), since iscoms can be formulated with purified
components from Quil A which proved innocuous in toxicological studies.

REFERENCES

Åkerblom, L., Nara, P., Dunlop, N. and Morein, B., 1991, Cross-neutralizing
    antibodies to HIV-1 in mice after immunization with gp160 iscoms.
    Dissection of the immune response Aids res. and human retroviruses,
    7:621.
Allison, A.C. Antigens and adjuvants for a new generation of vaccines,

1989. in: Immunological Adjuvants and Vaccines, eds., Gregoriadis, G., Allison, A.C. and Poste, G.

Eriksson, S., 1992, Immunogenicity of a gp120/Sf20iscom conjugate in comparison with other adjuvant systems. in: Modern Approaches to New Vaccines. Cold Spring Harbor.

Fekadu, M., Shaddock, J.H., Ekstrom, J., Osterhaus, A., Sanderlin, D.W., Sundquist, B. and Morein, B., 1992, An immune stimulating complex (iscom) subunit rabies vaccine protects dogs and mice against street rabies challenge, Vaccine, 10:192.

Heeg, K., Kuon, W. and Wagner, H., 1991, Vaccination of class I major histocompatibility complex (MHC)-restricted murine CD8+ cytotoxic T lymphocytes towards soluble antigens: immunostimulating ovalbumin complexes enter and allow sensitization against the immunodominant peptide, Eur.J.Immunol., 21:1523.

Höglund, S., Dalsgaard, K., Lövgren, K., Sundquist, B., Osterhaus, A. and Morein, B., 1989, Iscoms and immunostimulation with viral antigens. in: Subcellar biochemistry 15:39-68, ed. J.R. Harris, Plenum Publishing Corp.

Jaraquemada. D., Marti, M. and Long. E.O., 1990, An endogenous processing pathway in vaccinia virus-infected cells for presentation of cytoplasmic antigens to class II-restricted T cells, J.Ex.Med., 172, 947.

Jones, P.D., Tha Hla, R., Morein, B., Lövgren, K. and Ada, G.L., 1988, Cellular immune responses in the murine lung to local immunization with influenza A virus glyocproteins in micelles and immuno-stimulating complexes (iscoms), Scand.J.Immunol., 27:645.

Kelly, A., Powis, S.H., Kerr, L-A., Mockridge, I., Elliot, T., Bastin, J., Uchanska-Ziegler, B., Ziegler, A., Trowsdale, J. and Townsend, A., 1992, Assembly and function of the two ABC transporter proteins encoded in the human major histocompatibility complex, Nature, 355:641.

LaRosa, G.J., Davide, J.P., Weinhold, K., Waterbury, J.A., Profy, A.T., Lewis, J.A., Langlois, A.J., Dreesman, G.R., Boswell, R.N., Shaddock, P., Holley, L.H., Karplus, M., Bolognesi, D.P., Matthews, T.J., Emini, E.A., Putney, S.D. 1990, Conserved sequence and structural elements in the HIV-1 principal neutralizing determinant, Science, 249:932.

Larsson, M., Lövgren, K. and Morein, B., Immunopotentiation of synthetic oligopeptides by chemical conjugation to iscoms. Submitted.

Lövgren, K., 1988, The serum antibody responses distributed in subclasses and isotypes following intranasal and subcutaneous immunization with influenza virus iscoms, Scand.J.Immunol., 2:241.

Lövgren, K., Kaberg, H. and Morein, B., 1990, An experimental influenza subunit vaccine (ISCOM) - induction of protective immunity to challenge infection in mice after intranasal or subcutaneous administration, Clin.Exp.Immunol., 82:435.

Maloy, K.J., Donachie, A.M. and Mowat, A. Mcl., 1992, Iscoms as adjuvants for induction of class I MHC-restricted T cells and secretory immunity by oral immunisation, in: Abstracts, 8th Intern. Congress of Immunology, Budapest, Hungary, Springer Hungarica.

Monaco, J.J., 1992, A recent HLA Nomenclature Committee meeting resolved that the names Tap-1 and Tap-2 (for transporter associated with antigen processing) be used for Ham-1 and Ham-2 and for their homologs in rats and humans, Immunol.Today, 13, 173.

Morein, B. and Höglund, S., 1990, Subunit vaccines against infection by enveloped viruses, In Viral Vaccines, 69.

Morein, B., Helenius, A., Simons, K., Petterson, R., Kaariainen, L. and Schirrmacher, V., 1978, Effective subunit vaccines against enveloped animal virus, Nature, 276.

Morein, B. and Simons, K., 1985, Subunit vaccines enveloped viruses: virosomes, micelles and other protein complexes, Vaccine, 3:83.

Morein, B. and Åkerblom, L., 1992, The iscom - an approach to subunit vaccines, in: Recombinant DNA Vaccines, ed. R.I. Isaacson.

Morein, B., Lövgren, K. and Hoglund, S., 1989, Immunostimulating complex (iscom), in: Immunological adjuvant and vaccines. pp 153-161, eds Gregoriadis, G., Allison, A.C. and Poste, G. Plenum publising Corp.

Morgan, A., Finerty, S., Lövgren, K., Scullion, F.T. and Morein, B., 1988, Prevention of Epstein-Barr (EB) virus-induced lymphoma in cottontop tamarins by vaccination with EB virus envelope glycoprotein gp340 incorporated into immune-stimulating complexes, J.Gen.Virol., 69:2093.

Mowat, A. and Donachie, A., 1991, Iscoms - a novel strategy for mucosal immunization, Immunol,Today 12:383.

Nuchtern, J.G., Biddison, W.E. and Klausner, R.D., 1990, Class II MHC molecules can use endogenous pathway of antigen presentation, Nature 343: 74.

Osterhaus, A., de Vries, P. and Heeney, J., 1992, Scientific correspondence, Nature, 355:684.

Powis, S.J., Deverson, E.V., Coadwell, W.J., Ciruela, A., Huskinsson, N.S., Smith, H., Butcher, G.W. and Howard, J.C., 1992, Effect of polymorphism of and MHC-linked transporter on the peptides assembled in a class I molecule, Nature, 357:211.

Putkonen, P., Byorling, E., Åkerblom, L., Thorstensson, R., Benthin, L., Lovgren, K., Biberfeld, G., Morein, B., Norrby, E., Wigzell, H., 1992. Protection of macaques against HIV-2 with an HIV-2 iscom vaccine. Submitted.

Pyle, S., Morein, B., Bess, J., Åkerblom, L., Nara, P., Nigida, S., Lerche, N., Fishinger, P. and Arthur, L., 1989, Immune response to immunostimulatory complexes (iscoms) prepared from human immunodeficiency virus type 1 (HIV-1) or the HIV-1 external envelope glycoprotein, Vaccine, 7:465.

Rudensky, A.Y., Preston-Hurlburt, P., Hong, S-C., Barlow, A. and Janeway, Jr, C.A., 1991, Sequence analysis of peptides bound to MHC class II molecules, Nature, 353:622.

Speijers, G.P.A., Danse, L.H.J.C., Beuvry, E.C., Strik, J.J.T.W.A. and Vos, J.G., 1988, Local reactions of the saponin Quil A containing iscom measles vaccine after intramuscular injection of rats: A comparison with the effect of DPT-Polio vaccine, Fundamental and Applied Toxicology, 10:425.

Takahashi, H., Takeshita, T., Morein, B., Putney, S., Germain, R.N. and Berzofsky, J., 1990, Induction of CD8+ cytotoxic T cells by immunization with purified HIV-1 envelope proteins in iscoms, Nature, 344:873.

Van Binnendijk, R.S., van Baalen, C.A., Poelen, M.C:M., de Vries, P., Boes, J., Cerulundo, V., Osterhaus, A.D.M.E. and UytdeHaag, F.G.C.M., 1992, Measles virus transmembrane fusion protein synthesized de novo or presented in endogenously processed for HLA class I and class II-restricted cytotoxic T-cell recognition, J.Exp.Med., 176, 119.

Villacres-Eriksson, M. and Morein, B., Macrophages, B cells and dendritic cells as antigen presenting cells for iscom borne antigen. Submitted.

Villacres-Eriksson, M., Bergstrom-Mollaogly, B., Kaberg, H., Lövgren, K. and Morein, B., 1993, The induction of cell-associated and secreted interleukin-1 by iscoms, matrix or micelles in murine splenic cells, Clin.Exp.Immunol., in press.

Villacres-Eriksson, M., Kaberg, H., Mollaoglu, M. and Morein, B., 1992, Involvement of Interleukin-2 and interferon-gamma in the immune response induced by influenza virus iscoms, Scand.J.Immunol., 36:421.

Watson, D., Watson, N., Fossum, C., Lövgren, K. and Morein, B., 1989, Inflammatory response and antigen localization following immunization with influenza virus iscoms, Inflammation, 13:641.

Watson, D.L., Watson N.A., Fossum, C., Lövgren, K. and Morein, B., 1992, Interactions between immune-stimulating complexes (ISCOMs) and peritoneal mononuclear leucocytes, Microbiol.Immunol., 36:199.

NANOPARTICLES AS POTENT ADJUVANTS FOR VACCINES

Jorg Kreuter

Institut für Pharmazeutische Technologie
J.W. Goethe-Universitat
D-6000 Frankfurt/Main, Germany

INTRODUCTION

Many antigens, especially smaller peptides, virus subunits, and antigens produced by genetic engeneering procedures are weak antigens and produce none or a low protection. For this reason, it may be necessary to add adjuvants to render these antigens potent enough to be useful for vaccines.

A large number of adjuvants have been used including minerals, emulsions, peptides, lipids, and surfactants. Unfortunately, many of these adjuvants cause adverse toxicological effects or they are difficult to manufacture in a reproducable manner (Kreuter and Haenzel, 1978). It has been shown previously (Kreuter et al., 1986; Grafe, 1971) that the particle size of the adjuvant can significantly influence the immune response. With emulsions, however, the size of the emulsion droplets may change from preparation to preparation. Moreover, the particle size that will be relevant for the immune response may be further altered during and after injection. The consistency of the tissue in which the vaccine is injected may yield a smaller droplet size due to fricton between the vaccine liquid and the tissue. On the other hand, for instance, fatty tissue may induce a coalescence of the droplets resulting in a larger particle size.

The structure and the properties of aluminium compounds - the most commonly used adjuvants for commercial vaccines - may change significantly with slight alterations in production conditions and with ageing (N. Kerkhof et al., 1975; Nail et al., 1976a-c). This can lead to significant differences in the immune response obtained (Kreuter et al., 1986; Haas and Thommsen, 1960; Jolles and Paraf, 1973; Wunderli, 1950; Muggleton and Hilton, 1967; Pyl, 1953). This is demonstrated by the fact that the adsorption properties of different qualities of aluminium hydroxides showed no correlation with the adjuvant effect obtained (Pyl, 1953).

For this reason, adjuvants on the basis of submicron biodegradable polymeric particles - so-called nanoparticles - were developed (Kreuter and Speiser, 1976). These particles can be prepared in a physico-chemically reproducable manner within narrow limits (Kreuter, 1983a; Berg et al., 1986). The most promising polymer - poly(methyl methacrylate) - has been used in surgery for over 50 years and was shown to be slowly biodegradable in the form of nanoparticles (Kreuter et al., 1983). The observed slow degradation rate of about 30% to 40% of the nanoparticulate polymer per year seems to be very promising for vaccines, because a prolongation of the contact of the antigen with the immuno-competent cells of the organism in most instances is required in order to maintain a long immunity. Additionally, preliminary results showed that the local tissue response of this adjuvant was comparable to the fluid antigen without adjuvant and much less severe than the response after using aluminium hydroxide (Kreuter et al., 1976). For this reason, these polyacrylic nanoparticles hold promise for use as adjuvants for vaccines.

Definition of Nanoparticles

Nanoparticles are solid colloidal particles ranging in size from 10 to 100 nm (1 μm). They consist of macromolecular materials in which the active principle (drug or biologically active material) is dissolved, entrapped, encapsulated and/or to which the active principle is adsorbed or attached (Kreuter, 1983b).

This definition includes not only particles with a solid or porous matrix-type interior, but also nanocapsules with a shell-like wall or "microspheres" if they are below 1 μm in size. Nanoparticles for use in humans or animals have to be biocompatible and biodegradable.

Preparation of Nanoparticles

Polyacrylic materials so-far have been shown to hold most promise for use as adjuvants for vaccines. For this reason, the preparation methods of nanoparticles made of these types of polymers are shortly re-viewed. Polyacrylic nanoparticles are produced by emulsion polymerization. This name is misleading because in the case of the most promising adjuvant polymer - poly(methyl methacrylate) - no emulsifier is required. As discussed later, an emulsifier even may reduce the adjuvant effect since it will reduce the hydrophobicity of the adjuvant surface.

The production of poly(methyl methacrylate) nanoparticles is very simple. Monomeric methyl methacrylate purified from polymerization inhibitors (Kreuter and Speiser, 1976; Berg et al., 1986) is added to pure distilled water or to a desired buffer solution up to amounts of 1.5% (vol./vol.). Normally amounts of 0.5 to 1.0% are used. The polymerization is then carried out by gamma-irradiation (500 krad) (Kreuter and Zehnder, 1978). Alternatively, the polymerization can be carried out by heating to 65-85°C and addition of a polymerization initiator such as potassium peroxodisulfate or ammonium peroxodisulfate in concentrations between $3x10^{-4}$ to $5x10^{-2}$ mol. After the polymerization, the resulting preparation may be stored as such or may be freeze-dried.

In the case of the polyalkylcyanoacrylates, the addition of surfactants or protective colloids is required at monomer concentrations exceeding 0.1% because these materials possess a much lower glass transition temperature (T_g). For this reason, the polymerizing particles would coagulate without protective surfactants or colloids. Cyanoacrylates are much more reactive than methyl methacrylate. As a result, they can be polymerized at room or even at refrigerator temperature without gamma-irradiation. Because of their high reactivity in neutral and alkaline media the pH has to be lower than 3 (Douglas et al., 1984).

The polymerization can be carried out in the presence of the antigens. In this case, the antigen will be entrapped in the particles (Kreuter and Speiser, 1976). However, polymerization in the presence of

Table 1. Physico-chemical characterization of poly(methyl methacrylate) nanoparticles (Kreuter (in press), reprinted with permission of the copyright holder)

Parameter	Method	
Particle size	Transmission electron microscopy	130 ± 28 nm
	Photon correlation	125 ± 21 nm
Specific surface area	Nitrogen adsorption (BET)	52.8 ± 2.3 m²/g
Density	Helium pycnometer	1.06 g/cm³
x-diffraction	Guinier-De Wolff Camera	amorphous
Surface charge	Electrophoresis	-2.76 ± 0.57 μm·cm/s·V
Wetability	Contact angle	73° ± 7°
Molecular weight	Gel permeation chromatography	20.000[a] 130.000[b]

a μ-irradiation initiated polymerization
b heat induced polymerization

the antigen using heat polymerization can be used only for heat-stable antigens.

Alternatively, the antigens may be added after polymerization leading to the adsoption of the antigens on the polymer surface.

Physico-Chemical Characterization

The physico-chemical characterization of nanoparticles is carried out using scanning and transmission microscopy, photon-correlation spectroscopy, BET-surface area analysis, helium pycnometry, X-ray diffraction, surface charge, wetability (Kreuter, 1983a), and gel permeation chromatography (Bentele et al., 1983). The physico-chemical parameters for poly-(methyl methacrylate) nanoparticles are given in Table 1.

Influence of Physico-Chemical Parameters on the Adjuvant Effect

A. Particle Size

The influence of the particle size of poly(methyl methacrylate) and of polystyrene nanoparticles on the adjuvant effect of influenza (Kreuter and Haenzel, 1978) and of bovine serum albumin (Kreuter et al., 1986) was investigated after intramuscular injection. The particle sizes of the nanoparticles in these experiments ranged from 62 nm to 10 μm. Smaller particles below about 350 nm exhibited a significant adjuvant effect (Kreuter et al., 1986). This adjuvant effect was better than aluminium hydroxide at sizes below 150 nm. The adjuvant effect of particles of a size of 62 nm was lower than that of those with a size of 103 nm and 132 nm. However, the latter result was not statistically significant and may be caused by the biological variations. For this reason, at present, it is not clear if a particle size optimum exists arround 100 nm or if particle sizes below this size yield equal or better adjuvant effects.

B. Hydrophobicity

The influence of hydrophobicity on the adjuvant effect using bovine serum albumin or influenza subunits as antigens was investigated by determination of the antibody response against bovine serum albumin and of the protection against infection with live influenza virus in mice (Kreuter et al., 1988). The hydrophobicity of the nanoparticles was altered by insertion of hydroxyl groups into the polymer or by the exchange of a methyl group of the acrylates against a cyano group and by variation in the length of the ester side chain. The hydrophobicity of the resulting polymers was investigated by the measurement of the water contact angles. Both the antibody determination as well as the mouse protection experiments led to the conclusion that the adjuvant effect increased with increasing hydrophobicity (Kreuter et al., 1988).

Adjuvant Effects with Nanoparticles using Influenza or HIV-1 and HIV-2 Virus Vaccines

A. Influenza

The adjuvant effect of whole virus and of subunit influenza vaccines were tested in mice and guinea-pigs. The antibody response using the chicken cell hemagglutination inhibition test (Kreuter and Haenzel, 1978; Kreuter and Speiser, 1976; Kreuter et al., 1976; Kreuter and Liehl, 1981) as well as the protection against infection by a challenge with live mouse-adapted influenza virus using a dose of 50 LD_{50} were determinated (Kreuter and Liehl, 1978; 1981). Both types of nanoparticle vaccine preparations, incorporation of the whole virus or of virus subunits into the nanoparticles in presence of the antigens or adsoption of the antigens on empty nanoparticles, were investigated.

The experiments showed that an optimal adjuvant effect was obtained 4 weeks afterprimary immunization in mice and guinea-pigs at a nanoparticle concentration of 0.5% relative to the injection volume (Kreuter and Speiser, 1976; Kreuter and Liehl, 1981). This optimum was observable with the incorporated and with the adsorbed antigens. The optimum after adsorption, however, was much less pronounced than after incorporation. After incorporation, the antibody response with higher polymer concentrations of about 2% decreased significantly below that of the fluid vaccine, indicating that so much antigen surface was covered by the polymer that only a very minimal immune response was exhibited. After prolonged time periods and boosting, the differences in antibody response caused by the variation of the polymer content decreased and high polymer content vaccines yielded equal or sometimes even better responses than the 0.5% concentration (Kreuter and Speiser, 1976; Kreuter and Liehl, 1981). This may be an indication that additional antigen surface becomes accessible to the immunocompetent cells probably due to degradation of the nanoparticle polymer.

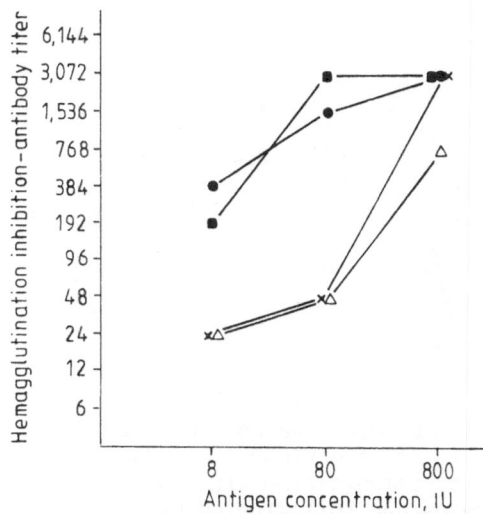

Fig. 1. Dependence of the antibody response on the antigen
concentration after 3 weeks. Sera were pooled before
titration. ■ = 0.5% PMMA + 0.5% PAA, polymerized in the
presence of antigens; ● = 0.8% PMMA, polymerized in the
absence of antigens; × = 0.1 Al(OH); △ = fluid vaccine.
(Sera were titrated individually) (Kreuter et al, 1976;
reproduced with permission of the copyright holder)

With whole virus vaccines substantially higher antibody responses
were obtained after incorporation of the virions into the nanoparticles
used at an optimal concentration of 0.5% poly(methyl methacrylate) in
comparison to the adsorbed vaccine and to aluminium hydroxide (Kreuter and
Speiser, 1976). The latter two vaccines yielded significantly higher antibody responses than the fluid vaccines. The protection obtained with these
three adjuvants, however, was equal when the adjuvants were added in the
above mentioned optimal concentration (Kreuter and Liehl, 1978). The adjuvant effect disappeared when the adjuvants were diluted, together with
the antigen, down to very low adjuvant concentrations. The protection experiments, on the other hand, indicated that both nanoparticle preparations
were superior to aluminium hydroxide at low antigen concentrations and
optimal adjuvant concentrations.

These observations were confirmed by the antibody determination in
guinea-pigs and mice after immunization with influenza subunits (Kreuter et
al., 1976). Both nanoparticle preparations, incorporated as well as adsorbed product, were significantly superior to aluminium hydroxide and to
the fluid vaccine. Again by far the highest differences were observed at
low antigen concentrations with the nanoparticles in comparison to the
latter two vaccines (Fig. 1).

The antibody response and the protection with the nanoparticle
adjuvants in comparison to the aluminium and to the fluid vaccine were
especially pronounced after longer time periods (Kreuter and Liehl, 1981;
Fig. 2).

In order to determine the heat stability of nanoparticle influenza
vaccines, vaccines with a low antigen content of 21 IU/ml were stored at
40°C for 0, 15, 30, 60, 120, and 240 hours and then administered intramuscularly to mice. After 4 weeks, the mice were challenged. The heat
storage did not reduce significantly the protection obtained with both

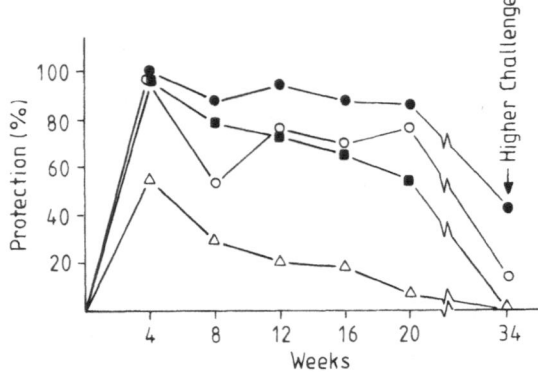

Fig. 2. Protection of mice against infection with a dose
of 50 LD_{50} of infectious mice adapted influenza virus
after s.c. immunization with 20 IU of A PR 8 whole
virus using the following adjuvants: ● incorporation
into 0.5% poly(methyl methacrylate); ○ adsorption onto
0.5% poly(methyl methacrylate); ■ adsorption onto 0.2%
aluminium hydroxide; △ fluid vaccine without adjuvant.
Higher challenge = 250 LD_{50}

nanoparticle preparations. A significant drop in the protection was observed with aluminium hydroxide, and the protection disappeared totally
after storage of the fluid vaccine for 60 hours.

B. HIV

Enormously higher adjuvant effects in comparison to the results
obtained with influenza were observed with HIV-1 and especially with HIV-2
(Stieneker et al., 1991 and (in press). The antibody responses against
these antigens grown on Molt-4 clone 8 cell cultures were determined by
ELISA after a single subcutaneous injection of 5 to 50 ug antigen/0.5 ml to
mice. The virus was purified by sucrose gradient centrifugation, split
with Tween 80/ether, and inactivated with formalin 0.5% for 72 hours. Only
nanoparticle preparations produced by adsorption were used in these experiments and compared to 0.2% aluminium hydroxide and to the fluid antigen
without adjuvant. The antibody response with HIV-2 was 10 to 200-fold
higher than with aluminium hydroxide (Fig. 3).

The latter adjuvant in turn yielded 10 to 100-fold higher responses
than the fluid vaccine (Steineker et al., 1991). With HIV-1 a delayed
antibody response was observed. Significantly high antibody titres were
reached after 10 weeks whereas HIV-2 yielded high titres already after 4
weeks (Fig. 3). Although the HIV-1 antibody titers were about 5 to 10
times lower than with HIV-2, they still were about 10 times higher than
those with aluminium hydroxide or with the fluid vaccine, even in comparison to HIV-2. Again, the titres obtained with aluminium hydroxide were
generally statistically higher than those of the fluid vaccine.

In a final experiment using a single injection of 5 µg antigen and a
single time point determination after 10 weeks, poly(methyl methacrylate)
nanoparticles were compared to 24 different adjuvants including different

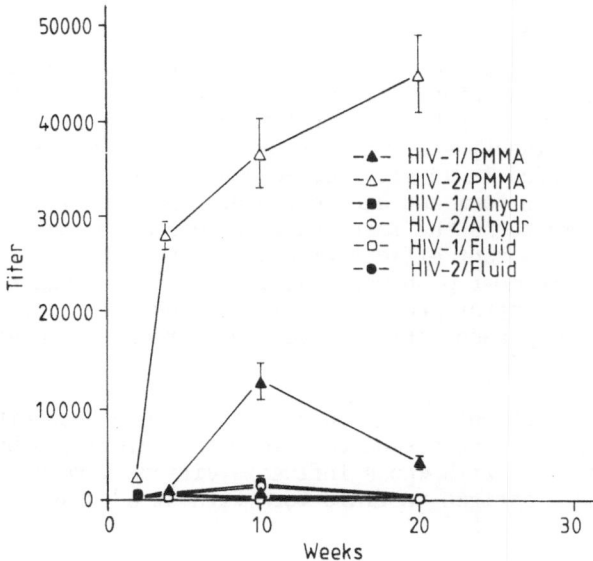

Fig. 3. Antibody response (ELISA) of mice after immunization
with HIV-1 and HIV-2 vaccines (10 µg antigen/mouse)
containing different adjuvants. PMMA = poly(methyl
methacrylate); Alhydr = aluminium hydroxide; Fluid =
fluid vaccine without adjuvant

aluminium compounds, Freund's complete and incomplete adjuvant, different
liposomes, surfactants Iscoms, and muramyl peptides. Again, the nanoparticles were by far the superior adjuvant (Kreuter, 1992).

Body Distribution and Elimination of Poly(methyl Methacrylate) Adjuvant Nanoparticles

After subcutaneous injection of poly(methyl methacrylate) nanoparticles labelled in the polymer chain with ^{14}C to rats almost all of the
radioactivity stayed at the injection site (Kreuter et al., 1979; 1983).
Less than 1% of the injected dose was found in the residual body. After an
initial urinary and fecal excretion rate of about 1% of the administered
dose per day, the elimination rate dropped rapidly to very low levels of
about 0.005% per day via the feces and 0.0005% per day via the urine.
Between about 200 and 280 days, the excretion rate in the feces and the
concentration in the residual body increased by a factor of 100 to 300.
The urinary excretion as well as the organ to blood radioactivity ratio
remained constant. The residual radioactivity at the injection site after
287 days amounted to 60-70% of the initial dose (Kreuter et al., 1983).
The local response at the injection site one year after intramuscular
injection of poly(methyl methacrylate) nanoparticles with entrapped whole
influenza virions was not different from the pure antigen (Kreuter et al.,
1976). Eosinophiles and giant cells were observed in both cases whereas
severe granuloma formation was observed with aluminium hydroxide. These
results strongly indicate that poly(methyl methacrylate) nanoparticles are
slowly biodegradable. Excretion of the biodegradation products seems to
occur primarily via the bile and the feces. Residual nanoparticles at the
injection site exhibited only mild tissue reactions.

CONCLUSION

Poly(methyl methacrylate) nanoparticles are powerful adjuvants for a number of antigens including influenza whole virions and subunits, bovine serum albumin, rabies (unpublished preliminary observations), and especially HIV-1 and HIV-2 inactivated whole virion subunit antigens. This polymer material has been used in surgery for 50 years. Due to its biodegradability in nanoparticulate form and due to its hydrophobicity poly(methyl methacrylate) seems to be the most promising polymer material for immunological adjuvants. Adjuvant effects with this material in comparison to other adjuvants were most pronounced at low antigen concentrations, with weaker antigens, and after prolonged time periods. In addition, poly-(methyl methacrylate) nanoparticles seem to increase the heat stability of antigens.

The tissue reactions at the injection site one year after injection exhibited much less severe tissue reactions than aluminium hydroxide. These tissue reactions with whole influenza virions were not different from those with the same antigen in fluid form without adjuvant.

ACKNOWLEDGEMENTS

The preceding chapter includes extracts from the chapter "Nanoparticles" in the book "Colloidal Drug Delivery Systems" edited by J. Kreuter and published by Marcel Dekker, New York, in press.

REFERENCES

Bentele, V., Berg, U.E. and Kreuter, J., 1983, Molecular weights of poly(methyl methacrylate) nanoparticles, Int.J.Pharm., 13:109.

Berg, U.E., Kreuter, J., Speiser, P.P. and Soliva, M., 1986, Herstellung und in vitro-Prüfung von polymeren Adjuvantien für Impfstoffe, Pharm.Ind., 48:75.

Douglas, S.J., Illum, L., Davis, S.S. and Kreuter, J., 1984, Particle size and size distribution of poly(butyl-2-cyanoacrylate) nanoparticles. I. Influence of physico-chemical factors, J.Colloid Interface Sci., 101:149.

Grafe, A., 1971, Hochdisperses δ-Al$_2$O$_2$-Aerosol als Adjuvans bei der Herstellung inaktiver Virus- oder Impfstoffe, Arzneim.-Forsch., 21:903

Haas, R. and Thomssen, R., 1960, Über den Entwicklungsstand der in der Immunbiologie gebräuchlichen Adjuvantien, Ergeb.Mikrobiol. Immunitätsforsch,Exp., 34:27.

Jollès, P. and Paraf, A., 1973, Substances exhibiting an adjuvant effect: Preparation of 'crude material', in: "Chemical and Biological Basis of Adjuvants: Molecular Biology, Biochemistry, and Biophysics", A. Kleinzeller, Springer, G.F. and Wittmann, H.G., eds., Springer-Verlag, Berlin, 1973, pp. 5-18.

Kerkhof, N.J., White, J.L. and Hem, S.L., 1975, Effect of dilution on reactivity and structure of aluminium hydroxide gel, J.Pharm.Sci., 64:940.

Kreuter, J., 1983a, Physico-chemical characterization of polyacrylic nanoparticles, Int.J.Pharm., 14:43.

Kreuter, J., 1983b, Evaluation of nanoparticles as drug delivery systems I: Preparation methods, Pharm.Acta Helv., 58:196.

Kreuter, J., (1992), Physico-chemical characterization of nanoparticles and their potential for vaccine preparation, Vaccine Res., 1:93.

Kreuter, J. and Haenzel, I., 1978, Mode of action of immunological adjuvants: Some physico-chemical factors influencing the effectivity of polyacrylic adjuvants, Infect.Immunity, 19:667.

Kreuter, J. and Liehl, E., 1978, Protection induced by inactivated influenza vaccines with poly(methyl methacrylate) adjuvants, Med.Microbiol.Immunol., 165:111.

Kreuter, J. and Liehl, E., 1981, Long-term studies of microencapsulated and adsorbed influenza vaccine nanoparticles, J.Pharm.Sci., 70:367.

Kreuter, J. and Speiser, P.P., 1976, New adjuvants on a polymethylacrylate base, Infect. Immunity, 13:204.

Kreuter, J. and Zehnder, H.J., 1978, The use of ^{60}Co-γ-irradiation for the production of vaccines, Radiation Effects, 35:161.

Kreuter, J., Mauler, R., Gruschkau, H. and Speiser, P.P., 1976, The use of new poly(methylmethacrylate) adjuvants for split influenza vaccines, Exp.Cell Biol., 44:12.

Kreuter, J., Täuber, U. and Illi, V., 1979, Distribution and elimination of poly(methyl-2-^{14}C-methacrylate) nanoparticle radioactivity after injection in rats and mice, J.Pharm.Sci., 68:1443.

Kreuter, J., Nefzger, M., Liehl, E., Czok, R. and Voges, R., 1983, Distribution and elimination of poly(methyl methacrylate) nanoparticles after subcutaneous administration to rats, J.Pharm.Sci., 72:1146.

Kreuter, J., Berg, U., Liehl, E., Soliva, M. and Speiser, P.P., 1986, Influence of the particle size on the adjuvant effect of particulate polymeric adjuvants, Vaccine, 4:125.

Kreuter, J., Liehl, E., Berg, U., Soliva, M. and Speiser, P.P., 1988, Influence of the hydrophobicity on the adjuvant effect of particulate polymeric adjuvants, Vaccine, 6:253.

Muggleton, P.W. and Hilton, M.L., 1967, Some studies on a range of adjuvant systems for bacterial vaccines, in: "International Symposium on Adjuvants of Immunity, Utrecht 1966", R.E. Regamey, Hennessen, W., Ikic, D. and Ungar, J., eds., Symposia Series Immunobiological Standardization, vol. 6., Karger, Basel, pp. 29-38.

Nail, S.L., White, J.L. and Hem. S.L., 1976a, Structure of aluminium hydroxide gel. I. Initial precipitate, J.Pharm.Sci., 65:1188.

Nail, S.L., White, J.L. and Hem, S.L., 1976b, Structure of aluminium hydroxide gel. II. Ageing mechanism, J.Pharm.Sci., 65:1192.

Nail, S.L., White, J.L. and Hem, S.L., 1976c, Structure of aluminium hydroxide gel. III. Mechanism of stabilization by sorbitol, J.Pharm. Sci., 65:1195.

Pyl, G., 1953, Die Prüfung von Aluminiumhydroxid auf seine Eignung für die Maul- und Klauenseuchevakzine, Arch.Exp.Veterinärmed., 7:9.

Stieneker, F., Kreuter, J. and Löwer, J., 1991, High antibody titres in mice with poly(methylmethacrylate) nanoparticles as adjuvants for HIV vaccines, AIDS, 5:431.

Stieneker, F., Kreuter, J. and Löwer, J., Different kinetics of the humoral immune response to inactivated HIV-1 and HIV-2 in mice: Modulation by PMMA nanoparticle adjuvant, Vaccine Res., (in press).

Wunderli, H.K., 1950, Untersuchungen über die Adsorption des Maul- und Klauenseuchevirus an Aluminiumhydroxid, Z.Hyg.Infektionskr., 132:1.

OPTIMIZATION OF CARRIERS AND ADJUVANTS

A MODEL STUDY USING SEMLIKI FOREST VIRUS INFECTION OF MICE

A. Snijders, I.M. Fernandez, C.A. Kraaijeveld and
H. Snippe

Eijkman-Winkler Laboratory for Medical Microbiology
Utrecht University, Heidelberglaan 100
3584 CX Utrecht, The Netherlands

INTRODUCTION

MODERN APPROACHES TO VACCINATION

Advances in molecular biology and chemistry, like genetic engineering (recombinant DNA technique), sequence analysis of DNA and the automated chemical synthesis of oligopeptides, have made it possible to develop alternative vaccines (Table 1). Combined with immunological methods, these techniques allow the prior identification of protein(s) and determinants on proteins (epitopes) that are effective in triggering a protective immune response against the whole pathogen, as well as the large scale production of the protein or peptide identified as a potential vaccine. These developments are further supported by the search for suitable adjuvants, that will enhance the immunogenicity of nonreplicating vaccines without being toxic themselves.

Genetically engineered vaccines

Genetic engineering opened a number of possibilities for vaccine production: (I) The production of extensively attenuated live vaccines by the introduction of controlled mutations (deletions) in virulent viruses resulting in vaccine strains, that are less liable to revert back to wild type. (II) The production of subunit vaccines consisting of one or more proteins of the pathogenic organism by inserting the gene(s) for these protein(s) into a vector (plasmid or virus) followed by its expression in and purification from bacteria, yeast or some insect or mammalian cell line. (III) The production of live recombinant vaccines (carrier vaccines) by introduction of genes encoding proteins of a pathogenic virus into the genome of another non-pathogenic virus (vector) in such a way that the foreign protein will become an integral part of a new recombinant virus. (IV) The production of recombinant vaccines consisting of live recombinant bacteria that express whole viral proteins or isolated determinants [Aldovini and Young, 1991; Agterberg and Tommassen, 1991].

Much research is done on the development of viral carrier vaccines. A number of nonpathogenic viruses that will replicate in the human host, have been engineered to act as vector (carrier) for the expression of protein from pathogenic viruses [reviewed in Bostock, 1990]. Live carrier

Table 1. Viral vaccine types

Live attenuated virus	- virus modified by conventional method [*] - virus modified by genetic engineering
Inactivated virus	- (whole virulent) virus inactivated by chemical treatment [*]
Viral subunit	- natural protein from the virus [*] - recombinant viral protein [*]
Live recombinant virus	- viral protein carried by an unrelated live virus
Live recombinant bacteria	- viral protein carried by live bacterium
Synthetic peptides	- peptide conjugated to carrier protein - free peptides

[*] vaccine type in use for humans

vaccines have a number of advantages: Their effectiveness is usually similar to other live viral vaccines. They contain only little genetic material from the pathogenic virus. They can be used to induce protective immunity to more than one viral disease. However, carrier vaccines still share the disadvantage of being harmful to immunocompromised individuals with the other live vaccine types. For veterinary applications this is usually no problem but it is a serious drawback for any vaccine intended for use in humans. Another problem encountered in the study of carrier and subunit vaccines is that for some viruses, like lymphocytic chorio-meningitis virus (LCMV) and respiratory syncytial virus (RSV), the immune reaction evoked by a single viral protein can cause more severe disease [Oehen et al, 1991; Porterfield, 1986]. In LCMV and RSV infection cytotoxic T cells are probably responsible for the aggravation of the disease [Oehen et al, 1991; Cannon et al, 1988]. Antibodies induced by vaccination can enhance subsequent infection and disease by viruses that infect macrophages [Portefield, 1986], like dengue virus [Halstead, 1988] and feline infectious peritonitis virus [Vennema et al, 1990].

In the production of recombinant subunit vaccines one must be careful that the correct conformation of the original protein is preserved: a bacterial protein expressed in another bacterium will fold correctly, but the expression in bacteria of proteins from viruses that normally infect mammals will often lead to unsatisfactory products, especially because posttranslational modifications like glycosylation do not occur in bacteria, but do influence protein folding [Vidal et al, 1989; Wright et al, 1989]. To date, the only recombinant vaccine that is available for human use is the hepatitis B vaccine produced in yeast. The recombinant viral protein was folded correctly, fortunately without being glycosylated by the yeast cell [Valenzuela et al, 1982]. A yeast type of glycosylation would result in strong antibody responses against the yeast sugar-groups (that differ considerably from those of glycosylated proteins produced in human cells) and would prevent antibody responses to the hepatitis B protein. Expression of viral proteins in mammalian cells will give the correct products, but usually gives very low yield, requires large-scale culture of cells that can easily be infected with other pathogenic viruses and is relatively expensive, making it rather unpractical for vaccine purposes.

Synthetic peptide vaccines

The development of synthetic peptide vaccines started with the hope that it would be possible to trigger the immune system specifically with certain isolated determinants or epitopes without triggering any unnecessary responses. Therefore such vaccines were postulated to be especially safe. In contrast to conventional vaccines, synthetic peptides are relatively easy and cheap to produce, stable for long periods without the need of refrigeration, while scale-up of production and purification is easy [Steward et al, 1991]. This type of vaccine would therefore be especially suitable for use in third-world countries.

To synthesize a peptide which can elicit antibodies reacting with the coat proteins of viruses, it is necessary to determine the sequence of amino acids that comprise an epitope. In cases where the nucleic acid sequence of the gene coding for the relevant viral protein is known, the amino acid sequence can be deduced and a number of predictive approaches [reviewed in Van Regenmortel and Daney de Marcillac, 1988] and experimental approaches [Geysen et al, 1984; Thole et al, 1988] can be employed to identify potential immunogenic determinants. The number of viruses for which the amino acid sequence of one or more of their proteins is known is already large and increases rapidly.

Peptides, like any other antigen, must contain both a B- and a T_h-cell epitope in order to evoke antibody responses. Although peptides that contain only a B-cell epitope can be bound by antibodies (and by the immuno-globulin receptor on B cells), they are non-immunogenic. These peptides are rendered immunogenic by coupling to a large carrier protein (that contains many T_h-cell epitopes) or by incorporating an identified T_h-cell epitope into the peptide [Borras-Cuesta et al, 1987].

The efficacy of synthetic peptides in triggering antibody responses with the correct specificity is debated. The antibody binding sites (B-cell epitopes) on protein antigens are usually conformational structures on the surface of the protein that can not be represented by small linear peptides [Benjamin et al, 19984; Laver et al, 1990]. The conformational nature of B-cell epitopes was deduced from the observation that denaturation (unfolding) or proteins results in a dramatic reduction of their reactivity with antibodies raised to the native protein [Benjamin et al, 1984]. The epitopes that are dependent on the steric conformation of the protein are termed 'conformational' epitopes. Such epitopes contain non-adjacent residues (in the amino acid sequence of the protein chain) brought together by folding of the protein. The determinants that are still recognized in spite of protein denaturation are termed 'linear' or 'sequential' epitopes and represent small continuous stretches of the protein chain. The question is whether antibodies raised against linear epitopes (in the form of synthetic peptides) will have sufficient affinity for the whole virus to mediate protection.

An example of a naturally occurring linear B-cell epitope is an epitope of foot-and-mouth disease (FMDV) that is a flexible loop sticking out from the surface of the virion and resembling a peptide [Acharya et al, 1989]. Immunization of experimental animals with peptides containing this epitope coupled to an appropriate T-helper epitope results in virus-neutralizing antibodies [Francis et al, 1987]. Another example of success in raising virus-neutralizing antibodies with synthetic peptides is for a linear B-cell epitope of human immunodeficiency virus (HIV) [Ho et al, 1987; Palker et al, 1988; Rusche et al, 1988; Goudsmit et al, 1988] presumed also to be a projecting loop in the virus [Javaherian et al, 1989]. The antibodies evoked by this loop are type-specific and, due to the lack of a suitable test animal, it is not known if these antibodies that neutralize HIV in vitro are able to confer in vivo protection.

Full protection of experimental animals after vaccination with synthetic peptides (coupled to carrier proteins) was demonstrated for two viral diseases, i.e. for FMDV in guinea pigs [Bittle et al, 1982] and for hepatitis B virus (HBV) in chimpanzees [Itoh et al, 1986; Thornton et al, 1987]. These results already indicate that for some viral diseases it is possible to induce protective immunity with synthetic peptides.

In contrast to their generally poor effectiveness in triggering the relevant protective antibody response, peptides are very effective in triggering T-cell responses, presumably because they resemble natural T-cell epitopes. The finding of Milich [Milich et al, 1987], that a synthetic peptide containing a HBV T_h-cell epitope could prime experimental animals with the whole (inactivated) virus, boosted a second wave of interest in the potential of peptide vaccines. The validity of Milich's observation was confirmed with a number of other viruses like rabies virus [Ertl et al, 1989], HIV [Hosmalin et al, 1991] and human T cell lymphotropic virus type 1 (HTLV-1) [Kurata et al, 1989].

SEMLIKI FOREST VIRUS AS MODEL ANTIGEN

The choice for Semliki Forest Virus (SFV) infection of mice as a model to evaluate the vaccine efficacy of synthetic peptides is based on the following considerations:

-   SFV causes lethal encephalitis in mice, allowing protection to be monitored.

-   The lethal encephalitis by SFV can be prevented completely by antibodies [Boere et al, 1983]. Protective antisera and a panel of protective monoclonal antibodies are available.

-   SFV is a relatively small and simple virus, containing only 5 structural proteins, facilitating the choice of antigens to be studied.

-   The amino acid sequence, derived from the nucleotide sequence [Garoff et al, 1980], of the structural proteins of SFV is known making it possible to synthesize peptide representing regions of these proteins.

Objectives of this study

1. The identification of linear B-cell epitopes with vaccine potential on one of the structural proteins of SFV, the E2 membrane protein.
2. The identification of T_h-cell epitopes on the structural proteins of SFV.
3. To determine in mice the immunogenicity and vaccine efficacy of different peptides, each containing a combination of an identified B- and T_h-cell epitope.
4. Improvement of the immunogenicity of these peptides by adjuvants.
5. Evaluation of the immune mechanisms, that provide protection of peptide-immunized mice against SFV.

Table 2. Reactivities of antipeptide sera

Antiserum against peptide [*]	Elisa antibody titres (Log_{10}) against		Neutralization titre (Log_{10})	Survival (n)
	Homologous peptide	SFV-infected L cells		
KLH	<2.0	<2.0	<0.5	0% (16)
126 to 141	4.5	<2.0	<0.5	0% (6)
178 to 186	4.0	2.0	<0.5	20% (6)
240 to 255	5.0	3.5	<0.5	90% (15)

BALB/c mice were injected subcutaneously with a mixture of 50 µg peptide-KLH conjugate and 50 µg Quil A and received a booster injection 5 weeks later. Sera were obtained at week 7. The immune reactivity of the pooled sera of each group was tested. The mice were challenged intraperitoneally (i.p.) with $10LD_{50}$ of SFV at week 10. (n, number of mice)
[*] Numbers indicate the amino acid position in the SFV E2 protein.

Table 3. Delayed type hypersensitivity to fragments of the SFV E2 protein

Antigen used for primary immunization	Antigen used for footpad challenge	Mean footpad swelling (± SEM) 24 h after challenge
PBS	SFV	0.2 ±0.4
SFV	SFV	12.8[*] ±1.2
PBS	ß-Gal	1.9 ±0.4
SFV	ß-Gal	2.3 ±0.6
PBS	ß-Gal-E2 115 to 151	1.9 ±0.2
SFV	ß-Gal-E2 115 to 151	6.9[*] ±1.2
PBS	ß-Gal-E2 1 to 350	2.2 ±0.4
SFV	ß-Gal-E2 1 to 350	7.1[*] ±0.6
ß-Gal	SFV	0.1 ±0.3
ß-Gal-E2 115 to 151	SFV	3.6[*] ±0.5
ß-Gal-E2 1 to 350	SFV	3.6[*] ±1.0
ß-Gal-E2 169 to 406	SFV	0.4 ±0.6

Groups of 5 mice were immunized intracutaneously (i.c.) with either 0.8 µg UV-inactivated SFV or 1.0 µg of the indicated protein mixed with dimethyl dioctadecyl ammonium bromide (DDA, 100 µg per animal). Six days after immunization the mice were challenged with either 0.4 µg inactivated SFV or 0.5 µg hybrid protein. Footpad swelling is expressed in 0.1 mm units.
[*] Swelling significantly higher than control (p <0.05 in Student's t test).

The identification of linear epitopes with vaccine potential on the E2 membrane protein of SFV was based on the binding of SFV-specific antibodies to a set of overlapping synthetic hexapeptides (Pepscan) representing the complete E2 amino acid sequence. The 14 available E2-specific monoclonal antibodies which were protective in vivo proved to be unsuitable for the identification of linear epitopes because they recognized only conformational epitopes, as indicated by their lack of reactivity with unfolded, reduced E2 protein on immunoblots. Three epitopes were detected with polyclonal anti-SFV serum at amino acid positions 135 to 141, 177 to 185 and 240 to 246 of the E2 protein [Snijders et al, 1991]. Synthetic peptides containing these epitopes were coupled to a carrier protein and tested as a vaccine (Table 2). Mice immunized with the peptide containing amino acids 240 to 255 of protein E2 were protected against a challenge with virulent SFV but protection of mice with the peptides containing amino acids 126 to 141 or 178 to 186 was only marginally better than that of controls. The prechallenge sera of most peptide-immunized mice reacted with SFV-infected cells but none of these sera neutralized the virus in vitro. However, protection of individual mice correlated well with SFV-specific antibody titre, suggesting antibody-mediated protection.

IDENTIFICATION OF A DTH-INDUCING T-CELL EPITOPE

Mapping of T-cell epitopes on the structural proteins of SFV was performed by measuring the ability of recombinant SFV protein fragments to induce SFV-specific delayed type hyper-sensitivity (DTH) and to prime mice for SFV-specific in vitro proliferation of T-cells. The SFV protein fragments were expressed as hybrid proteins with cro-β-galactosidase in E. coli from constructed recombinant plasmids. DTH reactions were measured, as footpad swelling, in BALB/c mice after immunization with whole, UV-inactivated SFV and challenge with the hybrid proteins and vice versa, using the adjuvant dimethyl dioctadecyl ammoniumbromide to enhance DTH (Table 3). Two of the tested hybrid proteins induced DTH, and these DTH reactions were equally strong. The largest DTH-inducing hybrid protein contained the N-terminal 350 amino acids of E2 and part of E3, the smallest contained only the region from amino acid residue 115 to 151 of the E2 membrane protein without any other SFV protein parts. It was concluded

Table 4. SFV-specific in vitro proliferation of LN cells from SFV-protein fragment immunized mice

Antigen used for immunization (1 µg)	DTH	Mean proliferation [*]	
		Medium	SFV
ß-Gal	-	250 ± 40	225 ± 10
ß-Gal - E2 115 to 151	+	530 ± 380	2 000[□] ± 400
ß-Gal - E2 1 to 350	+	660 ± 230	18 000[□] ± 4000
ß-Gal - E2 169 to 406	-	170 ± 10	570 ± 380

[*] Lymph node cells (10^5 per well) from the indicated group of immunized BALB/c mice were incubated for 4 days in nonsupplemented medium or medium containing UV-inactivated SFV (24 µg/ml). ^{3}H-Thymidine (0.5 µCi per well) was added for the last 20 hours. Results are expressed in mean cpm ±SEM of three wells.
[□] Proliferation significantly higher than control (p < 0.05 in Student's t test).

Table 5. Amino acid sequences of the peptides used in this study

Peptide	Designation	Amino acid sequence *	Epitope orientation	Anti-SFV IgG**
1	125-141	^{125}CRIQYHHDPQPV**GREKF**^{141}	-B	-
2	115-141	^{115}IQDTRNAVRACRIQYHHDPQPV**GREFK**^{141}	T-B	-
3	127-151	^{127}IQYHHDPQPV**GREKF**TIRPHYGKEI^{151}	B-T	-
4	240-255	CGG^{240}**PFVPRADE**PARKGKVH^{255}	-B	-
5	115-129/ 240-255	^{115}IQDTRNAVRACRIQY^{129}------^{240}**PFVPRADE**PARKGKVH^{255}	T-B	+
6	137-151/ 240-255	^{137}GREKFTIRPHYGKEI^{151}_^{240}**PFVPRADE**PARKGKVH^{255}	T-B	+
7	240-255/ 137-151	^{240}**PFVPRADE**PARKGKVH^{255}__^{137}GREKFTIRPHYGKEI^{151}	B-T	-
8	110-120/ 240-255	^{110}SFERFEIFPKE^{120}__^{240}**PFVPRADE**PARKGKVH^{255}	T-B	+

* Bold type: B-cell epitope of SFV; underlined: T$_h$-cell epitope of either SFV (peptides 1 to 7) or influenza virus (peptide 8); numbers: amino acid positions in the SFV E2 membrane protein or the influenza virus haemagglutinin (peptide 8).

** Non-neutralizing, SFV-reactive antibodies in serum of peptide-immunized BALB/c mice (determined by ELISA); -, no detectable secondary antibody reponse induced; +, detectable secondary antibody reponse (titre ≥ 2.0).

Table 6. DTH to SFV E2 protein-derived peptides

Antigen used for primary immunization	Antigen used for footpad challenge	Mean footpad swelling (± SEM) 24 h after challenge
SFV	ß-Gal-E2 115 to 151	4.6* ±0.7
SFV	ß-Gal	1.3 ±0.2
SFV	ß-Gal- peptide 2	1.9 ±0.4
SFV	ß-Gal- peptide 3	4.0* ±0.5
SFV	ß-Gal- peptide 5	1.6 ±0.3
SFV	ß-Gal- peptide 6	2.5* ±0.4

Groups of 5 BALB/c mice were immunized i.c. with 0.8 µg UV-inactivated SFV mixed with DDA (100µg per animal). Six days after immunization the mice were challenged with 0.5 µg hybrid protein. Footpad swelling is expressed in 0.1 mm units.

* Swelling significantly higher than control (p < 0.05 in Student's t test).

that the segment between amino acid residues 115 and 151 of the E2 membrane protein of SFV was responsible for the observed DTH, and thus, contains a T-cell epitope. This conclusion was confirmed by the SFV-specific in vitro proliferation of T cells from mice that were immunized with recombinant SFV protein fragments (Table 4).

For more precise mapping, a set of peptides (Table 5) was synthesized containing different overlapping subregions of the identified DTH-inducing region. The peptides either were conjugated to ß -galactosidase of E. coli and tested for their ability to evoke DTH (Table 6) or tested for their ability to stimulate SFV-primed T cells in vitro (Table 7). The results

Table 7. Peptide-specific in vitro proliferation of LN cells from SFV-infected mice

Antigen used for in vitro stimulation	Dose in µg	Mean proliferation of control cells	Mean proliferation of SFV-primed cells
none		120 ±10	150 ± 26
peptide 2 : 115-141	0.08	100 ±21	130 ± 16
	8	95 ±22	450* ± 50
peptide 3 : 127-151	0.08	130 ±33	2900* ± 850
	8	260 ±96	2900* ±1160
peptide 5 : 115-129 / 240-255	0.08	130 ±31	200 ± 63
	8	140 ±37	2400* ±1240
peptide 6 : 137-151 / 240-255	0.08	120 ±10	1140 ± 870
	8	123 ±11	2500* ± 690

LN cells were obtained from BALB/c mice infected with 2000 PFU of SFV, or corresponding control mice, and were incubated for 4 days with the indicated peptide. Results are expressed in mean cpm ±SEM of three wells.
* Proliferation significantly higher than control (p < 0.05 in Student's t test).

Table 8. Induction of antiviral IgG and survival of peptide 6 -immunized mice

Mouse strain	Log$_{10}$ IgG titre to SFV	Survival (n)
BALB/c	4.0	70% (34)
DBA/2	4.5	100% (8)
C3H	<2.0	NT
C57BL/6J	<2.0	NT

Groups of 6 or more mice were immunized i.c. twice with 50 μg of the indicated peptide mixed with the adjuvant Quil A.
NT, not tested.

indicate that the DTH-inducing T-cell epitope was located between amino acids 137 and 151 of the SFV E2 membrane protein and that a second, non-DTH-inducing T-cell epitope was present between amino acids 115 to 129.

THE DTH-INDUCING T-CELL EPITOPE PROVIDES T HELPER ACTIVITY FOR ANTIBODY PRODUCTION

The ability of the identified DTH-inducing T-cell epitope to induce a specific T helper response in mice was evaluated using synthetic peptides (Table 5) that contained combinations of the DTH-inducing region and two different previously identified linear B-cell epitopes of E2. These peptides (in combination with the adjuvant Quil A) proved able to induce an antipeptide IgG response in H-2^{d} mice [Snijders et al, 1992].

IMMUNOGENICITY AND VACCINE EFFICACY OF SYNTHETIC PEPTIDES CONTAINING SFV B-AND T-CELL EPITOPES

One of the peptides (peptide 6) was able to induce high titres of SFV-reactive IgG. Although the peptide-induced antibodies did not neutralize SFV in vitro, 70-100% of the peptide-immunized H-2^{d} mice were protected against lethal SFV infection (Table 8), even when the viral challenge was given 4 months after the immunizations (Table 9). The protection could be transferred by antipeptide serum (Table 10) indicating that antibodies were responsible for the protection. When the T-cell epitope of this protective peptide was replaced by an influenza virus T$_{h}$-cell epitope or by another SFV T$_{h}$-cell epitope or by another SFV T$_{h}$-cell epitope (Table 5, peptide 8 and 5), the resulting peptides induced lower non-neutralizing SFV-reactive antibody titres and protected a correspondingly lower percentage of mice (respectively 50% and 30%). A peptide with the same T$_{h}$-cell epitope as the best protective peptide but a less effective SFV B-cell epitope (peptide 3) only protected 33% of the mice. These results indicate that protection against SFV by a synthetic peptide is primarily dependent on its ability to induce adequate amounts of antibodies with a relevant specificity and sufficient affinity, while the ability to induce a relevant (SFV-specific) T memory response plays only a minor role in protection.

91

Table 9. Survival of peptide-immunized BALB/c mice challenged 4 months post immunization

Peptide used for immunization	Anti-SFV IgG titre at month *			Survival (n)	Survival time in days
	1	2	4		
None, adjuvant Quil A only	<2.0	<2.0	<2.0	10% (8)	7.3
Peptide 6 : 137-151 / 240-255	4.0	3.5	3.5	90% (7)	12.0
Peptide 8 : 110-120 / 240-255	3.5	3.0	3.0	10% (7)	6.3

* Titres (Log_{10}) of pooled mouse sera.

EFFECT OF ADJUVANTS AND ORIENTATION OF B-CELL EPITOPES ON THE IMMUNO-GENICITY OF SYNTHETIC PEPTIDES

The orientation of the epitopes in peptide 6 is T-B. In peptide 7 the orientation is reversed (B-T). Both peptides were compared for their immunogenicity in BALB/c mice using 6 different adjuvants: the Nonionic Block Polymer L180.5, L180.5 in a squalene-in-water emulsion referred to as W/O/W emulsion, Quil A, Freunds Complete Adjuvant and Q Vac. Assessed by antipeptide IgG titre, all peptide-adjuvant mixtures were highly immuno-genic (Table 11), but SFV-specific antibodies were only induced by mixtures containing peptide 6. These titre differences indicate the importance of the orientation of T- and B-cell epitopes in the antigen. The isotype distribution of the evoked antibodies depended on the adjuvant used for immunization (Figure 1). Although none of the peptide-adjuvant mixtures evoked virus neutralizing antibodies, several groups of peptide 6-mice were protected against lethal SFV infection (not shown).

DISCUSSION

The difficulties in the development of synthetic peptide vaccines intended to evoke protective antibodies are indicated by the low number of

Table 10. Protection of mice by passively transferred antipeptide serum

Specificity of donor serum	Anti-SFV IgG titre in recipient mice	Survival (n)
untreated control	-	10% (12)
peptide 6 : 137-151 / 240-255	3.0	50% (12)
peptide 8 : 110-120 / 240-255	2.5	20% (12)

Groups of 6 mice received i.v. 0.5 ml pooled donor serum (anti-SFV IgG titre 3.5) and were challenged i.p. 24h later with $10LD_{50}$ of SFV.

Table 11. Effect of adjuvants on the immunogenicity of synthetic peptides

Adjuvant	Immunization with			
	Peptide 6 : 137-151/240-255		Peptide 7 : 240-255/137-151	
	Log_{10} serum titre		Log_{10} serum titre	
	On peptide 6	On SFV-infected cells	On peptide 7	On SFV-infected cells
L 180.5	5.0	2.6	3.5	<2.0
W/O/W	5.5	3.5	5.5	<2.0
Quil A	5.5	4.0	5.0	<2.0
FCA	5.5	2.6	5.5	2.0
Montanide	5.5	2.6	5.5	2.0
Q-VAC	5.5	4.0	5.5	<2.0

BALB/c mice were immunized i.c. with 50 µg peptide and the indicated adjuvant. A booster injection was given 8 weeks later. Sera were obtained 2 weeks after the last immunization.

reports on effective in vivo protection by peptides in spite of a wealth of research [reviewed in Milich, 1989]. Full protection of test animals immunized with synthetic peptides, coupled to heterologous carrier proteins, was reported for foot-and-mouth disease virus (FMDV) in guinea pigs [Bittle et al, 1982], for hepatitis B virus (HBV) in chimpanzees [Itoh et al, 1986; Thornton et al, 1987], for SFV in mice [Snijders et al, 1991] and for rotavirus in suckling mice [Ijaz et al, 1991]. Up till now, there are only four reports of protection of test animals immunized withfully

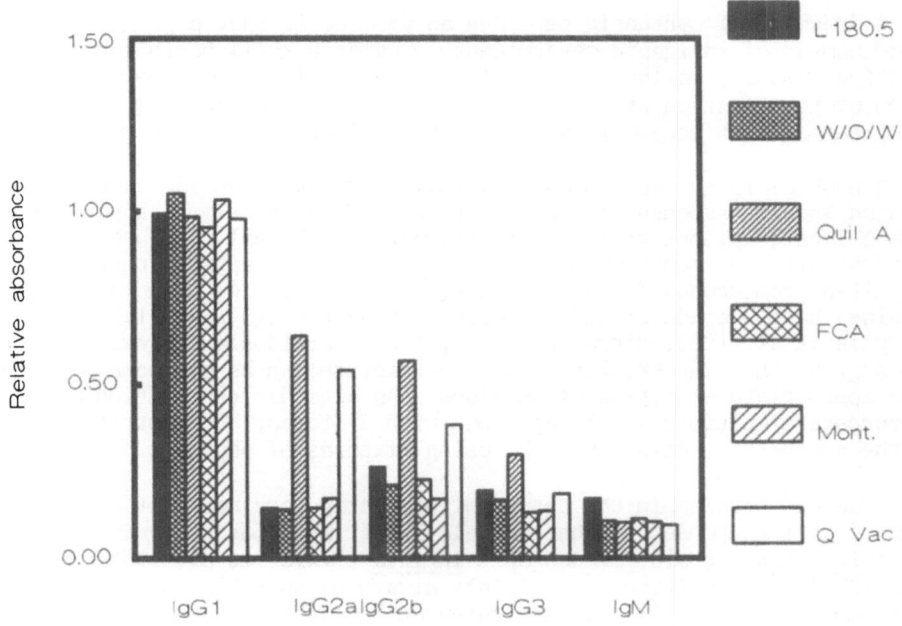

Fig. 1. Ab subclass determination peptide 6. Effect of adjuvants on the immunogenicity of synthetic peptides, see Table 11.

synthetic peptides containing a B- and a T_h-cell epitope, i.e. for rabies virus [Dietzschold et al, 1990], mouse hepatitis virus (MHV) [Koolen et al, 1990], venezuelan equine encephalomyelitis virus (VEEV) [Hunt et al, 1990] and for SFV [Snijders et al, 1992]. In all four cases the test animal were mice, the animal species most commonly used for such tests. In the case of MHV, VEEV and SFV, only non-neutralizing antibodies were evoked by the synthetic peptides. Still, protection could be mediated by high titres of non-neutralizing antibodies [Snijders et al, 1991; Snijders et al, 1992], showing that peptides that induce only non-neutralizing antibodies can be effective as a vaccine. Although immunization with these peptides did not prevent infection, it prevented disease, the main aim of vaccination. However, peptides that evoke only non-neutralizing antibodies are totally unsuitable as vaccine against any virus that can replicate in cells of the monocytemacrophage lineage, because non-neutralizing antibodies can give enhancement of infection of monocytes and macrophages by these viruses [Porterfield, 1986]. Furthermore, it is conceivable that a number of viruses will not contain linear B-cell epitopes that can evoke protective antibodies, because vireal epitopes are generally conformational [Snijders et al, 1991]. In our opinion, it will only be possible to induce protective immunity with synthetic peptides for some viral deseases and these cases are the exception not the rule.

In contrast ot the above reports on protection, a fully synthetic peptide with an FMDV B-cell epitope evoked neutralizing antibodies in cattle, the animals at risk for FMDV, but protection was inconsistent [Steward and Howard, 1987; Dimarchi et al, 1986; Doel et al, 1990], suggesting that protection against FMDV requires a high titre of neutralizing antibodies. The results with FMDV show further that <u>in vitro</u> neutralizing antibodies are sometimes insufficient for protection <u>in vivo</u> and that protection of test animals (like guinea pigs) is not always predictive for protection of the natural host (cattle). Vaccine efficacy should therefore also be evaluated in the natural host of the virus.

A drawback of synthetic peptides as vaccine is that peptides with a limited number of antigenic determinants provide a small basis for protective immunity, allowing the infectious organism to escape neutralization of inactivation by only a single mutation. The chances for the development of such escape mutants must be evaluated carefully.

A problem in the development of fully synthetic peptide vaccines that depend on T-cell responses is formed by the MHC restriction. A peptide with one T-cell epitope will usually trigger only T-cell responses in individual with the same MHC haplotype. A number of peptides that contain a so-called promiscuous T_h-cell epitope (that can combine with multiple MHC molecules) have been described recently [Panina-Bordignon et al, 1989; Sinigaglia et al, 1988; Berzofsky et al, 1988; Partidos and Steward, 1990], which suggest that the problem of MHC restriction can be overcome sometimes by the appropriate selection of peptides. An alternative solution may be to incorporate multiple T-cell epitopes with different MHC restrictions into the synthetic vaccine, e.g. by using mixtures of peptides.

A second problem in the design of synthetic vaccines that must trigger T-cell responses is that not all T-cell epitopes recognized as antigen <u>in vitro</u> are also immunogenic <u>in vivo</u> [Bixler et al, 1985; Gammon et al, 1987]. In addition, it will not always be possible to predict the T_h-cell epitopes that will trigger relevant T-cell responses in humans on the basis of responses in experimental animals, since only some human MHC haplotypes have homologues in animals [Bontrop et al, 1990; Tokunaga et al, 1988; Lawrance et al, 1989; Klein, 1989]. However, some T_h-cell epitopes are recognized both in humans and in mice [Sinigaglia et al, 1988;

Berzofsky et al, 1988; Partidos and Steward, 1990; Celis et al, 1988; Celis et al, 1989].

Antibodies raised to the KLH-conjugated peptides representing amino acid residue 134 to 141 and 240 to 255 of the SFV E2 protein were shown to bind to glutaraldehyde-fixed SFV-infected cells in ELISA (Table 2) and to intact SFV-infected cell in ADCMC [Snijders et al, 1991], indicating that at least part of these E2 sequences are at the surface of the spikes on SFV-infected cells. Immune serum evoked by KLH-coupled peptide 240 to 255, when injected together with a high dose of virulent SFV, caused a lower viraemia compared to control mice who received the same dose of virus combined with mock mouse serum (results not shown). This finding suggests that the antibodies to KLH-coupled peptide 240-255 were also able to bind intact infectious virions and that the 240-255 region is exposed on the virion. The exposure of (part of) the 240-255 region of E2 at the surface of the virion is also suggested by the presence of mutations conferring resistance to neutralization by monoclonal antibodies (mar-mutants) at position 246-251 of E2 of another alphavirus, Ross River virus (RRV) [Vrati et al, 1988], that is closely-related to SFV [Dalgarno et al, 1983].

Recombinant SFV fragments produced in E.coli as hybrid proteins with cro-β-galactosidase were used for the mapping of one linear B-cell epitope on the E1 membrane protein (not shown) and one on the E2 membrane protein [Snijders et al, 1991] and of T_h-cell epitopes on all structural proteins of SFV (Tables 3 and 4). As reported by others, these prokaryotic expression products are incorrectly folded products [Wahlberg et al, 1989; de Geus et al, 1987], that are, after SDS and 2-mercaptoethanol treatment and blotting, equivalent to denatured protein fragments [Kusters et al, 1989].

In general, the recombinant fragments were suitable for the mapping of T_h-cell epitopes. In spite of the presence of cro-β-galactosidase and certain contaminants like bacterial cell wall components in the protein preparation, the SFV-galactosidase hybrid proteins could be used to prime and to elicit SFV-specific DTH (Table 3) and to prime LN-cells in vivo for SFV-specific proliferation in vitro (Table 4). These hybrid proteins were however unsuitable for the in vitro stimulation of SFV-primed LN-cells, because the preparation of cro-β-galactosidase alone already stimulated LN cells strongly, probably due to the presence of LPS which is mitogenic for B cells. These results show that bacterial expression products are unsuitable for the mapping of T cell epitopes by in vitro stimulation of cell populations containing other cells besides the primed T cells, while they are suitable for in vitro stimulation of T cell clones [Thole et al, 1988] and for in vivo priming of T cells.

Synthetic peptides containing the identified B- and T_h-cell epitopes of the E2 membrane protein of SFV are immunogenic in BALB/c mice. For an effective immunization the presence of an adjuvant was required. Initially, Quil A was used in our studies but due to its hemolytic activity, a search for alternative adjuvants was initiated. The ultimate goal is the definition of a safe adjuvant formulation for application in humans, which helps the antigen to evoke, after one single injection, antibodies with the desired antigen specificity, affinity and isotype(s), resulting finally in a life-long protection against the viral disease.

SUMMARY

In this chapter we evaluate the efficacy of synthetic peptides as vaccines using Semliki Forest Virus (SFV) infection of mice as a model. We describe the identification and localization of three linear B-cell epitopes and two linear T_h-cell epitopes on the E2 membrane protein of SFV. Different peptides, each representing a combination of one of the identified B-cell epitopes and one of the identified T_h-cell epitopes, were immunogenic in mice. One of these peptides evoked high titres of non-neutralizing SFV-reactive antibodies and was effectively protective in mice. The orientation of the T- and B-cell epitope (T-B versus B-T) and the adjuvant used for immunization influenced the vaccine efficacy of the peptide profoundly.

REFERENCES

Acharya, R., Fry, E., Stuart, D., Fox, G., Rowlands, D. and Brown, F., 1989, The three-dimensional structure of foot-and-mouth disease virus at 2.9 A resolution, Nature, 337:709.

Agterberg, M. and Tommassen, J., 1991, Outer membrane protein PhoE as a carrier for the exposure of foreign antigenic determinants at the bacterial cell surface, Antoine van Leeuwenhoek, 59:249.

Aldovini, A. and Young, R.A., 1991, Humoral and cell-mediated immune responses to live recombinant BCG-HIV vaccines, Nature (London), 351:479.

Benjamin, D.C., Berzofsky, J.A., East, I.J., Gurd, F.R., Hannum, C., Leach, S.J., Margoliash, E., Michael, J.G., Miller, A., Prager, E.M., Reichlin, M., Sercarz, E.E., Smith-Gill, S.J., Todd, P.E. and Wilson, A.C., 1984, The antigenic structure of proteins: a reappraisal, Annu.Rev,Immunol., 2:67.

Berzofsky, J.A., Bensussan, A., Cease, K.B., Bourge, J.F., Cheynier, R., Lurhuma, Z., Salaun, J.J., Gallo, R.C., Shearer, G.M. and Zagury, D., 1988, Antigenic peptides recognized by T lymphocytes from AIDS viral envelope-immune humans, Nature, (London), 334:706.

Bittle, J.L., Houghten, R.A., Alexander, J., Shinnick, T.M., Sutcliffe, J.G., Lerner, R.A., Rowlands, D.J. and Brown, F., 1982, Protection against foot-and-mouth disease by immunization with a chemically synthesized peptide predicted from the viral nucleotide sequence, Nature (London), 298:30.

Bixler, G.S., Yoshida, T. and Atassi, M.Z., 1985, Antigen presentation of lysozyme: T-cell recognition of peptide and intact protein after priming with synthetic overlapping peptides comprising the entire protein chain, Immunology, 56:103.

Boere, W.A.M., Benaissa-Trouw, B.J., Harmsen, M., Kraaijeveld, C.A. and Snippe, H., 1983, Neutralizing and non-neutralizing monoclonal antibodies to the E2 glycoprotein of Semliki Forest Virus can protect mice from lethal encephalitis, J. Gen.Virol., 64:1405.

Bontrop, R.E., Elferink, D.G., Otting, N., Jonker, M. and de Vries, R.R., 1990, Major histocompatibility complex class II-restricted antigen presentation across a species barrier: conservation of restriction determinants in evolution, J.Exp.Med., 172:53.

Borras-Cuesta, F., Petit-Camurdan, A. and Fedon, Y., 1987, Engineering of immunogenic peptides by co-linear synthesis of determinants recognized by B and T cells, Eur.J.Immunol., 17:1213.

Bostock, C.J., 1990, Viruses as vectors, Vet.Microbiol., 23:55.

Cannon, M.J., Openshaw, P.J. and Askonas, B.A., 1988, Cytotoxic T cells clear virus but augment lung pathology in mice infected with respiratory syncytial virus, J.Exp,Med., 168:1.

Celis, E., Karr, R.W., Dietzschold, B., Wunner, W.H. and Koprowski, H., 1988, Genetic restriction and fine specificity of human T cell clones reactive with rabies virus, J,Immunol., 141:2721.

Celis, E., Ou, D., Dietzschold, B., Otvos, L.J. and Koprowski, H., 1989, Rabies virus-specific T cell hybridomas: identification of class II MHC-restricted T-cell epitopes using synthetic peptides, Hybridoma., 8:263.

Dalgarno, L., Rice, C.M. and Strauss, J.H., 1983, Ross River virus 26 s RNA: complete nucleotide sequence and deduced sequence of the encoded structural proteins, Virology, 129:170.

de Geus, P., van den Bergh, C.J., Kuipers, O., Verheij, H.M., Hoekstra, W.P. and de Haas, G.H., 1987, Expression of porcine pancreatic phospholipase A2. Generation of active enzyme by sequence-specific cleavage of a hybrid protein from Escherichia coli, Nucleic Acids Res., 15:3743.

Dietzschold, B., Gore, M., Marchadier, D., Niu, H.S., Bunschoten, H.M., Otvos, L.J., Wunner, W.H., Ertl, H.C., Osterhaus, A.D. and Koprowski, H., 1990, Structural and immunological characterization of a linear virus-neutralizing epitope of the rabies virus glycoprotein and its possible use in a synthetic vaccine, J.Virol., 64:3804.

Dimarchi, R., Brooke, G., Gale, C., Cracknell, V., Doel, T. and Mowat, N., 1986, Protection of cattle against foot-and-mouth disease by a synthetic peptide, Science, 232:639.

Doel, T.R., Gale, C., Do Amaral, C.M. Mulcahy, G. and Dimarchi, R., 1990, Heterotypic protection induced by synthetic peptides corresponding to three serotypes of foot-and-mouth disease virus, J.Virol., 64:260.

Ertl, H.C., Dietzschold, B., Gore, M., Otvos, L.J., Larson, J.K., Wunner, W.H. and Koprowski, H., 1989, Induction of rabies virus-specific T-helper cells by synthetic peptides that carry dominant T-helper cell epitopes of the viral ribon-ucleoprotein, J.Virol., 63:2885.

Francis, M.J., Hastings, G.Z., Syred, A.D., McGinn, B., Brown, F. and Rowlands, D.J., 1987, Non-responsiveness to a foot-and-mouth disease virus peptide overcome by addition of foreign helper T-cell determinants, Nature (London), 300:168.

Gammon, G., Shastri, N., Cogswell, J., Wilbur, S., Sadegh Nasseri, S., Krzych, U., Miller, A. and Sercarz, E., 1987, The choice of T-cell epitopes utilized on a protein antigen depends on multiple factors distant from, as well as at the determinant site, Immunol.Rev., 98:53.

Garoff, H., Frischauf, A.M., Simons, K., Lerach, H. and Delius, H., 1980, Nucleotide sequence of cDNA coding for Semliki Forest Virus membrane proteins, Nature (London), 288:236.

Geysen, H.M., Meloen, R.H. and Barteling, S.J., 1984, Use of peptide synthesis to probe viral antigens for epitopes to a resolution of a single amino acid, Proc.Natl.Acad.Sci., USA, 81:3998.

Goudsmit, J., Debouck, C., Meloen, R.H., Smit, L., Bakker, M., Asher, D.M., Wolff, A.V., Gibbs, C.J.J. and Gajdusek, D.C., 1988, Human immunodeficiency virus type 1 neutralization epitope with conserved architecture elicits early type-specific antibodies in experimentally infected chimpanzees, Proc.Natl,Acad,Sci., USA, 85:4478.

Halstead, S.B., 1988, Pathogenisis of dengue: challenges to molecular biology, Science, 239:476.

Ho, D.D., Sarngadharan, M.G., Hirsch, M.S., Schooley., R.T., Rota, T.R., Kennedy, R.C., Chanh, T.C and Sato, V.L., 1987, Human immunodeficiency virys neutralizing antibodies recognize several conserved domains on the envelope glycoproteins, J.Virol., 61:2024.

Hosmalin, A., Nara, P.L., Zweig, M., Lerche, N.W., Cease, K.B., Gard, E.A., Markham, P.D., Putney, S.D., Daniel, M.D., Desrosiers, R.C. and Berzofsky, J.A., 1991, Priming with T helper cell epitope peptides enhances the antibody response to the envelope glycoprotein of HIV-1 in primates, J.Immunol., 146:1667.

Hunt, A.R., Johnson, A.J. and Roehrig, J.T., 1990, Synthetic peptides of
    Venezuelan equine encephalomyelitis virus E2 glycoprotein. I.
    Immunogenic analysis and identification of a protective peptide,
    Virology, 179:701.
Ijaz, M.K., Attah Poku, S.K., Redmond, M.J., Parker, M.D., Sabara, M.I. and
    Babiuk, L.A., 1991, Heterotypic passive protection induced by
    synthetic peptides corresponding to VP7 and VP4 of bovine
    rotavirus, J.Virol., 65:3106.
Itoh, Y., Takai, E., Ohnuma, H., Kitajima, K., Tsuda, F., Machidam A.,
    Mishiro, S., Nakamura, T., Miyakawa, Y. and Mayumi, M., 1986, A
    synthetic petpide vaccine involving the product of the pre-S(2)
    region of hepatitis B virus DNA: protective efficacy in
    chimpanzees, Proc.Natl,Acad.Sci. USA, 83:9174.
Javaherian, K., Langlois, A.J., McDanal, C., Ross, K.L., Eckler, L.I.,
    Jellis, C.L., Profy, A.T., Rusche, J.R., Bolognesi, D.P., Putney,
    S.D. and Matthews, T.J., 1989, Principal neutralizing domain of the
    human immunodeficiency virus type 1 envelope protein,
    Proc.Natl.Acad,Sci., USA, 86:6768.
Klein, J., 1989, The MHC trans-species hypothesis: in the discussion
    period, Immunol.Suppl., 2:36.
Koolen, M.J., Borst, M.A., Horzinek, M.C. and Spaan, W.J., 1990,
    Immunogenic peptide comprising a mouse hepatitis virus A59 B-cell
    epitope and an influenza virus T-cell epitope protects against
    lethal infection, J.Virol., 64:6270.
Kurata, A., Palker, T.J., Streilein, R.D., Scearce, R.M., Haynes, B.F. and
    Berzofsky, J.A., 1989, Immunodominant sites of human T cell
    lymphotropic virus type 1 envelope protein for murine helper T
    cells, J,Immunol., 143:2024.
Kusters, J.G., Jager, E.J., Lenstra, J.A., Koch, G., Posthumus, W.P.,
    Meloen, R.H. and van der Zeijst, B.A., 1989, Analysis of an
    immunodominant region of infectious bronchitis virus, J,Immunol.,
    143:2692.
Laver, W.G., Air, G.M., Webster, R.G. and Smith Gill, S.J., 1990, Epitopes
    on protein antigens: misconceptions and realities, Cell, 61:553.
Lawrance, S.K., Karlsson, L., Price, J., Quaranta, V., Ron, Y., Sprent, J.
    and Peterson, P.A., 1989, Transgenic HLA-DR alpha faithfully
    reconstitutes IE-controlled immune functions and induces cross-
    tolerance to E alpha in E alpha O mutant mice, Cell, 58:583
Milich, D.R., McLachlan, A., Thornton, G.B. and Hughes, J.L., 1987,
    Antibody production to the nucleocapsid and envelope of the
    hepatitis B virus primed by a single synthetic T cell site, Nature
    (London), 329:547.
Milich, D.R., 1989, Synthetic T and B cell recognition sites: implications
    for vaccine development, Adv.Immunol., 45:195.
Oehen, S., Hengartner, H. and Zinkernagel, R.M., 1991, Vaccination for
    disease, Science, 251:195.
Palker, T.J., Clark, M.E., Langlois, A.J., Matthews, T.J., Weinhold, K.J.,
    Randall, R.R., Bolognesi, D.P. and Haynes, B.F., 1988, Type-
    specific neutralization of the human immunodeficiency virus with
    antibodies to env-encoded synthetic peptides, Proc,Natl.Acad.Sci.
    USA, 85:1932.
Panina-Bordignon, P., Tan, A., Termijtelen, A., Demotz, S., Corradin, G.
    and Lanzavecchia, A., 1989, Universally immunogenic T cell
    epitopes: promiscuous binding to human MHC class II and promiscuous
    recognition by T cells, Eur.J.Immunol., 19:2237.
Partidos, C.D. and Steward, M.W., 1990, Prediction and identification of a
    T cell epitope in the fusion protein of measles virus immuno-
    dominant in mice and humans, J.Gen.Virol., 71:2099.
Porterfield, J.S., 1986, Antibody-dependent enhancement of viral
    infectivity, Adv,Virus Res., 31:335.
Rusche, J.R., Javeherian, K., McDanal, C., Petro, J., Lynn, D.L., Grimalia,

R., Langlois, A., Gallo, R.C., Arthur, L.O., Fischinger, P.J.,
Bolognesi, D.P., Putney, S.D. and Matthews, T.J., 1988, Antibodies
that inhibit fusion of human immunodeficiency virus-infected cells
bind a 24-amino acid sequence of the viral envelope, gp120,
Proc.Natl.Acad.Sci., USA, 85:3198.

Sinigaglia, F., Guttinger, M., Kilgus, J., Doran, D.M., Matile, H.,
Etlinger, H., Trzeciak, A., Gillessen, D. and Pink, J.R., 1988, A
malaria T-cell epitope recongized in association with most mouse
and human MHC class II molecules, Nature (London), 336:778.

Snijders, A., Benaissa-Trouw, B.J., Oosting, J.D., Snippe, H. and
Kraaijeveld, C.A., 1989, Identification of a DTH-inducing T-cell
epitope on the E2 membrane protein of Semliki Forest Virus,
Cell.Immunol., 123:23.

Snijders, A., Benaissa-Trouw, B.J., Oosterlaken, T.A.M., Puijk, W.C.,
Posthumus, W.P., Meloen, R.H., Boerce, W.A.M., Oosting, J.D.,
Kraaijeveld, C.A. and Snippe, H., 1991, Identification of linear
epitopes on Semliki Forest Virus E2 membrane protein and their
effectiveness as a synthetic peptide vaccine, J.Gen.Virol., 72:557.

Snijders, A., Benaissa-Trouw, B.J., Visser-Vernooy, H.J., Fernandez, I.,
Snippe, H. and Kraaijeveld, C.A. (1992): A DTH-inducing T-cell
epitope of Semliki Forest Virus mediates effective T helper
activity for antibody production. Immunology, 72:99.

Steward, M.W., Stanley, C.M., Dimarchi, R., Mulcahy, G. and Doel, T.R.,
1991, High-affinity antibody induced by immunization with a
synthetic peptide is associated with protection of cattle against
foot-and-mouth disease, Immunology, 72:99.

Steward, M.W. and Howard, C.R., 1987, Synthetic peptides: a next generation
of vaccines? Immunol.Today, 8:51.

Thole, J.E., van Schooten, W.C., Keulen, W.J., Hermans, P.W., Janson, A.A.,
de Vries, R.R.,Kold, A.H. and van Embden, J.D., 1988, Use of
recombinant antigens expressed in Escherichia coli K-12 to map B-
cell and T-cell epitopes on the immunodominant 65-kilodalton
protein of Mycobacterium bovis BCG, Infect.Immun., 56:1633.

Thornton, G.B., Milich, D.R., Chisary, F., Mitamura, K., Kent, S.B.,
Neurath, R., Purcell, R. and Gerin, J., 1987, Immune responses in
primates to the pre-S2 region of hepatitis B surface antigen:
identification of a protective determinant, in: "Vaccines, '87",
R.M. Chanock, R.A. Lerner, F. Brown and H. Ginsberg, eds. Cold
Spring Harbor Laboratory, New York. 77.

Tokunaga, K., Saueracker, G., Kay, P.H., Christiansen, F.T., Anand, R. and
Dawkins, R.L., 1988, Extensive deletions and insertions in
different MHC supratypes detected by pulsed field gel
electrophoresis, J.Exp.Med., 168:933.

Valenzuela, P., Medina, A., Rutter, W.J., Ammerer, G. and Hall, B.D., 1982,
Synthesis and assembly of hepatitis B virus surface antigen
particles in yeast, Nature (London), 298:347.

Van Regenmortel, M.H. and Daney de Marcillac, G., 1988, An assessment of
prediction methods for locating continuous epitopes in proteins,
Immunol.Lett., 17:95.

Vennema, H., de Groot, R.J., Harbour, D.A., Dalderup, M., Gruffydd Jones,
T., Horzinek, M.C. and Spaan, W.J., 1990, Early death after feline
infectious peritonitis virus challenge due to recombinant vaccinia
virus immunization, J.Virol., 64:1407.

Vidal, S., Mottet, G., Kolakofsky, D. and Roux, L., 1989, Addition of high-
mannose sugars must precede disulfide bond formation for proper
folding of Sendai virus glycoproteins, J.Virol., 63:892.

Vrati, S., Fernon, C.A., Dalgarno, L. and Weir, R.C., 1988, Location of a
major antigenic site involved in Ross River virus neutralization,
Virology, 162:346.

Wahlberg, J.M., Boere, W.A. and Garoff, H., 1989, The heterodimeric
association between the membrane proteins of Semliki Forest Virus

changes its sensitivity to low pH during virus maturation, J.Virol., 63:4991.

Wright, K.E., Salvato, M.S. and Buchmeier, M.J., 1989, Neutralizing epitopes of lymphocytic choriomeningitis virus are conformational and require both glycosylation and disulfide bonds for expression, Virology, 171:417.

IMMUNOTARGETING AS AN ADJUVANT-INDEPENDENT SUBUNIT VACCINE DESIGN OPTION

Danna L Skea and Brian H Barber

Department of Immunology, Medical Sciences Building
University of Toronto
Toronto, Canada M5S 1A8

INTRODUCTION

Developments over the past decade in the concepts and techniques of molecular biology, synthetic peptide chemistry, and immunology, have placed within our grasp the potential to construct a new generation of defined subunit vaccine agents. The ability to clone and express in quantity the product of specific genes from virtually any infectious agent of interest, now provides an array of previously inaccessible antigens for consideration as vaccine candidates. In some cases, the generation of pathogen-neutralizing monoclonal antibodies has identified the key gene products on which to focus in order to block these infectious agents. Additionally, developments in the methods for identifying and synthesizing specific B- and T-cell epitopes now offer the potential to induce disease-neutralizing immune responses with completely synthetic structures. Although these advances have fostered much research into new vaccine candidates, and numerous studies have been carried out in different animal models, significant hurdles yet remain in terms of translating these findings into a new set of human vaccine agents.

One of these hurdles is the lack of safe and effective alternatives to the use of alum as a vaccine adjuvant. At the moment, alum is the only adjuvant licensed for general use in human vaccines [Warren et al., 1986]. Although effective in certain circumstances, in general, alum fails to augment the immunogenicity of defined immunogens as well as many other experimental adjuvants [Edelman, 1980]. The long-standing "gold standard" with respect to the enhancement of the immunogenicity of purified protein and peptide antigens is the water-in-oil emulsion formed with aqueous antigen in Freund's complete adjuvant [Warren et al., 1986]. However, for a variety of reasons, Freund's adjuvant cannot be used for either human or veterinary vaccine applications [Weidemann et al., 1991]. As a result, considerable effort has been invested in the search for safe and effective alternatives which would be acceptable to the vaccine regulatory authorities.

As an alternative to the search for new adjuvants, we have been exploring the possibility of an adjuvant-independent means of enhancing the immunogenicity of purified protein and peptide antigens. Our approach, which we refer to as immunotargeting, involves the conjugation of the antigen to a monoclonal antibody specific for a determinant expressed in

Table 1. Immunotargeting avidin to class II MHC in $(H-2^k \times H-2^b)$F1 mice

Antibody-Avidin Conjugate	Targeting Antibody Isotype	Targeting‡ Specificity	anti-avidin IgG* µg/ml
10-3.6.2 - avidin	mouse IgG2a	I-A^k	17.3
H16-L10-4R5 - avidin	mouse IgG2a	flu NP	0.42

* - Geometric mean response for 5 mice immunized subcutaneously with immunoconjugate containing 10 µg of avidin, and then boosted once intraperitoneally 4 weeks later with 10 µg of avidin only, in saline. Indicated are serum anti-avidin IgG levels one week post-boost as determined by ELISA, relative to a monoclonal anti-avidin IgG standard.

‡ - References for the monoclonal antibodies are as follows: 10-3.6.2 - Oi, et al, 1978; H16-L10-4R5 - Yewdell et al, 1981.

vivo on the surface of cells of the immune system. Because the antigen-antibody conjugate is injected in saline, the immunization is adjuvant-independent. Thus, any enhancement of immunogenicity observed can be attributed to the physical properties of the conjugate itself, presumably dominated by the binding specificity of the antibody and its ability to localize antigen at the surface of specific target cells.

IMMUNOTARGETING ANTIGEN TO CLASS II MHC

Our initial examination of the concept of immunotargeting was focussed on the use of class II major histocompatibility complex (MHC) gene products as potential in vivo target structures. In the mouse, class II gene products are constitutively expressed on B-cells, and also on certain cells in the monocyte-macrophage lineage specialized for the presentation of antigen to T-helper cells (e.g. dendritic cells). Given that in vivo priming for a good secondary antibody response requires both B-cell and T-cell recognition, it was reasoned that delivery of antigen to class II MHC-bearing cells could be an effective means of promoting the immunogenicity of the conjugated antigens. Using avidin as a model protein antigen able to bind with high affinity to biotinylated monoclonal antibodies, we found that anti-class II MHC antibodies were able to augment the immunogenicity of the bound protein [Carayanniotis and Barber, 1987].

Results from a representative anti-class II MHC immunotargeting experiment are depicted in Table 1. The anti-avidin IgG response in (H-2^{k}xH-2^b)F1 mice immunized with avidin conjugated to an anti-I-A^k monoclonal antibody (MAb) is compared with that observed for avidin bound to an isotype-matched control anti-influenza nucleoprotein MAb. Avidin bound to the anti-class II MAb is approximately 40-fold more immunogenic than avidin conjugated to the anti-NP MAb, a control antibody which would not be expected to react with murine cells (other than possibly through Fc receptors specific for IgG$_{2a}$). These results are in accord with our previous findings which indicated that directing avidin to class II MHC determinants augmented the immunogenicity of the targeted protein without requiring any adjuvant [Carayanniotis and Barber, 1987; 1990]. Using different strains of mice, including intra H-2 recombinants, we previously demonstrated that the targeting effect was dependent upon the presence in the recipient mouse of the class II allele for which the antibody was specific [Carayanniotis and Barber, 1990], and that enhancement of immunogenicity was only observed when the avidin was bound to the targeting antibody, and not when merely co-injected with the non-biotinylated antibody [Carayanniotis and Barber, 1987].

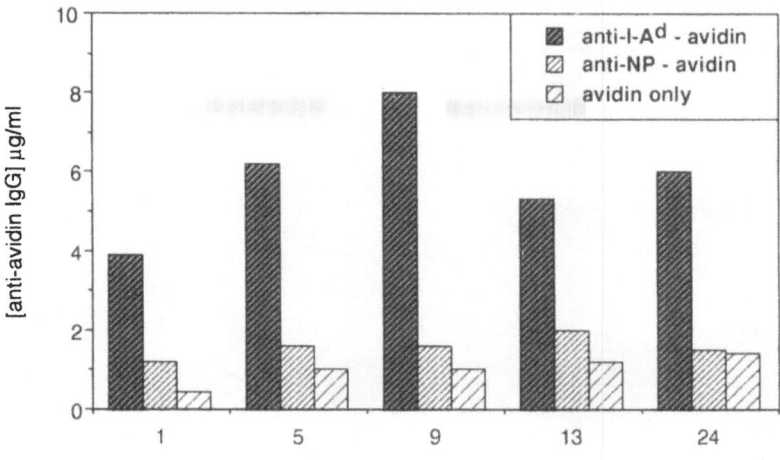

weeks after boost

Fig. 1. Long-lived antibody response in BALB/c mice primed by immunotargeting. Groups of 5 mice each were immunized subcutaneously with immunoconjugates (containing 25 μg of avidin)
or with 25 ug of avidin only. Four weeks later each mouse was
boosted intraperitoneally with 10 ug of avidin only. Serum
samples were collected at the indicated times. The data
represent the geometric mean level of serum IgG anti-avidin
antibody as measured by ELISA, relative to a monoclonal anti-
avidin IgG standard.

Previous examination of the kinetics of the primary response to the
anti-class II immunoconjugates indicates an early IgG response which peaked
10–15 days post-immunization [Carayanniotis and Barber, 1990]. A strong
memory response to the initial priming with immunoconjugate could be
induced by challenge with avidin alone at 3 weeks post-priming. In order
to see how long this secondary IgG response would last, mice immunized with
anti-class II MHC-avidin conjugates on day 0, and boosted on day 28 with
avidin only, were followed by periodic serum sampling for the anti-avidin
IgG level for a number of months post-immunization. As indicated by the
data presented in Figure 1, the anti-avidin response persisted for at least
6 months post-immunization. In other experiments of a similar nature, it
was observed that even at 6 months post-priming, the level of anti-avidin
IgG could be boosted more than 10-fold by re-exposure to avidin alone,
indicating that memory can be long-lived in mice immunized via
immunotargeting [Skea and Barber, manuscript in preparation].

The establishment of persistent antibody production and the potential
for a long-lived memory response, are two particularly important features
of any immunization strategy relevent to vaccine design. Therefore, it was
encouraging to see that immunotargeting could establish these manipulations
of the immune system in the absence of adjuvant. It is also interesting,
in the context of recent speculation concerning the molecular basis of
immunological memory [Gray et al., 1991], that anti-class II MHC
immunoconjugates seem able to "deliver" antigen to the in vivo cellular
locations required to achieve the maintenance of long-term memory.

A further point of interest with respect to the antibody response
induced by immunotargeting to class II MHC relates to the subclass

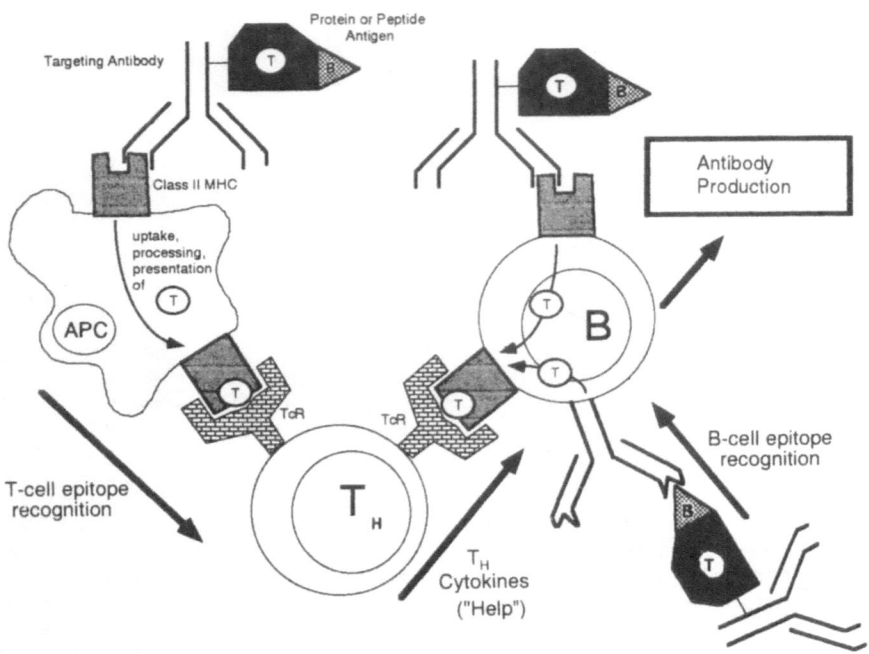

Fig. 2. Schematic representation of the proposed mechanism for the
        enhancement of immunogenicity by immunotargeting. T = T-helper
        cell epitope; B = B cell epitope; TcR = T cell receptor; APC =
        antigen presenting cell; B = B cell; TH = T-HELPER cell

distribution of the antigen-specific IgG. As expected, when mice are
immunized with avidin adsorbed to alum, the avidin-specific IgG response is
almost exlusively IgG1. In contrast, the response to avidin emulsified in
FCA contains a significant fraction (approx. 20-30%) of the response as
IgG_{2a}. It has been argued that the IgG_{2a} class of antibody in the mouse
represents the most effective class of antibody with respect to anti-viral
activity [Allison, 1989]. The IgG subclass distribution of antibody
response to avidin targeted to mouse class II MHC was very similar to the
profile observed for avidin in FCA [Skea and Barber, manuscript in
preparation], suggesting that the T-helper cells activated by immunotargeting, and the combination of lymphokines they produce, must also be
similar. Efforts are presently underway to determine the balance of TH1
versus TH2 helper T-cell activation involved by immunizations with
different adjuvants in comparison to immunotargeting.

RATIONALE FOR THE ENHANCEMENT OF IMMUNOGENICITY BY IMMUNOTARGETING

A schematic representation of the events which we believe could
account for the enhancement of immunogenicity observed for antigens bound
to anti-class II MHC MAbs is depicted in Figure 2. Cells bearing class II
MHC have the potential to act as antigen presenting cells (APC), implying
that the uptake, proteolytic processing and association of the released T-
cell epitope(s) (i.e., peptides) with class II MHC can result in the
activation of antigen-specific T-cell help [Berzofsky et al., 1989]. As a
result of subcutaneous immunization, anti-class II MHC immunoconjugates
could encounter class II determinants on specialized APCs such as dendritic
or Langerhans cells, or alternatively, interact with the class II MHC on B-

cells. Any of these events could result in the processing and presentation of antigen-specific T-cell determinants, which would in turn lead to the activation of class II MHC-restricted T-helper (TH) cells capable of providing the cytokines necessary for B-cell differentiation and expansion (i.e. "help") [Berzofsky et al., 1989]. The only B-cells capable of responding effectively to this "help" (often referred to as signal 2 [Cambier and Ransom, 1987]) would be those which had already received the traditional signal 1 [Jelinek and Lipsky, 1987], induced by engagement of their antigen-specific receptors with B-cell epitopes on the antigen. In this model, the enhancement of immunogenicity achieved by immunotargeting can be attributed to the significantly increased efficiency of antigen uptake into APCs mediated by the interaction of the immunoconjugate with class II MHC-bearing cells. Using an in vitro cell culture system, Casten et al., [1988] have shown that the efficiency of presentation of a defined T-cell epitope by class II bearing APCs can be enhanced by as much as three orders of magnitude by conjugation of this peptide to an anti-class II MAb. Snider et al., [1990] have also shown that targeting to class II MHC in vivo can serve to augment the immunogenicity of antigens carried in a cross-linked antibody complex.

SEARCH FOR THE OPTIMAL TARGET STRUCTURE IN VIVO

Although targeting antigens to class II MHC determinants in vivo has been demonstrated by us [Carayanniotis and Barber, 1987, 1990; Carayanniotis et al., 1988] and others [Pierce and Casten, 1988; Snider et al., 1990] to enhance the immunogenicity of the delivered antigen in the absence of adjuvant, it remained to be established whether or not this was the optimal "entry point" for the promotion of antibody responses. For this reason, we have investigated a number of monoclonal antibodies re-cognizing different surface markers on cells of the immune system for their ability to prime for an adjuvant-independent response to bound avidin. The results for one such set of rat MAbs recognizing differentiation antigens in the Balb/c mouse are depicted in Table 2. In this case, the control antibody was a rat anti-human class I MHC antibody. The immunoconjugate of avidin with this MAb provides a basal level of response which is not significantly different from avidin alone. The anti-class II antibody provides the best response, with the others ranging from relatively ineffective (e.g. the Langerhans cell specific MAb NLDC145) to moderately effective (e.g. the anti-CD4 and CD45 MAbs) in their ability to prime for a secondary IgG anti-avidin response. Previously we reported that avidin conjugates with the 33D1 anti-dentritic cell antibody were significantly more immunogenic than anti-I-A^k conjugates (with the 10-3.6.2 MAb) in H-2^k bearing mice [Carayanniotis et al., 1991]. In the current comparison, the response to the 33D1 conjugate was approximately one-third that of the anti-class II conjugate. We have also observed a range of responses to targeting with this antibody in other mouse experiments with avidin conjugates [Skea and Barber, manuscript in preparation]. This may reflect the acknowledged sensitivity of this antibody to inactivation by chemical modification of key amino acid side chains in the binding site [Nussenzweig et al., 1982], or alternatively, heterogeneity with respect to the functional capacity of the 33D1 cells encountered from one immunization to another.

It is important to recognize that the response observed for target-ing to a particular cell surface determinant is likely to be dependent upon a number of variables; including the affinity of the MAb for the determinant in question, the molecular nature of the target structure (e.g. its susceptibility to antibody-mediated cross-linking and internalization), and the antigen presenting properties of the cell bearing the target antigen. Thus, for example, the lack of response for NLDC-145 targeting to

Table 2. Immunotargeting avidin to different cell surface determinants in BALb/c mice

Antibody-Avidin Conjugate	Targeting‡ Specificity	anti-avidin IgG* µg/ml
B21-2 - avidin	I-A^{b,d}	26.0
33D1 - avidin	dendritic cells	8.1
NLDC 145 - avidin	Langerhans cells	2.9
GK 1.5 - avidin	CD4	17.2
M1/9.3.4.HL.2 - avidin	CD45	11.3
SFR8-B6 - avidin	HLA-Bw6	1.7
avidin only	-	1.4

* - Geometric mean response for 5 mice, one week post-boost, immunized as indicated in the legend to Table 1.

‡ - References for the monoclonal antibodies are as follows: B21-2 - Steinman, et al, 1980; 33D1 - Nussenzweig, et al, 1982; NLDC 145 - Kraal, et al, 1986; GK 1.5 - Wilde, et al, 1983; M1/9.3.4.HL.2 - Springer, et al, 1978; SFR8-B6 - Radka, et al, 1982.

Langerhans cells may reflect a lower affinity antibody, the nature of the surface antigen being recognized, or some property of the cells themselves.

CD45 is a marker which is expressed in various forms on most cells in the hematopoietic lineage [Johnson et al., 1989]. M1/9.3.4.HL.2 is an antibody which sees a determinant on CD45 common to all isoforms of the molecule and therefore would deliver antigen to B-cells, T-cells and various cells in the monocyte-macrophage differentiation pathway. Thus it is not surprising that immunotargeting to CD45 results in a significant response over background. Much less expected was the positive response to avidin targeted on GK1.5 to CD4. A level of response comparable to, or even significantly greater than that observed with anti-class II targeting has been observed in a number of experiments [Skea and Barber, manuscript in preparation]. Although human activated T-cells bear class II MHC antigens [Gerrard and Volkman, 1986], their presence on the surface of mouse T-cells is much more controversial [Graf et al., 1985]. A lack of class II MHC could make it difficult to explain CD4 T-cells acting as conventional APCs, but there may be other mechanisms whereby "help" can be delivered in vivo. Possibly antigen displayed on anti-CD4 antibodies serves to "focus" antigen-specific B-cells onto activated CD4+T-cells. Alternatively, the key event may be focusing antigen to a CD4+ dendritic cell population [Crowley et al., 1990] (rather than to T-cells), which is able to efficiently present T-cell determinants.

IMMUNOTARGETING INFLUENZA HA IN MICE

In an effort to explore immunotargeting in the context of viral components which represent potential vaccine candidates, we compared the adjuvant-independent response to anti-class II MHC immunotargeted influenza hemagglutinin (specifically the large water soluble bromelin fragment, designated BHA) with the response observed in mice using aluminum phosphate (i.e. alum), the adjuvant licensed for use in man. The result is depicted in Figure 3. As can be clearly seen, alum is not a good adjuvant for BHA in mice and the post-boost IgG response obtained was only marginally above the BHA in PBS control. The anti-I-A^k immunoconjugate with BHA, which was formed using the avidin bridge technique described previously

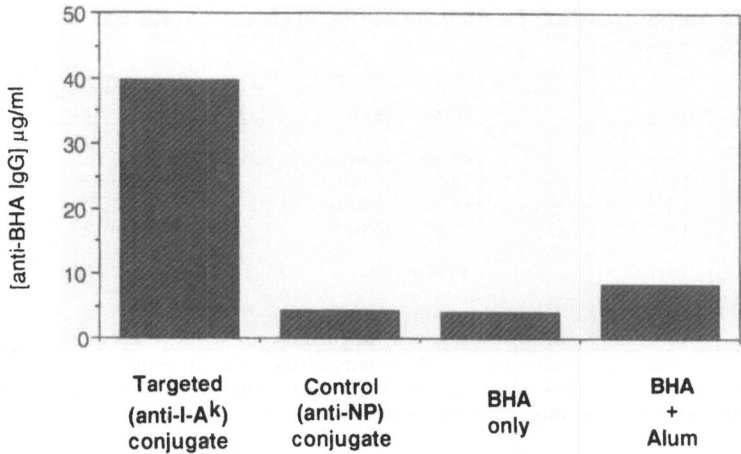

Fig. 3. Mouse antibody responses to immunotargeted BHA. Groups
of 5 mice each were immunized subcutaneously with MAb-
avidin-BHA conjugates (containing 25 ug of BHA), or with
25 µg of BHA mixed with alum, or with 25 µg of BHA alone.
Four weeks later, each mouse was boosted intraperitoneally
with 10 ug of BHA only. Serum samples were collected one
week after the boost. The data represent the geometric mean
level of serum IgG anti-BHA antibody as measured by ELISA,
relative to a monoclonal anti-BHA IgG standard

[Carayanniotis and Barber, 1990], gave a response which was more than 4-
fold greater than that with alum. The control conjugate, formed in the
same way with the anti-influenza nucleoprotein MAb, gave a background
response, confirming the ability to target antigenic complexes of
significantly greater size than avidin alone.

IMMUNOTARGETING IN NON-MURINE SPECIES

     We initially chose to examine the immunotargeting approach to
immunization in mice because of the potential to use MHC allele specific
MAbs and defined inbred strains of mice to determine whether or not any
enhancement of immunogenicity could be attributed to immunotargeting. This
system has certainly proven to be useful in establishing that targeting to
class II MHC determinants can be an effective means of augmenting immunogenicity [Carayanniotis and Barber, 1987, 1990]. Having identified this as
a useful target structure, we then wanted to determine whether or not this
approach was applicable to outbred species such as man, for which the MHC
genes represent highly polymorphic loci. One means of addressing this
problem is to identify anti-class II MHC MAbs which recognize framework
(i.e. non-polymorphic) regions of the molecule. One such antibody is
designated as 44H10. Generated by immunization of a Balb/c mouse with the
HOON human leukemic cell line [Dubiski et al., 1988], this antibody has
been shown to react with all human class II MHC HLA-DR alleles. In
addition, it recognizes a conserved framework determinant on the class II
molecules of a number of other species including rabbits, ferrets, and
macaques [Skea et al., 1992]. We have utilized this antibody to test the
immunogenicity of 44H10-avidin conjugates in the rabbit in an effort to
determine the efficacy of targeting to conserved class II MHC determinants.
In Table 3, the anti-avidin response for the individual rabbits immunized
with those immunoconjugates is compared with the response to avidin alone,

Table 3. A comparison of immunotargeting with immunization using FCA, in the rabbit

Rabbit	Immunogen	anti-avidin IgG* (arbitrary units/ml)
13	44H10 - avidin	3,776
15	44H10 - avidin	6,239
14	44H10 - avidin	7,179
16	44H10 - avidin	10,309
11	avidin only	225
17	avidin + FCA	67,154
18	avidin + FCA	82,180

* - Responses for individual rabbits immunized subcutaneously with the immunogens indicated (100 µg of avidin), boosted 4 weeks later with avidin only (80 µg), and bled one week post-boost.

or avidin in Freunds' complete adjuvant. Although the response achieved to avidin conjugated to the anti-class II MAb was clearly less than with Freunds' (approximately 10%), it was still markedly enhanced (approximately 30-fold) over the avidin only control. This response was achieved with no negative pathology at the site of injection, in contrast with the marked granuloma formation associated with the immunization in Freunds [Skea et al., 1992].

Thus the potential does exist to effectively utilize anti-class II MHC framework MAbs to immunotarget antigens in species other than inbred mice. Although it remains to be determined how effective this target site might be in general, it does provide a viable option for direct application of the immunotargeting technology to the development of candidate subunit vaccines for veterinary animals and man.

ACKNOWLEDGEMENTS

D.L.S. is the recipient of a Medical Research Council of Canada Postdoctoral Fellowship. This research was supported by Connaught Laboratories Limited and the Province of Ontario Technology Fund.

REFERENCES

Allison, A.C., 1989, Antigens and adjuvants for a new generation of vaccines, in: "Immunological Adjuvants and Vaccines" (G. Gregoriadis, A.C. Allison and G. Poste, eds) pp 1-12, Plenum Press, New York.

Berzofsky, J.A., Kurata, A., Takahashi, H., Brett, S.J. and McKean, D.J., 1989, Molecular studies of antigen processing and presentation to T cells by class II MHC molecules, Cold Spring Harbor Symp. Quant. Biol., LIV: 417.

Cambier, J.C. and Ransom, J.T., 1987, Molecular mechanisms of transmembrane signalling in B Lymphocytes, Ann.Rev.Immunol., 5: 175.

Carayanniotis, G. and Barber, B.H., 1987, Adjuvant-free IgG responses induced with antigen coupled to antibodies against class II MHC, Nature, 327: 59.

Carayanniotis, G. and Barber, B.H., 1990, Characterization of the adjuvant-free serological response to protein antigens coupled to antibodies specific for class II MHC determinants, Vaccine, 8: 137.

Carayanniotis, G., Skea, D.L., Luscher, M.A. and Barber, B.H., 1991, Adjuvant-independent immunization by immunotargeting antigens to MHC and non-MHC determinants in vivo, Molec.Immunol., 28: 261.

Carayanniotis, G., Vizi, E., Parker, J.M.R., Hodges, R.S. and Barber, B.H.,
    1988, Delivery of synthetic peptides by anti-class II MHC mono-
    clonal antibodies induces specific adjuvant-free IgG responses in
    vivo, Molec.Immunol., 25: 907.
Casten, L.A., Kaumaya, P. and Pierce, S.K., 1988, Enhanced T cell responses
    to antigenic peptides targeted to B cell surface Ig, Ia or class I
    molecules, J.Exp.Med., 168: 171.
Crowley, M.T., Inaba, K., Witmer-Pack, M.D., Gezelter, S. and Steinman,
    R.M. , 1990, Use of the fluorescence-activated cell sorter to
    enrich dendritic cells from mouse spleen, J.Immunol.Meth., 133: 55.
Dubiski, S., Cinader, B., Chou, C.T. Charpentier, L. and Letarte, M., 1988,
    Cross-reaction of a monoclonal antibody to human MHC class II
    molecules with rabbit B cells, Mol.Immunol., 5: 713.
Edelman, R. 1980, Vaccine adjuvants, Res.Inf.Dis., 2: 370.
Gerrard, T.L.,Volkman, D.J., Jurgerson, C.H. and Fauci, A.S., 1986,
    Activated human T cells can present denatured antigen, Human
    Immunol., 17: 416.
Graf, L., Koch, N. and Schirrmacher, V., 1985, Expression of Ia antigens in
    a murine T-lymphoma variant, Molec.Immunol., 12: 1371.
Gray, D., Kosco, M. and Stockinger, B., 1991, Novel pathways of antigen
    presentation for the maintenance of memory, International Immunol.,
    3: 141.
Jelinek, D.F. and Lipsky, P.E., 1987, Regulation of human B lymphocyte
    activation, proliferation and differentiation, Adv.Immunol., 40: 1.
Johnson, P., Greenbaum, L., Bottomly, K. and Trowbridge, I.S., 1989,
    Identification of the alternatively spliced exons of murine CD45
    (T200) required for reactivity with B220 and other T200-restricted
    antibodies, J.Exp.Med., 169: 1179.
Kraal, G., Breel, M., Janse, M. and Bruin, G., 1986, Langerhans' cells,
    veiled cells and interdigitating cells in the mouse recognized by a
    monoclonal antibody, J.Exp.Med., 163: 981.
Nussenzweig, M.C., Steinman, R.M., Witmer, M.D. and Gutchinov, B., 1982, A
    monoclonal antibody specific for mouse dentritic cells,
    Proc.Natl.Acad.Sci., 79: 161.
Oi, V.T., Jones, P.P., Goding, J.W., Herzenberg, L.A. and Herzenberg, L.A.,
    1978, Properties of monoclonal antibodies to mouse Ig allotypes, H-
    2 and Ia antigens, Curr.Topics Microbiol.Immunol., 81: 115.
Pierce, S.K. and Casten, L.A., 1988, Soluble globular protein antigens
    covalently coupled to antibodies specific for B cell surface
    strutures are effective antigens both in vitro and in vivo,
    Prog.Leuk.Biol., 7: 259.
Radka, S.F., Kostyn, D.D. and Amos, D.B., 1982, A monoclonal antibody
    directed against the HLA-BW6 epitope, J.Immunol., 128: 2804.
Skea, D.L., Douglas, A.R., Skehel, J.J. and Barber, B.H., 1993, The
    immunotargeting approach to adjuvant-independent immunization with
    influenza hemagglutinin, Vaccine, in press.
Snider, D.P., Kaubisch, A. and Segal, D.M., 1990, Enhanced antigen
    immunogenicity induced by bispecific antibodies, J.Exp.Med., 171:
    1957.
Springer, T., Galfri, G., Secher, D.S. and Milstein, C., 1978, Monoclonal
    xenogeneic antibodies to murine cell surface antigens:
    Identification of novel leukocyte differentiation antigens,
    Eur.J.Immunol., 8: 539.
Steinman, R.M., Nogueira, N., Witmer, M.D., Tydings, J.D. and Millman,
    I.S., 1980, Lymphokine enhances the expression and synthesis of Ia
    antigens on cultured mouse peritoneal macrophages, J.Exp.Med., 152:
    1248.
Warren, H.S., Vogel, F.R. and Chedid, L.A., 1986, Current status of
    immunological adjuvants, Ann.Rev.Immunol., 4: 369.
Wiedemann, F., Link, R., Pump, K., Jacobshagen, U., Schaefer, H.E.,
    Weismuller, K.H., Hummel, R.P., Jung, G., Bessler, W. and Boltz,

T., 1991, Histopathological studies on the local reactions induced by complete Freund's adjuvant (CFA), bacterial lipopolysaccharides (LPS) and synthetic lipopeptide (P$_3$C) conjugates, J.Pathol., 164: 265.

Wilde, D.B., Marrack, P., Kappler, J., Dialynas, D.P. and Fitch, F.W., 1983, Evidence implicating L3T4 in class II MHC antigen reactivity; monoclonal antibody GK 1.5 (anti-L3T4a) blocks class II antigen-specific proliferation, release of lymphokines, and binding by cloned murine helper T lymphocyte lines, J.Immunol., 131: 2178.

Yewdell, J.W., Frank, E. and Gerhard, W., 1981, Expression of influenza A virus internal antigens on the surface of infected P815 cells, J.Immunol., 126: 1814.

BCG VACCINE: THE NEXT GENERATION OF TARGETED DRUG DELIVERY SYSTEMS

M.J. Groves, Yan Lou and M.E. Klegerman

Institut for Tuberculosis Research (M/C 964)
College of Pharmacy, University of Illinois at Chicago
840 West Taylor Street, Room 2014 SEL
Chicago, IL 60607-7019, USA

INTRODUCTION

The relatively recent recognition of the value of bacillus Calmette-Guérin (BCG) as a general immunostimulant and its introduction into clinical practice as a treatment for superficial bladder tumours (Morales et al, 1976; Brosman and Lamm, 1990; Soloway and Perry, 1987; Khanna et al, 1990) has resulted in a number of studies of the mode of action of this attenuated living bacterial suspension.

The mode of action of BCG against tumours on the bladder wall can be divided into three phases although not all details are known unambiguously. Nevertheless, for most purposes it can be said that the organisms are targeted to the superficial tumour cells, stimulate an immune response and are finally phagocytosed by the tumour cells. It is anticipated that these activities could be simulated by non-viable microparticulate systems, thereby avoiding some of he side effects associated with BCG therapy (Lamm et al, 1986).

TARGETING OF TUMOR CELLS BY BCG

In 1987 Ratliff and his colleagues postulated that the apparent targeting of BCG cell for superficial bladder tumours was mediated by a fibronectin-fibronectin receptor reaction. Details of this mechanism have become clearer over the past five years and it appears that fibronectin (or rather fibronectins since these high molecular weight proteins are ubiquitous throughout the body, Goodheart and Silverman, 1991) is excreted at or very close to the tumour cell surface. Teppema et al, (1992) have demonstrated that virtually no BCG is associated with normal intact urothelium in patients treated with BCG for bladder cancer. These workers admitted that some BCG cells were occasionally associated with coagulation lesions. They suggest that there was little evidence that there was direct contact between BCG and urothelial cells although there was strong in vivo evidence that a human T-24 carcinoma cell line was capable of adhering, ingesting and degrading BCG. This is in contrast with the work of Ratliff's group (Ratliff et al, 1987; 1988a, b; 1989; Abou-Zeid et al, 1988; Hudson et al, 1990, 1991; Godfrey et al, 1992; Kavoussi et al, 1990) who provided strong evidence for the direct interaction of BCG cells with fibronectin coated surfaces. The BCG fibronectin receptors have been identified as members of the group of the Antigen 85 protein complex (Godfrey et al, 1992), an observation confirmed by Öner et al, (in press). The Antigen 85 complex are the major components of proteins excreted by BCG in culture and have molecular weights in the order of 30-32 kDa. However, these proteins are apparently readily washed off the surface of the cell (Shi et al, 1989) and the actual attachment factors may be proteins in the 60-65 kDa size range that are more firmly attached to the cell wall (Öner, unpublished; Ratliff, private communication). All three members of the 85 complex have been reported to be present on the Copenhagen substrain of BCG (Wiker et al, 1986) although, in the Tice substrain, the A antigen predominated (Öner et al, in press).

Immunological Response Induced by BCG

One aspect of the antitumour response induced by BCG is apparently a local immune response and a delayed hypersensitivity response (Kavoussi et al, 1990). The phagocytes are undoubtably stimulated involving,in part, a phagocytic response, and a granulomatous reaction is generated in the bladder wall (Lage et al, 1986; Prescott et al, 1989). As reviewed by Prescott at al, (1989), immunological studies have demonstrated that T-lymphocytes, monocytes/macrophages and polymorphonuclear leukocytes are the main infiltrating cells in response to intravesical BCG treatment. There is also a class II MHC (HLA-DR) antigen expression on urothelial tumour cells which have the effect of altering their phenotype, with implications for at least one mechanism of BCG antitumour activity (Stefanini et al, 1989)

The local response induced by the mycrobacterial cells is therefore an important element of the overall antitumour activity.

Phagocytosis

Becich et al, (1991) reported that BCG cells were internalized and digested by both human bladder tumour (T-24) and murine bladder tumour (MBT-2) cell lines, and suggested that this observation was likely to occur in vivo in the clinical situation. It is true that a number of non-phagocytic cell lines are capable of digesting mycobacteria in vitro (Shepherd, 1957) and M. leprae cells are ingested by Schwann cells both in vitro and in vivo (Band et al, 1986). In our laboratory we have demonstrated phagocytosis of BCG cells by a murine sarcoma (Devados et al, 1993). Becich et al, (1991) commented that the subsequent fate of the tumour cells is by no means certain but demonstrated that the internalized BCG cells were completely destroyed. These authors found internalized and degraded BCG cells in urothelial cells obtained in bladder washers from patients undergoing BCG treatment. This latter observation has not been confirmed by Teppema et al, (1992) and there remains some controversy in this area.

One point needs to be made here since it is evident from our own published observations that BCG is capable of killing murine sarcoma cells after they have been internalized (Klegerman et al, 1991a, b). The possibility exits that BCG contains a stable cytotoxic component. We have observed that the antineoplastic activity of archived BCG samples remains the same over a 25 year period while cell viability declines over the same period (Klegerman et al, 1991b). We have isolated a complex lipopolysaccharide (termed PS1) from lyophilized BCG by hot water extraction that has in vivo anti-tumour activity (Lou et al, in press). The identity of this material is currently under investigation. It has some chemical affinity with the lipoarabinomannan (LAM), identified as the toxic component of M. tuberculosis (Brennan et al, 1990; Chatterjee et al, 1992), although unpublished evidence currently suggests that it is not identical. It should be pointed out that M. tuberculosis is virulent whereas BCG is not, an

112

Table 1. Relative avidities (calculated as Scatchard affinity constants, K_a) of BCG and gelatin microparticles for human plasma fibronectin bound on glass beads (data from Olson, 1992)

	$K_a \times 10^9$
BCG Tice substrain lot 105149 (n=5)	6.58 (se + 21.1%)
BCG Tice substrain lot 105178	7.16
Connaught substrain	6.65
Pasteur substrain	5.08
Gelatin microparticles, acid washed, 60 Bloom diam. 15.3um	0.13
Gelatin microparticles, acid washed, 300 Bloom diam. 6.8um	0.21
Gelatin microparticles, base washed, 60 Bloom diam. 16.3um	0.16
Gelatin microparticles, base washed, 225 Bloom diam. 3.1um	0.21

observation that might indicate these two compounds are not necessarily the same in terms of toxicity, at least to the host animal.

Mimicking the Action of BCG

The main side effects associated with the clinical use of BCG are due to the fact that the vaccine is viable (Lamm et al, 1986). There is some doubt as to whether or not a heat-killed bacterial suspension could be used (Rosenthal, 1980) but at least for the present a viable vaccine is employed clinically. For this reason, it would be of interest to attempt to mimic the biological activity of BCG vaccine with a system that avoided most, if not all, of the side effects associated with a viable bacterial suspension.

We have investigated microparticular drug delivery systems as possible alternatives to the vaccine.

(a) Targeting activity. Based on the observation that fibronectin contains domains specific for a number of molecules including collagen (Engvall and Ruoslahti, 1977; Owens and Barelle, 1986), it seemed evident that gelatin, a degraded form of collagen, might have an avidity for fibronectin coated onto glass beads. This has been confirmed by Olson (1992) who, by measuring Scatchard affinity plots, was able to compare the avidities of BCG and micron-range gelatin microspheres, Table 1

This suggested that gelatin microspheres were capable of targeting tumour cells expressing fibronectin or with fibronectin at or near the periphery of the tumour due to damage of underlying tissues. The avidity is approximately 1-2 orders of magnitude lower than that of BCG but remains strong enough to be of potential application to a specific fibronectin-bearing surface.

(b) <u>Immunological response and phagocytosis</u>. It is known that gelatin microspheres are capable of stimulating macrophages and are phago-cytosed (Tabati and Ikada, 1987; 1989a, b, c). In an <u>in vivo</u> test of gelatin microspheres containing PS1 Lou, (unpublished), has recently de-monstrated that the formulation is capable of preventing a murine sarcoma from growing and that it produces a granulomatous reaction. This indicates a local immune response and phagocytosis by the tumour which is basically similar to those reactions observed for BCG for the same tumour. The direct antitumour effect of the PS1 itself is marked (Lou et al, in press) but is without the granulatomous response, suggesting a direct involvement of the drug delivery system itself.

SUMMARY

It will be evident that the mode of action of BCG in superficial bladder cancer is multifaceted and complex. Nevertheless, evidence is beginning to accrue that a gelatin microparticulate system containing PS1 or, perhaps, oncolytic drugs may provide a non-viable alternative treatment for bladder cancer. Thus, based on the study of the mode of action of BCG, a new generation of targeted drug delivery systems is beginning to appear that has promise in the treatment of disease.

REFERENCES

Abou-Zeid, C., Ratliff, T.L., Wiker, H.G., Harboe, M., Bennedsen, J. and Rook, G.A.W. 1988, Characterization of fibronectin-binding antigen released by <u>Micobacterium tuberculosis</u> and <u>Mycobacterium bovis</u> BCG, <u>Inf.Immun.</u>, 56:3046.

Band, A.H., Chitamber, S.D., Bhattacharya, A. and Talwar, G.P. 1986, Mechanisms of phagocytosis of mycobacteria by Schwann cells and their comparison with macrophages, <u>Int.J.Leprosy</u>, 54:294.

Becich, M.J., Carroll, S. and Ratcliff, T.L., 1991. Internalization of bacillus Calmette-Guerin by bladder tumorcells, <u>J.Urol.</u> 145:1316.

Brennan, P.J., Hunter, S.W., McNeil, M., Chatterjee, D. and Daffe, M., 1990, Reappraisal of the chemistry of mycobacterial cell walls, with a view to understanding the role of individual entities in disease processes, <u>in</u>: Microbial Determinants of Virulence and Host Response, Am. Soc. Microbiol., Washington, D.C., pps. 55-75.

Brosman, S.A. and Lamm, D.L. 1990, The preparation, handling and use of intravesical bacillus Calmette-Guerin for the management of stage Ta, Ti carcinoma <u>in situ</u> and transitional cell cancer, <u>J.Urol.</u> 144:313.

Chatterjee, D., Roberts, A.D., Lowell, K., Brennan, P.J. and Orme, I.M. 1992, Structural basis of capacity of lipoarabinomannan to induce secretion of tumor necrosis factor, <u>Inf.andImmnl.</u> 60, (3), 1249.

Devados, P.O., Klegerman, M.E. and Groves, M.J., 1993, Phagocytosis of <u>Micobacterium bovis</u> BCG organisms by murine 5180 sarioma cells, <u>Cytobios</u> 74: 49.

Engvall, E. and Ruoslahti 1977, Binding of soluble form of fibroblast surface protein, fibronectin to collagen, <u>Int.J.Cancer</u>, 28:1.

Godfrey, H.P., Feng, Z., Mandy, S., Mandy, K., Huygan, K., deBrayn, J., Abou-Zeid, C., Wiker, H.G., Nagai, S. and Tasaka, H. 1992, Modulation of expression of delayed hyposensitivity by Mycobacterial Antigen 85 Fibronectin-binding proteins, <u>Inf.& Immunol.</u>, 60 (6),2522.

Goodheart, C.R. and Silverman, R.H. 1991, Cellular and plasma fibronectin, <u>SIM Ind.Microbiol. News</u>, 41:266.

Hudson, M.A., Ritchey, J.K., Catalona, W.J., Brown, E.J. and Ratliff, T.L. 1990, Comparison of the fibronectin-binding ability and antitumor efficacy of various mycobacteria, Cancer Res., 50:3843.

Hudson, M.A., Brown, E.J., Ritchey, J.K. and Ratliff, R.L. 1991, Modulation of fibronectin-mediated bacillus Calmette-Guerin attachment to murine bladder mucosa by drugs influencing the coagulation pathway, Cancer Res., 51:3726.

Kavoussi, L.R., Brown, E.J., Ritchey, J.K. and Ratliff, T.L. 1990, Fibronectin-mediated Calmette-Guerin bacillus attachment to murine bladder mucosa, J.Clin.Invest., 85:62.

Khanna, O.D., Son, D.L., Mazer, H., Read, J., Nugent, D., Cottone, R., Heeg, M., Rezvan, M., Viek, N., Uhlman, R. and Friedman, M. 1990, Multi-center study of superficial bladder cancer treated with intravesical bacillus Calmette-Guerin or adriamycin, Urology, 35 (2):101.

Klegerman, M.E., Uijainwala, L. and Zeunert, P. 1991a, High dose inhibition and low-dose enhancement of murine sarcoma growth exhibited by BCG vaccine, Canc.Let., 56:137.

Klegerman, M.E., Zeunert, P.L., Lajeune, J., Lou, Y. and Groves, M.J. 1991b, Relative tumour inhibitory and stimulatory activities of BCG vaccine preparations, lots and substrains in a quantitative mouse sarcoma bioassay, Anticancer Res., 11:1707.

Lage, J.M., Bauer, W.C., Kelley, D.R., Ratliff, T.L. and Catalona, W.J. 1986, Histological parameters and pitfalls in the interpretation of bladder biopsies in bacillus Calmette-Guerin treatment of superficial bladder cancer, J.Urol., 135:916.

Lamm, D.L., Stogdill, V., Stogdill, B. and Crispen, R. 1986, Complications of bacillus Calmette-Guerin immunotherapy in 1278 patients with bladder cancer, J.Urol., 135:272.

Morales, A., Eidinger, D. and Bruce, A.W. 1976, Intracavitary bacillus Calmette-Guerin in the treatment of superficial bladder tumors, J.Urol., 116:180.

Olson, W.P. 1992, Doctoral thesis, University of Illinois at Chicago.

Owens, R.J. and Barelle, F.E. 1986 Mapping the collagen-binding site of human fibronectin by expression in Escherichia coli, E.M. Bo. J., 5:2825.

Prescott, S., James, K., Busultil, A., Hargreave, T.B., Chisholm, G.D. and Smith, J.F. 1989, HLA-DR expression by high-grade superficial bladder cancer treated with BCG, Bri.J.Urol., 63:264.

Ratliff, T.L., Palmer, J.O., McGarr, J.A. and Brown, E.J. 1987, Intra-vesicular bacillus Calmette-Guerin therapy for murine bladder tumors: initiation of the response by fibronectin-mediated attachment of bacillus Calmette-Guerin, Cancer Res., 47:1762.

Ratliff, T.L., Kavoussi, L.R. and Catalona, W.J. 1988a, Role of fibronectin in intravesical BCG therapy for superficial bladder cancer, J.Urol., 410.

Ratliff, T.L., McGarr, J.A., Abou-Zeid, C., Rook, G.A.W., Stanford, J.L., Aslanzadeh, J. and Brown, E.J. 1988b, Attachment of mycobacteria to fibronectin-coated surfaces, J.Gen.Microbiol., 134:1307.

Ratliff. T.L. 1989, Mechanisms of action of intravesicular BCG for bladder cancer in: "Superficial Bladder Cancer", DeBruyne, F.M.J., Denis, L., van der Meijden, A.P.M. eds., Alan R. Liss New York, pps. 107.

Rosenthal, S.R. 1980, "BCG Vaccine: Tuberculosis - Cancer", PSG Publishing, Littleton, MA.

Shepherd, C.C. 1957, Growth characterisitics of tubercle bacilli and certain other mycobacteria in He La cells, J.Exp.Med., 105:39.

Shi, M., Klegerman, M.E. and Groves, M.J. 1989, The effect of washing on the surface charge of Mycobacterium bovis, BCG vaccine, Tice substrain, Microbios, 60:97.

Soleway, M.S. and Perry, A. 1987, Bacillus Calmette-Guerin for the treatment of superficial transitional cell carcinoma of the bladder in patients who have failed Thio-TEPA and/or Mitomycin C, J.Urol., 137:871.

Stefanini, G.F., Bercovich, E., Mazzeo, V., Grigloni, W.F., Emili, E.,

D'Errico, A., LoCigno, M., Tamagnini, N. and Mazeti, M. 1989, Class I and Class II HLA antigen expression by transitional cell carcinoma of the bladder: correlation with T-cell infiltration and BCG treatment, J.Urol., 141:1444.

Tabati, Y. and Ikada, Y. 1987, Macrophage activation through phagocytosis of muramyl dipeptide encapsulated in gelatin microspheres, J.Pharm-Pharmacol., 39:698.

Tabati, Y. and Ikada, Y. 1989a, In vivo effects of recombinant interferon alpha A/D incorporated in gelatin microspheres on murine tumor cell growth, Japa.J.Can.Res., 80:387.

Tabati, Y. and Ikada, Y. 1989b, Synthesis of gelatin microspheres containing interferon, Pharm.Res.,6 (5):422.

Tabati, Y. and Ikada, Y. 1989c, Protein precoating of polylactide microspheres containing a lipophilic immunopotentiator for enhancement of macrophage phagocytosis and activation, Pharm.Res., 6:296.

Teppema, J.S., deBoer, E.C., Steerenberg, P.A. and van der Meijden, A.P.M. 1992, Morphological aspects of the interactions of bacillus Calmette-Guerin with urothelial bladder cells in vivo and in vitro: relevance for antitumor activity? Urol.Res., 20:219.

Wiker, H.G., Harboe, M. and Lea, T. 1986, Purification and characterization of two protein antigens from the heterogenous BCG 85 complex in Mycobacterium bovis BCG, Int.Arch,Allergy & Appl.Immunol., 81:298.

SIGNIFICANCE OF VIRULENCE FACTORS AND IMMUNO-EVASION FOR THE DESIGN OF GENE-DELETED HERPESVIRUS MARKER VACCINES

Saul Kit

Baylor College of Medicine and NovaGene, Inc
Houston, TX, USA

INTRODUCTION

The family, HERPESVIRIDAE, consists of over 100 virus species that cause disease in animals from fish to man (Herpesvirus Study Group, 1992). This morning I will describe vaccines for two of these herpesviruses, namely, swine pseudorabies virus (PRV; Aujeszky's disease virus) and infectious bovine rhinotracheitis virus (IBRV; bovine herpesvirus-1). PRV and IBRV cause economically important diseases affecting the respiratory, reproductive and central nervous systems of the 500 million pigs and 1.28 billion cattle worldwide. PRV and IBRV also induce abortions in pregnant animals and lethal infections of newborn.

The genomes of the herpesviruses consist of linear, double-stranded DNA molecules, generally about 150 kbp in size. The viral DNAs are encased in an icosahedral capsid surrounded by a lipoprotein envelope containing 7 or more viral-encoded glycoproteins. The envelope glycoproteins are important for virus adsorption and penetration into host cells and they serve as antigenic determinants for the host's humoral and cellular immune responses.

The herpesvirus genomes encode about 70 proteins. These are often located at homologous locations on the viral DNAs and exhibit pronounced amino acid sequence homology. Genetic studies have revealed the astonishing fact that about one third of the herpesvirus genes are dispensable for virus replication in cell culture. In the case of the human herpes simplex virus (HSV), 25 or more of the 70 genes are nonessential for in vitro virus replication (Longnecker, 1987) (Table 1). Four of the 7 HSV glycoprotein genes and 4 of the 7 IBRV and PRV glycoprotein genes are dispensable. Only 3, that is HSV gB, gD, and gH and their PRV and IBRV homologs are essential for virus replication (Kit, S. and Kit, M., 1991) (Table 2). Likewise, the HSV, PRV and IBRV thymidine kinase (TK) genes are nonessential. Thirty years ago, we demonstrated that HSV mutants lacking the capacity to induce TK activity in TK-negative mouse fibroblast [LM (TK⁻)] cells replicated to high titers when grown in rapidly proliferating cells. Similar observations have been made with TK⁻ IBRV mutants.

The fact that so many herpesvirus genes are nonessential for replication in cultured cells does not, of course, signify that these genes

New Generation Vaccines, Edited by G. Gregoriadis
et al., Plenum Press, New York, 1993

Table 1. HSV genes nonessential for virus replication in cell culture

US2	US5	US9	US10	US11
UL4	UL10	UL16	TK	UL24
UL46	UL47	UL50	UL51	UL55
gG	IEP27	IEP47	gE	
UL2	UL3	gC	UL39	UL56
US3	(protein kinase)			

Table 2. Functions of homologous herpesvirus glycoproteins

HSV gH,gB,gD PRV gH,gII,gp50 IBRV gH,gI,gIV	ESSENTIAL: required for membrane fusion, virus penetration, and direct cell to cell spread. gp50 needed for fusion of virus envelope to cell membrane but not for fusion of infected cell membrane to uninfected cell membranes. All induce high titers of VN Ab; Induce MHC-restricted lysis, NK activity
HSV gE PRV gI IBRV ge	NONESSENTIAL: membrane fusion and cell to cell transmission; neurotrophic virulence factor; Induce VN Ab, MHC-restricted lysis
HSV gI PRV gp63 IBRV gi	NONESSENTIAL: forms complex with gE and homologs; gE/gI complex bind Fc portion of IgG and may help HSV escape immune cytolysis (not shown for homologs)
HSV gC PRV gIII(g92) IBRV gIII	NONESSENTIAL: Adsorption to heparan proteoglycan receptors; C3 binding activity; Virus spread by adsorption of released virus to neighboring cells Induce VN and MHC-restricted lysis
HSV gG PRV gX IBRV gg	NONESSENTIAL: function unknown Induce Ab

lack important functions related to the in vivo survival strategy of herpesviruses, but does have enormous practical significance for vaccine design. Through recombinant DNA techniques, nonessential genes coding for virus virulence factors and immuno-evasive factors can be deleted from the viral genomes. Nevertheless, the deletion mutants can replicate in cultured cells to high titers so that potent MLV vaccines can be produced at low cost. Furthermore, because at least 10 kb of foreign DNA can be inserted to replace deleted nonessential herpesvirus genes at several different locations on the viral genomes, the herpesviruses make excellent vectors for the expression of foreign genes, which can serve as: (i) immunogens to protect animals against both herpesvirus and nonherpesvirus diseases; (ii) cloned antigens to coat wells of diagnostic test kits; and (iii) vehicles for the delivery of enzymes, growth factors, cytokines, and neurotransmitters to nerve and/or lymphoid cells. So it is not surprising that HSV, varicella-zoster, Epstein-Barr, herpesvirus saimiri, PRV and IBRV have already been used as virus vectors to express foreign genes. In this report, only recombinant IBRV vectors expressing foreign genes will be described.

SAFETY OF ATTENUATED MLV VACCINES

For viral vaccines to be licensed, it is essential that they be safe, pure (free of adventitious agents), efficacious, and potent. These requisites have been achieved traditionally either by inactivating the infectivity of the viruses or by attenuating their pathogenicity, and, more recently, by the preparation of subunit and polypeptide vaccines (Kit et al, 1991a). In addition, for control and eradication programs, it is important that the vaccines should be "marker" vaccines, so that veterinarians can distinguish vaccinated from infected animals, and so that government approval for the interstate movement of vaccinated animals can be obtained.

For over 100 years, from 1881 to 1981, conventional killed and MLV vaccines have been prepared essentially by the empirical methods of Pasteur; that is, by treating micro-organisms with chemical or physical inactivating agents, or by cultivating them either in adverse environments or in unconventional hosts. Killed and MLV vaccines prepared by these procedures each have their advocates. However, despite vaccination of farm animals for over 30 years with conventional killed and MLV vaccines, FMD outbreaks continue to occur in many parts of the world and pseudorabies and IBR outbreaks have increased. The widespread prevalence of PRV and IBRV in particular, and their ability to establish life-long latent infections in nerve cells has favored their enzootic perpetuation in populations.

The killed FMD vaccines currently in use have been prepared by Pasteur's method of treating them with chemicals, such as formaldehyde, but this treatment has not always fully inactivated infectivity and recent FMD outbreaks have been attributed to virus "escape" from manufacturing facilities. To attenuate the rabies agent for dogs, Pasteur serially passed the rabies agent in rabbits. Most conventional virus vaccines prepared in this way harbor point mutations which are potentially reversible. Furthermore, the vaccines prepared by serially passing viruses in unconventional hosts may often be either underattenuated or overattenuated. Even though some of the conventional PRV vaccines are safe for pigs, they are not safe for other farm animals and have caused fatalities after they were accidentally inoculated into sheep and chickens. Recent studies also suggest that variants of a conventional modified live PRV vaccine used in Poland became established in the field and proved to be pathogenic (Christensen et al, 1992).

To attenuate anthrax, Pasteur grew anthrax at the marginal temperature of 43°C. It is fascinating that although not known to Pasteur, this high temperature treatment induced a **deletion** of an anthrax plasmid harboring a virulence factor, thereby foreshadowing our modern vaccine design. With the advent of recombinant DNA technology, new procedures have become available for the rational design of safe MLV vaccines based on an understanding of the pathobiology of virus infection. Safe MLV vaccines can be produced by identifying the specific viral genes contributing to **virulence** and to **immuno-evasion**. These genes can then be **deleted** so as to permanently and irreversibly attenuate the viruses.

Two such virulence genes are the pseudorabies gI and gIII genes (and their homologs in other herpes viruses) (Table 2). PRV gI contributes to both virulence and neurotropsin (Card et al, 1992), while PRV gIII, by serving as a complement C3 receptor, can protect from complement-mediated cell lysis and complement-mediated virus neutralization. The overall effect of the viral C3 function is that of enhancing susceptibility to both viral and secondary bacterial infections (Huemer et al, 1992). Recent studies have shown that conventional Bucharest and NIA4 PRV vaccines harbor "spontaneous" deletions of the gI gene, and that the Bartha PRV vaccine has deletions in gI and gp63 and a mutation in gIII which drastically reduces the expression of the gIII glycoprotein.

The most important herpesvirus virulence factor known is the **TK gene**. Even though TK⁻ herpesviruses replicate well in proliferating cells, they replicate very poorly in vitro in quiescent G1 phase cells, and in vivo in the nerve cells where they normally establish latent infections. Neurons are nonreplicating cells with very low levels of TK and the related enzymes required for DNA synthesis. These considerations led us in 1983 to engineer the deletion of the PRV TK gene, thereby producing the **OMNIVAC-PR** vaccine, the first recombinant DNA-derived MLV virus vaccine to be licensed in the world (Kit, S. and Kit, M., 1991). Numerous studies on PRV, HSV, IBRV and similar neurotropic herpesviruses have shown that TK⁻ mutants of these viruses can replicate at the site of infection and establish latency in sensory ganglia, but reactivate very poorly from latency (Kit, S. and Kit, M., 1991; Tenser, 1991). Hence, the shedding of vaccine virus to in contact controls does not occur or rarely occurs. Another important attribute of the OMNIVAC-PR vaccine is its lack of virulence for newborn 1 -3 day-old pigs. By contrast, a study by Moormann et al, (1990) showed that a TK⁺ gI-deleted PRV vaccine could be safely used with **10-week-old** seronegative pigs, but caused severe disease or death in **3-day-old** pigs. Deletion of the TK gene from the latter TK⁺gI⁻ virus reduced its virulence so that it was safe for both 3-day-old and 10-week-old pigs.

DELETION MUTANT MARKER VACCINES

The availability of deleted **marker vaccines** has fueled the optimism that national eradication programs are feasible and practical. Marker deleted vaccines permit herd clean-up through the identification and culling of vaccinates that have been infected with wild-type virus. Furthermore, the availability of marker vaccines enables governments to authorize the movement of vaccinated animals from breeders to finishers, from farrow to finish farms, and from farms to animal shows with the confidence that such traffic will not spread disease. Marker vaccines are produced by deleting a major virus glycoprotein gene from the viral DNA. In the case of our MLV PRV vaccine, we chose for deletion the most reliable marker available, that is, the nonessential PRV glycoprotein gIII(g92) gene (homologous to HSV gC and IBRV gIII) (Table 2). The licensed OMNIVAC-PR vaccine was the starting material for the deletion of the gIII gene. The second generation marker vaccine obtained by deleting the gIII gene was named OMNIMARK-PR (Kit, S. and Kit, M., 1991).

The rationale for the use of the gIII marker is simple. Pigs vaccinated with OMNIMARK-PR develop antibodies to all PRV proteins, **except gIII**. However, wild-type virus-infected pigs develop antibodies to all viral proteins, **including gIII**. Thus, the sera of infected animals can be distinguished from that of vaccinated animals with differential ELISA tests by the presence or absence of gIII antibodies. Those animals identified as having gIII-positive sera can then be culled from the herd and sent to slaughter.

After we constructed the OMNIMARK-PR vaccine, it was discovered that many commercial PRV vaccines had a "spontaneous" 3-4 kb deletion mutation in the unique short region of the PRV genome, which removed the nonessential glycoprotein gI, i.e., the neurotropic virulence factor. Even though these vaccines were TK^+, and hence, not as safe as TK^- vaccines, it was understandable that some vaccine producers chose to use gI as the marker for their vaccines. I will come back to the comparative merits of gIII versus gI marker vaccines later. For the moment, let me emphasize that OMNIMARK-PR and OMNIVAC-PR vaccines were derived by engineering deletions in a clone of a TK^+ PRV (Bucharest) vaccine strain (BUK-0) sent to us by D.P. Gustafson of Purdue University. He, of course, had received the TK^+ PRV (Bucharest) sample many years after its isolation in Europe. Thus, the original PRV (Bucharest) isolate probably underwent many genome variations over the years. Restriction nuclease analyses of the PRV(BUK-0) sample received in Houston and of seven clones prepared from it revealed that: (i) PRV(BUK-0) consisted of a mixture of at least 2 genotypes; and (ii) PRV(BUK-0) and all of its clones had the characteristic 3-4 kb deletion in the unique short region of the genome (KpnI-I and BamHI-7/12 fragments), typical of all PRV (Bucharest) vaccine strains. As a result of the KpnI-I(BamHI-7/12) deletion, the gI gene was deleted in our clone 3 PRV(BUK), but contrary to expectation, clones 5 and 7 and the derived OMNIVAC-PR and OMNIMARK-PR vaccines were gI^+. Thus, OMNIVAC-PR is TK^- gI^+gIII^+ and OMNIMARK-PR is $TK^-gI^+gIII^-$. It is interesting that recently, gI^+ variants of the PRV (Bucharest) (Suivac A) vaccine have been isolated from pigs on Polish farms. The original Suivac A vaccine and field isolates all induced antibodies to gI^- (Christensen et al, 1992), confirming that PRV Bucharest strains can be either gI^- or gI+. The history of the Polish isolates indicates that the gI^+ derivatives were virulent for pigs. The establishment and spread of gI^+ derivatives of the Suivac A vaccine and the absence of a glycoprotein marker are adverse features of this conventional vaccine.

At the time OMNIMARK-PR was engineered, we were concerned that removing the gIII gene might greatly impair the protective response elicited by OMNIMARK-PR, because PRV gIII is a major target for the induction of PRV antibodies and cellular immune responses. However, this did not occur. It may be recalled that PRV encodes 7 envelope glycoproteins (Table 2). Six glycoprotein genes and the genes for other structural and nonstructural PRV proteins were intact in OMNIMARK-PR. Thus, despite the deletion of the gIII gene, adequate levels of protective humoral and cellular immunity are induced. To illustrate, the virus neutralizing antibodies induced by OMNIVAC-PR and OMNIMARK-PR neutralize the infectivity of wild-type PRV as effectively as they neutralize the infectivity of the vaccine viruses and vice versa.

But the proof of efficacy of the OMNIMARK-PR vaccine does not depend upon theory but practice. The safety and efficacy of OMNIMARK-PR has been verified by extensive studies in the USA, New Zealand, and Japan (Kit, S. and Kit, M., 1991a). OMNIMARK-PR is safe for pregnant pigs in all stages of gestation, for newborn pigs 1-3 days of age, and for sheep and cattle, and protects these animals from lethal challenge doses of virulent virus.

Let me illustrate by citing the eradication program now in progress in Japan. Pseudorabies was introduced into Japan in 1975 by the importation of infected breeding sows from Holland. An outbreak of pseudorabies occurred in 1981 and disease spread rapidly to 19 of the 47 Japanese prefectures. On May 22, 1989, the Japanese Ministry of Agriculture established a "Control and eradication program for Aujeszky's disease". OMNIMARK-PR was licensed by the Ministry of Agriculture shortly afterwards for exclusive use in several prefectures. Studies over the last year have demonstrated that as little as 10^3 $TCID_{50}$ of OMNIMARK-PR protected newborn pigs from 100 times that amount of a virulent Japanese strain of pseudorabies virus, and protection was observed even when the pigs were exposed to virulent virus as early as 7 days after vaccination. Over a million doses of vaccine have been used in Japan. No vaccine-induced abortions or stillbirths have been reported. The litter sizes have increased from an average of 8 to 11 piglets per litter. The total number of newborn pigs has increased several percent and the overall health of the herds has improved to the great benefit of the farmers.

GENE-DELETED IBRV MARKER VACCINE

Besides a gene-deleted pseudorabies marker vaccine, we have genetically engineered a gene-deleted IBRV marker vaccine, designated IBRV(NG)dltkdlgIII. Several studies have shown that TK^- IBRV protects calves and pregnant cows from the abortifacient activity of virulent IBRV (Kit et al, 1986; Miller et al, 1991). As with the PRV vaccine, the IBRV(NG)dltkdlgIII vaccine was irreversibly attenuated by a deletion/insertion mutation in the TK gene (Kit et al, 1992a, b). The glycoprotein gIII gene was also deleted to provide a marker to distinguish vaccinated from infected cattle.

Collaborative studies recently carried out by Fernando Osorio and Eduardo Flores at the University of Nebraska have verified the safety and efficacy of the IBRV(NG)dltkdlgIII vaccine. These studies have shown that vaccinated SPF cattle were significantly protected from a very severe challenge with the virulent IBRV(USDA/Cooper) strain. The criteria for protection were the magnitude and extent of hyperthermia, the degree and extent of postchallenge nasal shedding, and the intensity of upper and lower respiratory tract symptoms. The vaccine induced virus neutralizing antibodies, and at 7 days after virulent virus challenge, an anamnestic response was seen. Experiments to evaluate the capacity of the IBRV vaccine to block the challenge virus from establishing latency are in progress.

DIFFERENTIAL gIII BLOCKING ELISA TEST

Sensitive and specific differential pseudorabies and IBR diagnostic tests have been developed for determining the disease status of swine and cattle herds, respectively. The US-licensed PRV gIII blocking ELISA test which accompanies the OMNIMARK-PR vaccine has several interesting features (Kit, S. and Kit M., 1991). First, the PRV gIII antigen used to coat microtiter test wells was prepared by cloning the PRV gIII gene in an IBRV expression vector (Kit et al, 1992a). Since PRV gIII was the only PRV antigen in the wells, there was no interference by other PRV antigens and the specificity of the test approached 100%. Second, undiluted test sera could be used in the test. Hence, time consuming and laborious dilutions of test sera prior to addition to the antigen-coated wells were not required. The test utilized a mouse anti-PRV monoclonal antibody

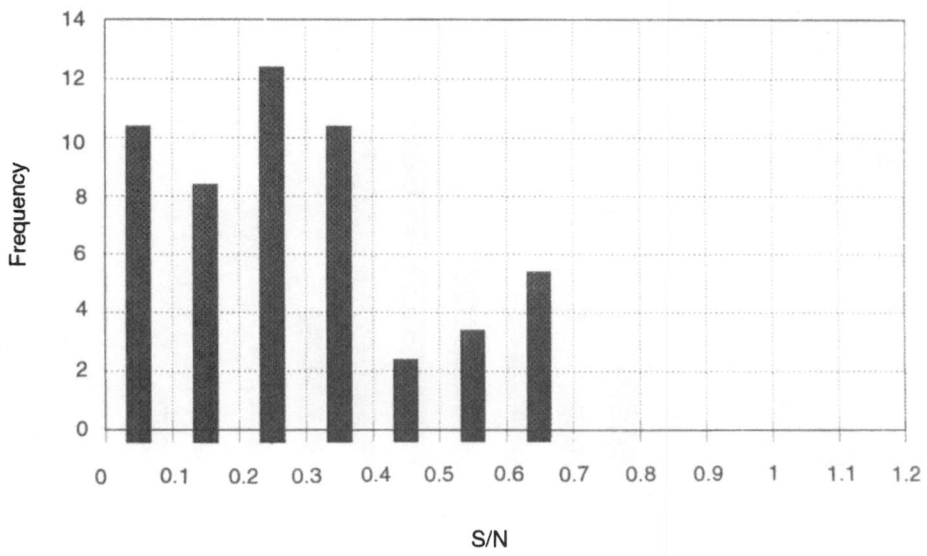

Fig. 1. IBRV gIII blocking ELISA; field sera VN=1:2 to 1:256

conjugated to horse radish peroxidase plus TMB substrate for color
development. The test results were expressed as S/N values, defined as the
optical density at 650 nm of the test sera/optical density at 650 nm of the
control sera. S/N values of >0.70 (less than 30% blocking of color
development) were considered negative, while S/N values of <0.70 were
considered positive. The PRV gIII blocking ELISA test was 4-16 times more
sensitive than the virus neutralization test and was comparable in
sensitivity to the screening ELISA and latex agglutination tests. gIII
antibodies were detected in sera of pigs infected with 200 PFU of virulent
PRV by 7 days post infection and vaccinated pigs infected with wild-type
PRV developed gIII$^+$ antibodies as early as 7-10 days post challenge.

Undiluted test sera were also used with the differential IBRV gIII
blocking ELISA test (Kit et al, 1992b). The specificity of the IBRV test
was demonstrated by showing that gnotobiotic cattle sera containing high
titers of heterologous antibodies did not interfere with the test. Control
experiments also showed that heterologous PRV and equine herpesvirus-1
antibodies gave negative test results (S/N >0.70), as expected. Likewise,
sera from cattle collected 30 days postvaccination and from rabbits twice
immunized with IBRV(NG)dltkdlgIII gave negative S/N values. However, the
cattle and rabbit sera became positive for gIII antibodies 7-14 days after
wild-type IBRV challenge. Figure 1 illustrates results obtained with
bovine field sera exhibiting positive virus neutralization titers of 1:2 to
1:256. All of the sera had positive S/N values <0.70. In addition, gIII
antibodies were detected in about a third of the bovine field sera showing
negative virus neutralization titers (VN <1:2 or <1:4) (Fig. 2),
illustrating the potential usefulness of this test for IBRV herd cleanup
and eradication programs.

RECOMBINANT IBRV VECTORS EXPRESSING FOREIGN GENES

Foreign genes inserted in nonessential herpesvirus genes can be
expressed by recombinant herpesvirus vectors. For example, recombinant
IBRV vectors in which the coding sequences of the "early" PRV TK gene, the
"late" PRV gIII gene, and the E.coli beta galactosidase gene were ligated

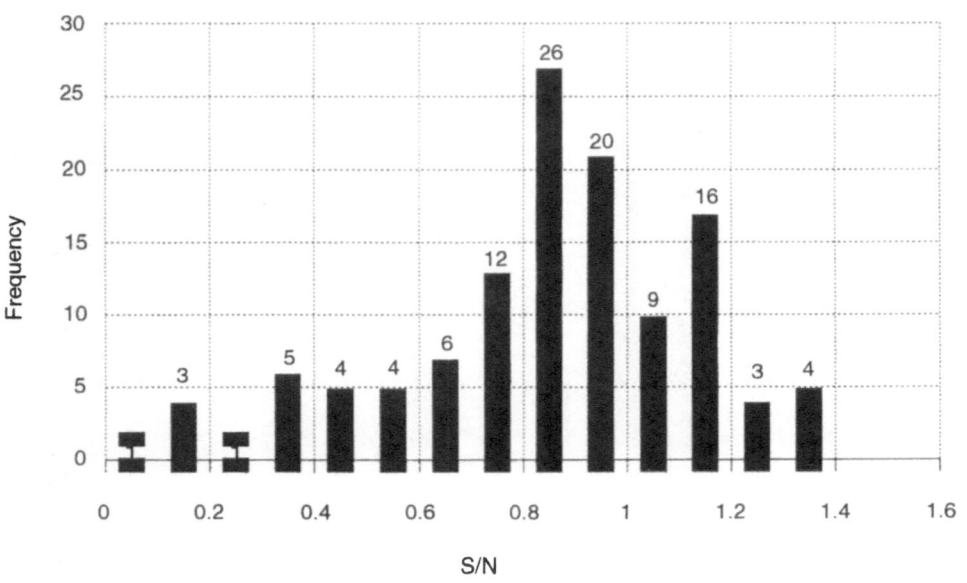

Fig. 2. IBRV gIII blocking ELISA; field sera VN <1:2 or <1:4

to the late IBRV gIII promoter and inserted in place of the IBRV gIII coding sequences, efficiently expressed these foreign genes (Kit et al, 1992a). Likewise, the PRV gIII gene inserted at a second site, i.e., that of the IBRV TK gene, was expressed efficiently in cells infected by the recombinant. The IBRV recombinant that expresses the E.coli beta galactosidase gene can be recognized by the colored plaques made in the presence of X-gal and has been useful as an intermediate in the construction of other recombinant viruses expressing foreign DNAs from the IBRV gIII promoter. The high levels of expression of PRV gIII by the recombinant designated IBRV(NG)dltkdlgIII(PRVg92) enabled us to utilize the PRV gIII thereby produced as an antigen to plate microtiter test wells in the sensitive and specific PRV gIII differential blocking ELISA test (Kit, S. and Kit, M., 1991). This recombinant may also have the potential for affording cattle protection from fatal PRV infection.

Reciprocal constructs, that is, recombinant PRV vectors in which the IBRV gIII gene has replaced the PRV gIII gene, have also been prepared. However, for reasons that are not entirely clear, the PRV recombinant vectors have not expressed foreign genes quite as efficiently as the IBRV vectors express foreign DNAs (Kit et al, 1992a).

One of the most interesting herpesvirus vectors prepared to date is an IBRV recombinant containing FMDV VP1 epitope sequences inserted at the amino terminal end of the IBRV gIII gene (Kit et al, 1992b). The foreign FMDV sequences are efficiently expressed as a fusion protein with IBRV gIII on the surface of recombinant IBRV/FMDV particles and on the surface of infected cells. Since the IBRV vector was attenuated by insertion/deletion mutations in the TK gene, the IBRV/FMDV recombinant can be classified as a modified-live SUBUNIT VACCINE, capable of protecting cattle from both IBR and FMD.

The rationale for the design of the recombinant is based upon data developed by many investigators showing that the major immunogenic regions providing protective immunity against FMD are located on the viral capsid protein VP1, at amino acids residues 140-160 and 200-213. One of our collaborators, R. DiMarchi, showed that these epitopes could be linked together through a tripeptide bridge to produce peptides that protected cattle against FMD. We synthesized the DNA sequence that encodes DiMarchi's FMD peptide and inserted it into the gIII gene of the attenuated TK⁻ IBRV(NG)dltk vector.

Two IBRV/FMDV recombinants have been prepared. The first was a IBRV/FMDV (Monomer) in which the IBRV gIII promoter was ligated to the bovine growth hormone signal sequence and then to the sequences encoding FMDV (O1K) VP1 amino acids 200-213, a pro-pro-ser spacer, and VP1 amino acids 141-158, and a second spacer. Nucleotides encoding the first 39 amino acids of the IBRV gIII gene were deleted by unidirectional exonuclease digestion to avoid redundant signal sequences, but gIII epitope sequences recognized by monoclonal antibodies were retained. The gIII DNA sequence were in phase with the FMDV epitope sequences. The second recombinant was an IBRV/FMDV (dimer) in which the FMDV epitope sequences were repeated in tandem. With the dimer recombinant, however, only the N-terminal 21 amino acids of the IBRV gIII signal sequence were deleted. The IBRV gIII gene was chosen as the gene to use in the construction of an FMDV-IBRV fusion protein for several reasons. IBRV gIII is a major glycoprotein of the IBRV envelope and is assembled into homodimers visible as 20-25 nm protruding spikes. It was anticipated that with the fusion protein, FMDV antigen presentation would be optimized.

The modified-live IBRV/FMDV vaccine has a number of advantages over previous peptide and subunit FMDV candidate vaccines. The IBRV/FMDV vaccine can be produced at low cost. Immunization with or without adjuvants, can protect cattle from both IBR and FMD. Delivery can be either by the convenient intramuscular route, or, if desired, by the more labor intensive intranasal route which would be expected to induce IgA immunoglobulins. The vaccine is by definition a "marker" vaccine. That is, any other FMDV protein can be used in a diagnostic test so that vaccinated animals can be distinguished serologically from animals infected with FMDV field strains. Administration of the live IBRV-FMDV vaccine allows in vivo amplification of the same chemically defined FMDV VP1 epitopes that are used in synthetic peptide vaccines (something synthetic peptides and biosynthetic peptide vaccines cannot do), and, as with the subunit vaccines, there is no infectious FMDV present to cause disease.

Experiments demonstrating the safety and efficacy of the IBRV/FMDV vaccine have been carried out in cattle. Suffice it to say that intramuscular or intravenous vaccination of calves with either of the recombinants induced IBRV neutralizing antibodies and protected cattle from challenge exposure to virulent IBRV(Cooper). Anti-FMDV ELISA antibodies and VN FMDV antibodies were also induced. Finally, pilot studies performed in Germany have shown that 2 of 5 cattle immunized once with IBRV-FMDV(monomer) and then challenged intradermolingually at 21 days post-vaccination with 10^4 $TCID_5$ of FMDV-O1K were solidly protected and had no lesions on any of the four feet. Clearly additional efficacy studies are required with both monomer and dimer recombinants. Nevertheless, the results illustrate the potential usefulness of modified-live herpesvirus vectors for immunizing against diseases caused by RNA viruses.

COMPARISON OF PRV gIII AND gI AS VACCINE MARKERS

The conventional PRV vaccines previously developed and some of the genetically engineered PRV vaccines have glycoprotein gI deletions as the marker to distinguish vaccinated from infected pigs. The government of the Netherlands has mandated that only vaccines with gI deletions may be sold in that country and other European nations may follow this lead. Recent studies, however, suggest that the gIII marker may have significant advantages over gI as a marker. To understand why, it is necessary to consider the functions of PRV glycoproteins in relationship to the replication strategy and the immuno-evasion tactics of PRV (Fitzpatrick and Bielefeldt-Ohmann, 1991).

The essential PRV glycoproteins, gH, gII, and gp50, have a role in virus penetration, a process that involves the fusion of the virus membrane with that of the cell (Table 2). In addition, gH and gII have a role in the direct cell to cell spread of viruses through fusion of infected with uninfected cell membranes. PRV gI also has an accessory role in the fusion of infected cell membranes with uninfected cell membranes, and hence, with direct cell to cell spread. By contrast, PRV gIII, like HSV gC and IBRV gIII, functions in virus adsorption to heparan proteoglycan receptors, and thus, virus spread by adsorption of released virus to receptors on neighboring cells. The fusion of herpesvirus-infected cells and neighboring uninfected cells has been recognized for many years as a key evasive mechanism which allows local spread of infection even in the presence of extracellular humoral defenses (Black and Melnick, 1955). This capability of spreading even in the presence of herpes antibodies has been demonstrated for simian herpesvirus-B and for HSV-1. Likewise, virus plaques can be formed by HSV(MP), a mutant that does not express HSV-1 gC, the homolog of PRV gIII (Hoggan and Roizman, 1959). Thus, it is most significant that Zsak et al, (1992) found that virus neutralizing antibodies greatly reduce the plaque size of a gI$^-$ PRV deletion mutant, while the plaques sizes of wild-type PRV and gIII$^-$ PRV deletion mutants were not affected. The gI$^+$gIII$^-$ mutant exhibited antibody insensitive cell to cell spread, while their gI$^-$gIII$^+$ mutant relied on the antibody-sensitive mode of spread through release from infected cells and heparan-dependent adsorption to uninfected cells.

These differences are extremely important for vaccination strategies. Current management practices favor vaccination of pregnant sows prior to farrowing to build up maternal antibodies. This passive immunization of the newborn with colostral antibodies protects the piglets from clinical signs of disease but not necessarily from infection and latency by virulent virus. The maternal antibodies decrease with a defined half-life. It is only at 10-14 weeks, after the maternal antibodies have waned that veterinarians immunize the piglets with MLV gI-deleted vaccines. In practice, the titers of residual antibodies are not known, unless expensive antibody titrations are performed. If vaccination with gI$^-$vaccine is performed too early, the material antibody interferes with active immunization. In addition, many producers use gI$^-$ vaccines that are TK$^+$. These vaccines may not be safe for 1 to 3 day-old piglets, again compelling producers to rely on maternal antibodies for protection for the first 2 months of the animal's life.

Recently, van Oirschot et al, (1991) demonstrated that even when the maternal antibodies did not have gI antibodies (derived from mothers vaccinated with gI-negative vaccines), these maternal antibodies interfered with the induction of gI antibodies after the pigs were experimentally infected with a gI$^+$ field strain. Thus, the presence of virulent field virus may not be detected by the differential gI test and field virus may continue to circulate in the herd.

Eloit et al, (1992) have also pointed out that animals repeatedly vaccinated with gI$^-$ vaccines may not develop detectable gI antibodies when they become infected with gI$^+$ field viruses. They state, "It is likely that gI antibodies will be more and more difficult to detect in mass screening of infected herds." In yet another study, vaccinated pregnant sows were inoculated into the arteria uterina at 70 days of gestation either with PRV-infected autologous blood mononuclear cells or with cell-free PRV (Nauwynck and Pensaert, 1992). PRV was shown to reach the uterine and fetal tissues via infected mononuclear cells even in the presence of the circulating antibodies induced by vaccination. This **cell-associated virus** spread led to abortion. **Cell-free virus** did not induce abortion under similar circumstances. Finally, Zsak et al, (1992) showed that mice passively immunized with PRV antibodies were much better protected from lethal doses of a gI⁻ PRV than from wild type PRV or gIII⁻ PRV; i.e., the gI⁻ virus was much more susceptible to antibodies than the gIII⁻ virus.

Unlike conventional TK⁺ gI⁻ vaccines, and even the genetically engineered TK⁻ gX⁻ Tolvid (Upjohn Corp) vaccine, the OMNIMARK TK⁻ gIII⁻ vaccine has been safely used to actively immunize 1-3 day old pigs. We have hypothesized that the gIII⁻ TK⁻ vaccine will induce protection in 1-3 day old pigs even in the presence of maternal antibodies, because TK⁻ gIII⁻ viruses exhibit immuno-evasive direct cell to cell spread. An experiment has been initiated to test this hypothesis. Two to 3-day-old piglets showing maternal antibody titers of 1:32 to 1:256 were vaccinated with either the TK⁻gIII⁻ OMNIMARK vaccine or with a TK⁻gI⁻ vaccine, or with neither. The 3 groups have been transported from Indiana to Texas and will be challenged with virulent PRV after the maternal antibodies decrease to low levels. It should be know by September, 1992 whether or not the gIII⁻ TK⁻ vaccinates are significantly better protected, as we predict, than the gI⁻ TK⁻ vaccinates. If so, this would be a strong incentive to rescind regulations requiring only gI⁻ vaccines and to permit producers to develop vaccination strategies entailing active immunization of newborn pigs.

SUMMARY

Using recombinant DNA technology, safe and efficacious modified-live pseudorabies and IBRV vaccines have been constructed with attenuating deletions in the thymidine kinase genes and a marker deletion in the gIII gene. Glycoprotein gIII binds complement C3, which suggest that its deletion may also serve to protect from complement-mediated cell lysis and virus neutralization. Although glycoprotein gIII is a major target for the induction of virus neutralizing antibodies, its deletion does not compromise the protective response, because adequate humoral and cellular immune responses are induced by the 6 other envelope glycoproteins and by viral structural and nonstructural proteins. Sensitive and specific anti-gIII differential blocking ELISA tests have been devised for use in conjunction with the gIII-deleted marker vaccine to distinguish sero-logically between vaccinated and field virus-infected animals. Glyco-protein gIII promotes the adsorption of virus particles to heparan proteo-glycan receptors on the host cell. Deletion of gIII from the vaccine promotes direct virus spread by fusion of membranes from infected to un-infected cells, a process that is relatively antibody insensitive. Hence, deletion of gIII from the vaccine may facilitate the effective immunization of newborn animals even in the presence of maternal antibodies. Finally, IBRV and PRV have been used as viral vectors for the expression of foreign genes. A recombinant IBRV virus vaccine expressing monomer and dimer forms of FMD capsid protein epitopes on the surface of hybrid virus particles and on the surface of infected cells has been constructed. These recombinants induce virus neutralizing antibodies against IBRV and FMDV and have the potential of protecting cattle from disease caused by either virus.

REFERENCES

Black, F.L. and Melnick, J.L., 1955, Micro-epidemology of poliomyelitis and herpes-B infections. Spread of the viruses within tissue cultures, J.Immunology, 74:236.

Card, J.P., Whealy, M.E., Robbins, A.K. and Enquist, L.W., 1992, Pseudorabies virus envelope glycoprotein gI influences both neurotropism and virulence during infection of the rat visual system, J.Virol., 66:3032.

Christensen, L.S., Medneczky, I., Strandbygaard, B.S. and Pejzak, Z., 1992, Characterization of field isolates of suid herpesvirus-1 (Aujeszky's disease virus) as derivatives of attenuated vaccine strains, Arch.Virol., 124:225.

Eloit, M., Vannier, P., Hutet, E. and Fournier, A., 1992, Correlation between gI, gII, gIII, and gp50 antibodies and virus excretion in vaccinated pigs infected with pseudorabies virus, Arch.Virol., 123:135.

Fitzpatrick, D.R. and Bielefeldt-Ohmann, H., 1991, Mechanisms of herpesvirus immuno-evasion, Microbiol.Pathogenesis, 10:253.

Herpesvirus study group of the International Committee on Taxonomy of viruses 1992. The family Herpesviridae: An update. Arch.Virol., 123:425.

Hoggan, M.D. and Roizman, B., 1959, The isolation and properties of a variant of herpes simplex producing multinucleated giant cells in monolayer cultures in the presence of antibody, Am.J.Hygiene, 70:208.

Huemer, H.P., Larcher, C. and Coe, N.E., 1992, Pseudorabies virus glycoprotein III derived from virions and infected cells binds to the third component of complement, Virus Research, 23:271.

Kit, S. and Kit, M., 1991, Genetically engineered herpesvirus vaccines. Accomplishments in pigs and prospects in humans, in: "Prog.Med.Virol.," J.L. Melnick, ed., Basel-Karger 38, 128, New York.

Kit, S., Kit, M. and McConnell, S., 1986, Intramuscular and intravaginal vaccination of pregnant cows with thymidine kinase-negative, temperature-resistant infectious bovine rhinotracheitis virus (bovine herpesvirus 1), Vaccine, 4:55.

Kit, S., Kit, M., DiMarchi, R., Little, S. and Gale, C., 1991a, Modified-live infectious bovine rhinotracheitis virus (IBRV) vaccine expressing foot-and-mouth disease virus (FMDV) capsid protein epitopes on surface of hybrid virus particles, in: "Immunology of Proteins and Peptides VI," M.Z. Atassi, ed., Plenum Press, N.Y. p211.

Kit, S., Kit, M., DiMarchi, R.D., Little, S.P. and Gale, C., 1991b, Modified-live infectious bovine rhinotracheitis virus vaccine expressing monomer and dimer forms of foot-and-mouth disease capsid protein epitopes on surface of hybrid virus particles, Arch.Virol., 120:1.

Kit, S., Otsuka, H. and Kit, M., 1992a, Expression of porcine pseudorabies virus genes by a bovine herpesvirus-1 (infectious bovine rhinotracheitis virus) vector, Arch.Virol., 124:1.

Kit, S., Otsuka, H. and Kit, M., 1992b, Blocking ELISA to distinguish infectious bovine rhinotracheitis virus (IBRV)-infected animals from those vaccinated with a gene-deleted marker vaccine, J.Virol.Methods, 40:45.

Longnecker, R., Chatterjee, S., Whitley, R.J. and Roizman, B., 1987, Identification of a herpes simplex virus 1 glycoprotein gene within a gene cluster dispensable for growth in cell culture, Proc.Natl.Acad,Sci.USA, 84:4303.

Miller, J.M., Whetstone, C.A., Bello, L.J. and Lawrence, W.C., 1991, Determination of ability of a thymidine kinase-negative deletion mutant of bovine herpesvirus-1 to cause abortion in cattle, Am.of Vet.Res., 52:1038.

Moormann, R.J.M., de Rover, T., Briaire, J., Peeters, B.P.H., Gielkens,

A.L.J. and van Oirschot, J.T., 1990, Inactivation of the thymidine kinase gene of a gI deletion mutant of pseudorabies virus generates a safe but still highly immunogenic vaccine strain, J.Gen.Virol., 71:1591.

Nauwynck, H.J. and Pensaert, M.B., 1992, Abortion induced by cell-associated pseudorabies virus in vaccinated sows, Am.J.Vet.Res., 53:489.

Tenser, R.B., 1991, Role of herpes simplex virus thymidine kinase expression in viral pathogenesis and latency, Intervirology, 32:76.

van Oirschot, J.T., Davs, F., Kimman, T.G. and Van Zaane, D., 1991, Antibody response to glycoprotein I in maternally immune pigs exposed to a mildly virulent strain of pseudorabies virus, Am.J.Vet.Res.,52:1788.

Zsak, L., Zuckermann, F., Sugg, N., and Ben-Porat, T., 1992, Glycoprotein gI of pseudorabies virus promotes cell fusion and virus spread via direct cell-to-cell transmission, J.Virol., 66:2316.

ERADICATION OF SYLVATIC RABIES USING A LIVE RECOMBINANT VACCINIA-RABIES

VACCINE

Marie Paule Kieny, Bernard Brochier* and
Paul-Pierre Pastoret*

Transgene S.A., 11 Rue de Molsheim, 67082 Strasbourg Cedex
France
*Service de Virologie-Immunologie, Faculte de Medecine
Veterinaire, Universite de Liege, 45 Rue des Veterinaires
B-1070 Bruxelles, Belgique

SUMMARY

　　　Rabies is prevalent in most parts of the world. The disease is
propagated extensively through wild animals where the fox remains a major
vector of the disease in North America and in Europe. Field trials of
vaccination of wild carnivores against rabies with attenuated rabies virus
have yielded in the early '80s promising results. This approach has been
extended by the introduction of a recombinant vaccinia virus expressing the
rabies glycoprotein to protect foxes against the infection. The
recombinant virus was found to be innocuous to foxes and to the numerous
non-target species tested. Administration of live recombinant virus to
foxes via the subcutaneous, intradermal or oral routes uniformly elicited
high titers of neutralizing antibodies and animals receiving the
recombinant virus (VVTGgRAB) in bait were protected against rabies.
Moreover, five campaigns of fox vaccination between 1989 and 1991 have
resulted in the near disappearance of rabies cases reported in domestic
animals (the best indicator of rabies prevalence) in a 10700 km^2 area of
Belgium. Because of the efficiency and inocuity of VVTGgRAB and of the
excellent quality of the baits, this recombinant rabies vaccine appears to
be the ideal vaccine to achieve global eradication of rabies in Europe.

INTRODUCTION

　　　Rabies is a viral disease which affects all warm-blooded animals and
is widespread in most countries of the world (Steck, 1982). Consequences
for Public Health are substantial in South America, Africa and Asia where
veterinary and sanitary structures are often lacking. However, rabies is
still a subject of great concern, even in Europe and North America due to
its propagation amongst wild animals which constitute a considerable
reservoir of the virus: foxes in Europe and foxes, skunks and raccoons in
North America (Debbie, 1983; Anderson et al, 1981). Dogs represent the
major vector in Africa and Asia whereas in Central and South America both
dogs and bats have been implicated, with the latter being responsible for
large economic losses in livestock.

New Generation Vaccines, Edited by G. Gregoriadis
et al., Plenum Press, New York, 1993

The disease is transmitted through the bite of an infected animal whose saliva contains large quantities of virus. The rabid animal undergoes behavioural changes during the final stages of the disease and aggressive behaviour facilitates transmission. As rabies is nearly always fatal, immune populations do not exist.

Prophylactic measures (other than the vaccination of domestic animals) aim to eliminate or reduce the population of the principal reservoir through, for example, poisoning or gassing (Andral and Blancou, 1982). These measures have proved to be moderately successful for stray dogs but are less effective for wild animals. In Europe, their application to foxes has reduced the number of rabies outbreaks but has not significantly contained the disease. The present upsurge of European rabies, which first appeared in Poland in 1935 in foxes and badgers (Toma and Andral, 1977), has since spread throughout Western Europe, presenting enzootic waves as in North America (Anderson et al, 1981). Vaccination of wild animals (particularly foxes) is now thought to present an alternative, and perhaps more effective, countermeasure.

Oral administration is the only appropriate route for the vaccination of large numbers of wild carnivores and this was first attempted by Baer et al, in North America (Baer et al, 1971) and by Mayr et al, in Europe (Mayr et al, 1972). In small-scale trials, live attenuated rabies virus introduced into various baits has been successfully used to vaccinate foxes (Blancou, 1979). Both in Switzerland (Kappeler et al, 1985) and in West Germany (Schneider et al, 1985), field trials have been successful in eradicating rabies cases from localized areas. However, attenuated viruses remain pathogenic to some rodents and can revert to virulence at a significant frequency. It is of note that inactivated rabies virus is ineffective when administered orally. For these reasons a safe and effective recombinant vaccine is certainly an improved alternative to attenuated rabies virus.

CONSTRUCTION OF THE VACCINIA-RABIES RECOMBINANT

Vaccinia virus (VV), a large (180 kb) double-stranded DNA orthopox virus, has been used extensively to control and eradicate smallpox in man (Behbehani, 1983). The relative innocuity of VV has stimulated its development as a live vector for viral antigens and derivatives expressing surface proteins from influenza, hepatitis B, herpes simplex and many other organisms have been used to confer protection against the respective diseases (Panicali et al, 1983; Smith et al, 1984).

Rabies virus (RV) is a rhabdovirus, an enveloped negative single-stranded RNA virus related to vesicular stomatitis virus. The glycoprotein (G) is presented at the exterior surface of the virion where it aggregates to form surface projections or spikes . G is the only protein capable of inducing or reacting with virus-neutralizing antibody (VNA) and appears to be the main viral protein capable of eliciting protection (Wunner et al, 1983; Kieny et al, 1986), although recent results have shown that vaccination with the nucleocapsid (N) protein can also confer protection against rabies (Fu et al, 1991; Summer et al, 1991).

We have developed a recombinant VV (VVTGgRAB) bearing the rabies G coding sequence and expressing the rabies surface antigen (Kieny et al, 1984). To do that, the rabies G cDNA was inserted on a plasmid vector downstream of the vaccinia virus P7.5K promoter into the non-essential vaccinia TK gene. Double reciprocal recombination in vivo between this plasmid and the VV genome permitted integration of the DNA insert into the viral genome. Infection of cell cultures with VVTGgRAB elicited the produc —

Table 1. Vaccination of animals with rabies vaccines

Vaccine	Route of inoculation	Fraction surviving
VVTGgRAB 10^8 pfu	intradermal	2/2
	subcutaneous	2/2
	oral	8/8
	oral (presentation in chicken-heads baits)	4/5*
Wild type VV 10^x pfu	intradermal	0/2
Control inactivated rabies vaccine	subcutaneous	2/2

* : 2/5 animals were observed to have ingested only part of the vaccine.

tion of a correctly processed rabies G which reacted strongly with rabies-neutralizing monoclonal antibodies.

Protection of mice and rabbits against rabies after inoculation with the live recombinant virus has already been described (Kieny et al, 1984; Lathe et al, 1985).

PRELIMINARY EXPERIMENTS IN FOXES

European foxes (Vulpes vulpes) captured and raised in captivity were inoculated or fed with live recombinant VVTGgRAB (Blancou et al, 1986). No generalized reaction to the virus nor impairment of digestive or alimentary function was observed. When the foxes were subsequently challenged by injection of rabies virus, all animals which had received 10^8 pfu of VVTGgRAB either orally or parenterally resisted challenge, including one animal exhibiting undetectable levels of rabies-neutralizing antibodies. Control animals injected subcutaneously with a commercial inactivated adjuvanted vaccine similarly resisted challenge. With such a vaccine, oral administration has previously been shown to be ineffective (Blancou et al, 1982).

Oral administration is the only route appropriate to the vaccination of wild animals. Accordingly, the vaccine must be presented in a form appropriate to ingestion. Capsules containing 108 pfu of VVTGgRAB were thus prepared, inserted into chicken heads and distributed to the test animals. These animals similarly produced high titers of VNA and resisted challenge (Blancou et al, 1986) with rabies virus (Table 1).

Horizontal transmission of the recombinant virus could have an important impact on the wild population. To address this question we examined whether vaccinated animals could transmit the virus to non-treated control animals. Four animals were vaccinated by direct application of 10^8 pfu of VVTGgRAB into the mouth. Each was subsequently housed in the same cage as an untreated animal of the opposite sex. Sera were analysed 28 days post-vaccination. All treated animals presented high titers of neutralizing antibodies. Surprisingly, one of the control (female) also possessed neutralizing activity in her serum (Blancou et al, 1986). It is note that this female shared a cage with a male exhibiting particular aggressive behaviour and was bitten by the latter immediately after the administration of the vaccine. This circumstance is likely to be rare in the wild.

Four animals were challenged one year after oral vaccination. At this time, two animals had no detectable neutralizing antibodies. Nevertheless, all animals survived challenge, attesting to the long duration of immunity conferred by the recombinant VVTGgRAB.

RACCOONS

Over the past 40 years, the raccoon (Procyon lotor), an especially ubiquitous mammalian carnivore, has become a prominent rabies reservoir in the USA (Centers for Disease Control, Atlanta, 1986), prompting laboratory trials with VVTGgRAB, since conventional attenuated rabies viruses had proved ineffective. Adult raccoons maintained in captivity received 10^8 pfu of VVTGgRAB contained in a 3 cm^3 polyurethane sponge coated with a beef tallow/paraffin wax mixture (Johnston and Lawson, 1987), by ingestion. All 20 raccoons given VVTGgRAB recombinant virus in sponge developed rabies neutralizing antibodies and 17 of 20 animals survived challenge by live rabies virus (Rupprecht and Kieny, 1988). As for foxes several experiments were conducted to investigate the potential roles of horizontal or vertical transmission of VVTGgRAB in raccoons. In cage trials with pairs of adult raccoons, 2 of 5 non-immunized contact animals developed low rabies virus neutralizing antibody levels and survived rabies challenge.

NON-TARGET SPECIES

In view of the possibility of conducting field-trials with VVTGgRAB containing baits, it was important to verify the absence of pathogenicity, excretion and transmission of the virus in non-target animal species as well as in target species. The potential pathogenicity of VVTGgRAB by various routes was evaluated in mice, rabbits, hamsters, ferrets, cattle, sheep, pigs, dogs and cats (Wiktor et al, 1988; Blancou et al, 1989; Desmettre et al, 1990; Wiktor et al, 1985; Soria Baltazar et al, 1987; Brochier et al, 1988). Except for intradermal inoculation, neither local or general reaction was observed in animals receiving 10^8 pfu or more of recombinant virus.

Several non-target wild species were also chosen for testing (Brochier et al, 1989) because of their presence in the areas where this vaccine may be distributed: wild boar, badger, wood mouse, several species of voles, several species of birds. Animals received VVTGgRAB orally and no sign of pathogenicity was detected, either by direct observation or after necropsy (Pastoret et al, 1992). The innocuity of VVTGgRAB vaccine seems thus to be established in all wild animals tested as well as in domestic and laboratory species.

FIELD TRIALS OF VACCINATION IN EUROPE

Taking into account all the data concerning the efficacy and safety of VVTGgRAB, limited field trials were authorized by the Belgian and French authorities in 1987 (6 km^2, Marche-en-Famenne, Belgium) and 1988/89 (25 ha, Mars La Tour, France). No adverse phenomenon was observed in wild life in the vaccinated area.

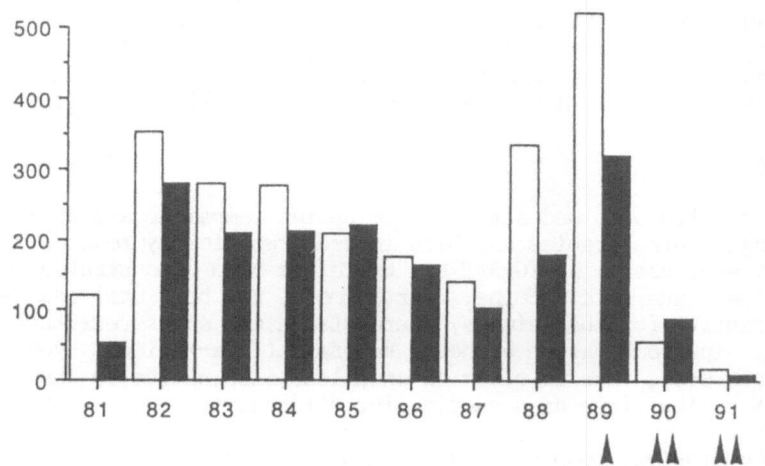

Fig. 1. Yearly evolution of rabies incidence in Belgium since 1981 in number of reported cases. Open bars: foxes; closed bars: domestic animals. (From Coppens et al, 1992).

The Belgian Government subsequently authorized in 1988 a 435 km^2 field trial in the Province of Luxembourg. The vaccine used consisted of 10^8 pfu of VVTGgRAB contained in a plastic bag (2.2 ml) and enclosed in a bait consisting of a mixture of fishmeal, fishoil and a synthetic polymer. The bait engineered for this purpose had the shape of parallelepiped (5x3x2 cm) and a weight of 40 g. It also contained tetracyclin (150 mg) as a bone biomarker of bait consumption.

The distribution was performed by hand in October-November 1988 and 222 animals were collected after the vaccination campaign for post-mortem examination. 61% of the collected foxes had tetracyclin incorporated in their bones, as did 47% of wild boars, 37% of stone martens, 27% of cats, 12% of wood mice and 10% of crows. No pox-like lesion could be detected in the animals, and no abnormal mortality or morbidity was noted, attesting to the safety of the vaccination campaign. Despite important variations in environmental temperature, VVTGgRAB titer remained constant in the baits for more than 3 months (Pastoret et al, 1992).

LARGE-SCALE VACCINATION CAMPAIGNS

After these confirmations of VVTGgRAB safety, five fox vaccination campaigns were carried out in Belgium in November 1989, April 1990, October 1990, May 1991 and October 1991. Baits were dispersed in the field at a mean density of 15 baits/km^2 of vaccinable area. The baits were distributed by helicopter or airplane dropping, and the total of the contaminated area of Belgium (10,700 km^2) was covered (the first 2 campaigns concerned only 2200 km^2 in the Province of Luxembourg, the rest of the area being initially vaccinated with the SAD B19 strain of attenuated rabies virus). Field controls of bait uptake, performed after the first 3 releases, have shown that 80 to 96 % of the baits were either consumed or removed by animals by day 30 (Pastoret et al, 1992). Tetracyclin was found in 74%, 62%, 74%, 62% and 79% of the bones of the foxes collected after the first, second, third and fifth campaigns of vaccination respectively (Pastoret et al, 1992; Brochier et al, 1991; Coppens et al, 1992).

As shown in Figure 1, rabies incidence has severely decreased in the treated area. It is worth noticing that only 6 cases of bovine rabies have been diagnosed in 1991 (Coppens et al, 1992). As in the previous trials, no poxvirus-like lesion has been observed in the collected specimens.

DISCUSSION

The results obtained after 5 vaccination campaigns are very encouraging. Animal rabies has been indeed dramatically reduced, if not completely eradicated. VVTGgRAB has therefore been demonstrated to be a very efficient and safe vaccine. Furthermore, the bait utilized has proven to be attractive for the animals, thermostable and shock-resistant (aerial dropping). The combination of both, Raboral[R] (Iffa-Merieux, Lyon, France) and the bait seems thus to offer an excellent alternative to the attenuated strains of rabies virus as a vaccine for wild carnivores.

In conclusion, complete eradication of fox rabies may be achieved after a limited number or vaccinations with a potent, safe and stable vaccine, an efficient bait system, and an effective method of bait distribution.

ACKNOWLEDGEMENTS

We would like to thank N. Monfrini for typing the manuscript.

REFERENCES

Anderson, R.M., Jackson, H.C., May, R.M., Smith, A.D.M., 1981, Population dynamics of fox rabies in Europe, Nature, 289:765.

Andral, L. and Blancou, J., 1982, La rage. Nouveaux developments en matiere de vaccination, Rev.Sci.Tech,Off.Int.Epiz. 1:927.

Baer, G.M., Abelsethm, M.K. and Debbie, J.G., 1971, Oral immunization of foxes against rabies, Amer.J.Epidemiol., 93:487.

Behbehani, A.M., 1983, The smallpox story: life and death of an old disease, Microbiol.Rev., 47:455.

Blancou, J., 1979, Prophylaxie medicale de la rage chez le renard, Rec.Med.Vet., 155:733.

Blancou, J., Kieny, M.P., Lathe, R., Lecocq, J.P., Pastoret, P.P., Soulebot, J.P. and Desmettre, P., 1986, Oral vaccination of the fox against rabies using a live recombinant vaccinia virus, Nature, 322:373.

Blancou, J., Andral, L., Aubert, M.F.A., Barrat, M.J., Cain, E. and Selve, M., 1982, Vaccination du renard contre la rage par voie orale, Bull.Acad.Vet.de France, 55:351.

Blancou, J., Artois, M., Brochier, B., Thomas, I., Pastoret, P.P., Desmettre, P., Languet, B. and Kieny, M.P., 1989, Innocuite et efficacite du virus recombinant vaccine-rage adminstre par voie orale chez le renard, le chien et le chat, Ann.Rech.Vet., 20:195.

Brochier, B., Languet, B., Blancou, J., Thomas, I., Kieny, M.P., Lecocq, J.P., Desmettre, P. and Pastoret, P.P., 1988, Innocuite du virus recombinant vaccine-rage chez quelques especes non-cibles, in: Vaccination to control rabies in foxes P.P. Pastoret, B. Brochier, I. Thomas, and J. Blancou, eds, Office for Official Publications of the European Communities, Brussels-Luxembourg, 118.

Brochier, B., Languet, B., Blancou, J., Thomas, I., Kieny, M.P., Costy, F., Desmettre, P. and Pastoret, P.P., 1989, Use of recombinant vaccinia-rabies virus for oral vaccination of wildlife against rabies: innocuity to several non-target bait consuming species, J.Widll.Dis., 25:540.

Brochier, B., Costy, F., Hallet, L., Duhaut, R., Peharpre, D., Afiademanyo, K., Bauduin, B. and Pastoret, P.P., 1991, Controle de la rage en Belgique. Resultats obtenus apres trois compagnes de vaccination du renard roux, Ann,Med.Vet., 135:191.

Centers for Disease Control, Atlanta, Rabies surveillance annual summary, 1986.

Coppens, P., Brochier, B., Costy, F., Peharpre, D., Marchal, A., Hallet, L., Duhaut, R., Bauduin, B., Afiademanyo, K., Libois, R. and Pastoret, P.P., 1992, Lutte contre la rage en Belgique. Bilan epidemiologique 1991 et strategie future, Ann.Med.Vet., 136:129.

Debbie, J.G., 1983, Rabies Control in Wildlife, in: "Veterinary Learning Systems", Princeton, Reports on Rabies: 23.

Desmettre, P., Languet, B., Chappuis, G., Brochier, B., Thomas, I., Lecocq, J.P., Kieny, M.P., Blancou, J., Aubert, M.F.A., Artois, M. and Pastoret, P.P., 1990, Use of a vaccinia rabies recombinant for oral vaccination of rabies vectors, Vet.Microbiol., 23:227.

Fu, Z.F., Dietzschold, B., Shumacher, C.L., Wunner, W.H., Ertl, H.C.J. and Koprowski, H., 1991, Rabies virus nucleoprotein expressed in and purified from insect cells is efficacious as a vaccine, Proc.Natl.Acad.Sci.USA., 88:2001.

Johnston, D.H. and Lawson, K., 1987, U.S. Patent No.4:650.

Kappeler, A., Wandeler, A.I. and Capt, S., 1985, Geographical barriers to wildlife rabies spread: the concept and its application in oral immunization of foxes against rabies, Rev.Ecol.(Terre Vie), 40:267.

Kieny, M.P., Desmettre, P., Soulebot, J.P. and Lathe, R., 1986, Rabies vaccines: traditional and novel approaches, Prog.Vet.Microbiol.Immun., 3.

Kieny, M.P., Lathe, R., Drillien, R., Spehner, D., Skory, S., Schmitt, D., Wiktor, T., Koprowski, H. and Lecocq, J.P., 1984, Expression of rabies virus glycoprotein from a recombinant vaccinia virus, Nature, 312.

Lathe, R.,Kieny, M.P., Lecocq, J.P.,Drillien, R., Wiktor, T. and Koprowski, H., 1985, Immunization against rabies using a vaccinia-rabies recombinant virus expressing the surface glycoprotein, in: Vaccines, R. Lerner, R. Chanock, F. Brown, eds, Cold Spring Harbor Laboratory, New York: 157.

Mayr, A., Kraft, H., Jaeger, O. and Haacke, H., 1972, Oral immunisierung von Fuchsen gegen Tollwut, Zentbl.Vet.Med., 19:615.

Panicali, D., Davis, S.W., Weinberg, R.L. and Paoletti, E., 1983, Construction of live vaccines by using genetically engineered poxviruses: biological activity of recombinant vaccinia virus expressing influenza virus haemaglutinin, Proc.Natl.Acad.Sci.USA, 80:5364.

Pastorett, P.P., Brochier, B., Blancou, J., Argois, M., Aubert, M., Kieny, M.P., Lecocq, J.P., Languet, B., Chappuis, G. and Desmettre, P, Development and deliberate release of a vaccinia-rabies recombinant virus for the oral vaccination of foxes against rabies, in: Recombinant Poxviruses, M. Mattheu, Binns and Geoffrey L. Smith, eds, CRC Press, New York, USA, 163.

Rupprecht, C.E. and Kieny, M.P., 1988, Development of a vaccinia-rabies glycoprotein recombinant virus vaccine, in: Rabies, J.B. Campbell, K.M. Charlton, eds, Kluwer Academic Publishers, Boston.

Schneider, L.G., Cox, J.H. and Muller, W.W., 1985, Field trials of oral immunization of wildlife animals against rabies in the Federal Republic of Germany: a mid course assessment, Rec.Ecol.(Terre Vie) 40:267.

Steck, F., 1982, Rabies in Wildlife, Symp.zool.Soc.Lond.,50:57.

Smith, G.L., Mackett, M. and Moss, B., 1985, Infection vaccinia virus recombinants that express hepatitis B virus surface antigen, Nature, 302:490.

Smith, G.L., Mackett, M. and Moss, B., 1984, Recombinant vaccinia viruses as new life vaccines, Biotechnol.Genet.Engng.Rev., 2:383.

Soria Baltazar, R., Blancou, J. and Artois, M., 1987, Resultats de l'administration par voie orale au mouton de deux vaccins contenant un virus de la rage modifie (SAD B19) ou un recombinant du virus de la vaccine et de la rage (187 XP), Ann.Med.Vet., 131:481.

Summer, J.W., Fekadu, M., Shaddock, J.H., Esposito, J.J. and Bellini, W.J., 1991, Protection of mice with vaccinia virus recombinants that express the rabies nucleoprotein, Virology, 183:703.

Toma, B. and Andral, L., 1977, Epidemiology of Fox Rabies, Adv.Virus Res., 21:1.

Wiktor, T.J., Mc Farlan, R., Reagan, K., Dietzschold, B., Curtis, P., Wunner, W., Kieny, M.P., Lathe, R., Lecocq, J.P., Mackett, M., Moss, B. and Koprowski, H., 1984, Protection from rabies by a vaccinia virus recombinant containing the rabies glycoprotein gene, Proc.Natl.Acad.Sci.USA, 81:7194.

Wiktor, T.J., Kieny, M.P. and Lathe, R., 1988, New generation of rabies vaccine. Vaccinia-rabies glycoprotein recombinant virus, in: Applied Virology Research, E. Kurstak, ed, vol. 1 Plenum Press, New York, 69.

Wiktor, T.J., Mac Farlan, R.I., Dietzschold, B., Ruprecht, C., Wunner, W.H., 1985, Immunogenic properties of vaccinia recombinant virus expressing the rabies glycoprotein, Ann.Inst.Pasteur Virol., 136E:405.

Wunner, W.H., Dietzschold, B., Curtis, P. and Wiktor, T.J., 1983, Rabies subunit vaccines, J.Gen.Virol., 64:1649.

VACCINATION OF WILDLIFE: THE ROLE OF RECOMBINANT POXVIRUS VACCINES

E. Paul J. Gibbs

College of Veterinary Medicine
University of Florida, Gainesville
Florida 32610, USA

INTRODUCTION

There are two major milestones in the history of vaccinology in the 18th and 19th centuries. The first milestone, the birth of vaccinology, occured in England. On May 14th 1796, Edward Jenner used cowpox virus taken from the wrist of a milkmaid, Sarah Nelmes, to vaccinate James Phipps; six weeks later Jenner inoculated Phipps with smallpox virus taken from a pustule of a patient with the disease. The result of this experiment is known to all associated with vaccines and led to one of the most significant medical achievements of the 20th century; the global eradication of smallpox in 1977. The second milestone occured in France on July 6th 1885 when Louis Pasteur treated Joseph Meister, who had been bitten by a rabid dog three days previously, by vaccinating him with attenuated strains of rabies virus grown in rabbit brain. Joseph Meister survived and the concept of vaccination was firmly established.

The science of vaccinology has made major advances since Pasteur's historic studies and today we are on the threshold of a new generation of vaccines, produced principally through recombinant technology as described in the proceedings of this NATO Advanced Scientific Institute. The prospect of a range of more effective vaccines is at hand. For use in wildlife, recombinant vaccines, using poxvirus as the vector, are particularly attractive. They offer the opportunity to use vaccine in baits since the vectors are thermostable and, in some cases, can initiate infection by the oral route. Interestingly, the first live-virus recombinant-vectored vaccine to be released for field studies (a further milestone in the history of vaccinology) is a vaccinia virus expressing the gene for the surface glycoprotein of rabies virus; a fusion of the science initiated by Jenner and Pasteur!

This recombinant is currently being used as an oral vaccine to control rabies in wildlife populations; red foxes (<u>Vulpes vulpes</u>) in Europe and raccoons (<u>Procyon lotor</u>) in the USA. It is recognized to be the paradigm for the development of other vaccines intended for use with wild-life. The reason for this vaccination campaign is principally to protect human health by reducing/eliminating a reservoir of infection. However, vaccination extends beyond the need to protect human health.

The United Nations "Earth Summit" in Rio de Janeiro in 1992 on

New Generation Vaccines, Edited by G. Gregoriadis
et al., Plenum Press, New York, 1993

protection of the environment emphasized conservation of animal species. Vaccines have, and will continue to play, an important role in protecting endangered species both in captivity and, to a lesser extent, in the wild. The protection of the black-footed ferret (Mustela nigripes) is a well known example of a species taken into captivity and protected from a disease that was threatening its existence (Thorne and Williams, 1988).

While some may wish it to be otherwise, wildlife species also have economic value as an important source of "fur and food" in developing countries and recreation in the form of hunting in industrialized countries. Setting aside the revenue generated by tourism in many countries of the world, the recreational value of wildlife is enormous. The "value" of wildlife in the USA can be used as an example. In 1991, sportsmen, bird-watchers, etc. spent $59.5 billion dollars in pursuit of their interests; this represented approximately 1% of the gross national product. Within this broad field of activity, there were 14 million hunters of which 10.7 million hunted big game. The revenue generated from the license fees paid by these hunters is then used to support conservation in general. The interest of these big game hunters is focussed on cloven-hoofed animals (deer, wild swine, etc.) the population of which in North America is estimated to be 32 million. The white-tailed deer (Odocoileus virginianus) population alone is estimated to be 19.6 million.

A precipitous decline in any wildlife population, whether endangered or not, or the existence/discovery of a zoonotic disease in a wildlife species is, therefore, likely to generate concern from many sectors of society. Two examples will illustrate this point. In 1988, approximately 18,000 common seals (Phoca vitulina) and some gray seals (Halichoerus grypus) were washed up around the North Sea coastline (Hall and Harwood, 1990). The cause of the deaths of these seals was identified as a morbillivirus similar to canine distemper virus. The public, which had expressed great concern over the death of approximately 50% of the seal population in the southern part of the North Sea, questioned why was it not possible to vaccinate the remaining seals. The second example concerns the American bison or buffalo (Bison bison). Bison in the Yellowstone National Park in the USA and in the Wood Buffalo National Park in Canada are infected with brucellosis. This disease has been largely eradicated from domestic cattle (economic and zoonotic justification) and the bison populations are seen as a reservoir of infection that must be eliminated. A recommendation to slaughter all the bison in the Wood Buffalo National Park has been rejected by the Canadian government and a policy allowing bison that stray from Yellowstone Park onto neighboring farmland to be shot is highly controversial (Thorne et al, 1991). Vaccination is seen as the only viable option.

But can vaccines be developed and used successfully in such situations? The answer is not simple. Notwithstanding the excitement and opportunities generated by the emergence of the new generation vaccines, vaccination of wildlife raises a number of practical and philosophical challenges. This review attempts to identify the advantages of the new-generation poxvirus-vectored vaccines and their potential role in wildlife vaccination. The examples quoted are frequently virus diseases, but the principles apply to infectious diseases in general. It is divided into two parts. The first part outlines why a seemingly attractive proposal to vaccinate a wildlife population must be carefully considered. Any decision to proceed with vaccination must be based on sound scientific data supported by society's appreciation of the complexity of the situation and financial backing for such action. Usually, it will be found that no suitable vaccine is available and a research program must be developed first. The second part gives examples where new-generation vaccines have been shown to be effective, hold considerable promise, or are needed for the control of disease in wildlife.

PART 1   VACCINATION OF WILDLIFE: A FRAMEWORK FOR DECISION MAKING[1]

Objectives of Vaccination. What are we attempting to do and can we justify our decision?

These questions can be answered most easily by focussing on those actions directed at the health of individuals and those concerned with the population as a whole. Vaccination of an individual wild animal brought to an animal hospital, rescue center, or breeding facility can be a life-saving procedure to protect it during its stay in confinement. The underlying concepts for considering a vaccination programme for such animals are no different from those faced when proposing to vaccinate domestic species. Vaccination of free-living wildlife populations (the major subject of this review) needs very different considerations.

Many diseases of domestic farm livestock and household pets infect wildlife species, but the involvement of wildlife is often incidental. When the specific disease is controlled in the domestic population the incidence of infection in wildlife falls and often "dies out"; the history of rinderpest in Africa is a good example. In contrast, there are some diseases, rabies being the paradigm, where wildlife species represent the reservoir of infection with disease "spilling over" into domestic species and humans. There is also an intermediate category where nidi of infection are discovered in wildlife species only when the incidence of infection in the domestic population falls to reveal transmission from a wildlife source. In these cases, such as tuberculosis in deer in North America, domestic species are believed to be the original source of infection for the wildlife species. Thus, the epidemiological history of disease in wildlife species indicates that disease can be categorized into one of three groups; sylvatic infections, short term infections with "die out", and long-term, but localized.

The examples quoted above have emerged through epidemiological associations over many years. The directional flow of the pathogen (is it being transmitted predominantly from wildlife to domesticated species/humans or vice versa?) is not always so easy to discern. Molecular analyses of the sequence divergence of selected genes of pathogens isolated from both domestic and wild animals is proving to be a powerful tool in establishing this relationship. For example, it has recently been shown that feline immunodeficiency virus (FIV) infects wild Felidae and domestic cats; phylogenetically, however, the isolates examined indicate that transmission is not a regular occurrence between the two populations (Olmsted et al, 1992).

Thus, before embarking on a vaccination program for free-living wildlife species, there are many questions that need to be asked. The first group of four seemingly simple questions relates to justification: Is the driving force to consider vaccination a) to control/eradicate a

---

[1] The following analysis embraces several of the concepts advanced by Ailsa Hall and John Harwood (1990) of the Sea Mammal Research Unit, Natural Environment Research Council, High Cross, Madingley Road, Cambridge, England, in a publication called the "Intervet Guidelines to Vaccinating Wildlife". The stimulus for developing these guidelines was the epidemic of phocine distemper in the North Sea in 1988. The authors present an excellent series of decision trees to aid the reader in deciding whether vaccination is appropriate and feasible.

disease/infection present in the population that represents a major public health risk or a risk to domestic animals; b) to control/eradicate an existing disease of economic importance in the wildlife species concerned; c) to control a disease epidemic that appears to have grave consequences for the survival of the wildlife population concerned if no action is taken, or d) to protect an endangered species that it is believed will possibly become extinct if exposed to an infection from other wildlife species, domestic species, or human even though there is no evidence that such an infection is currently present in the endangered species? In some circumstances, there may be more than one reason for vaccination.

Examining the questions in more detail, one will see that questions a) and b) center around economics and public health policy while questions c) and d) focus on ecological and conservational policies. In all cases, risk assessment is of paramount importance. On closer examination of the questions, within the context of a specific problem, one often begins to recognize that the four seemingly simple questions cannot always be answered with confidence. Unfortunately, there is often very little baseline information on the prevalence of infection/disease in a wildlife population. The information that is available may have been collected fortuitously in conjunction with other studies or during an epidemic and, as such, is biased or statistically unsound. If vaccination is contemplated when unnaturally high mortality is occurring what is "unnaturally high"?. This raises the philosophical question "do diseases regulate populations and if so, is it appropriate to interfere even when extinction is imminent"?

Using the information on FIV infections of wild Felidae mentioned earlier as an example, it is logical for most people to argue, on first hearing that the endangered Florida panther (<u>Felis concolor coryi</u>) is infected with FIV, that the population should be darted and vaccinated against FIV. On closer examination of the situation, one finds that there is no vaccine for protection of either domestic or wild Felidae currently available. Even if there were, a decision to vaccinate this subspecies would need to be justified on a) the presence of a disease that can be associated with the infection in the population, and b) would need to recognize that a vaccine developed for domestic cats would not necessarily be effective in the panther and conceivably could be dangerous. Currently, since no clinical disease has been associated with FIV infections in the Florida panther, expenditure of funds on developing a vaccine would be difficult to justify.

The above analysis presupposes that society wishes to preserve a wildlife species in a given area. This is certainly true for the Florida panther, but wildlife populations vary from endangered to pest proportions. A wildlife species may be indigenous or introduced, the latter often being a pest. Feral animals are frequently regarded as wildlife species and, while many wish to eradicate them because of the environmental damage they inflict, others value them as game species and as a source of license revenue. (For introduced wildlife species and feral animals of little hunting value, there is rarely any concern for protecting them; rather it is the converse and several attempts to eradicate unwanted species through introducing disease have been documented. The most famous of these was the attempt to control the European rabbit (<u>Oryctolagus cuniculus</u>) in Australia by infecting them with myxomatosis virus (Fenner and Ratcliffe, 1965)).

Criteria for Intervention. We want to vaccinate but can we?

This section examines the feasibility of vaccination. It initially assumes that use of a vaccine to control disease in wildlife can be justified and a vaccine is available for the control of this disease in domestic species.

It is axiomatic that successful vaccination breaks the cycle of transmission between an infected host and a susceptible host, by making the susceptible host resistant to infection. (Some vaccines, such as those against herpes infections, protect the susceptible host from developing disease, but do not prevent infection or subsequent excretion of virulent virus. Within the context of controlling/eradicating disease in a population such vaccines are less effective but, since the levels of virus excretion are lower when compared with non-vaccinated animals, their use is accepted). The key to successful vaccination lies in an effective vaccine used at the most appropriate time in the life cycle of the pathogen. To paraphrase, identify the "weakest link" in the transmission chain and "target" the vaccine. Identifying the weakest link depends upon knowledge of the epidemiology of the infectious agent and the behaviour of the host. The routes of transmission and the basic mechanisms by which pathogens are maintained in nature are many and varied; a detailed analysis is beyond the scope of this review, but is available (Fenner et al, 1992). This knowledge is based primarily on studies in domestic animals and humans, not wildlife. With wildlife diseases, one is frequently surprised how little detailed knowledge is available on the behaviour of the host; for example the size of the home range, the family unit, etc. Thus, similar to the situation for the prevalence of disease, as discussed in the previous section, there is often only limited information upon which to base a focussed vaccination program.

Although a vaccine may be available for the disease in domestic species, the vaccine, if live, may carry a high risk, particularly if used to protect an endangered species. This was tragically demonstrated when a canine distemper vaccine, albeit safe in domestic European ferrets (Mustela putorius fuvo), was given to North American black-footed ferrets (Mustela nigripes) and led to their death from distemper (Carpenter et al, 1976). Equally, concern has been expressed that any live vaccine released into a wildlife population may "spill over" into other species in which it may cause dramatic disease. Finally, within this category, we also need to assess whether animals can be vaccinated before they become exposed to infection, but at a time when material (passive) immunity is unlikely to interfere. This is a problem when vaccinating domestic species and is an even greater problem when faced with vaccinating wildlife of unknown age.

The assumption up to this point in the analysis is that a vaccine is available for use in wildlife and further, that it can be effectively administered. The fact is that there are very few vaccines that have been evaluated for use in wildlife and those that are available must generally be given by injection. Vaccination of wildlife, having either captured the animal first or by using a darting rifle, is extremely inefficient and expensive. For example, a proposal supported by the World Wildlife Fund to dart 50 endangered desert African elephants (Loxodonta africanus) in Namibia to protect them against an anthrax outbreak will cost an estimated $40,000 (World Wildlife Fund, 1992). Administering vaccine by this route may also cause more problems than it solves. The disturbance and stress may pressure potentially infected animals to move out of the area and infect susceptible populations elsewhere. The additional stress of vaccination in populations, already stressed by environmental pollutants, should also be considered. Thus, in cases where vaccines are being used to control or eradicate disease, the problem may be exacerbated.

Clearly, oral vaccination is the most attractive route of administration, although the concept of aerosol vaccination administered by low flying aircraft is also attractive for animals that congregate in large numbers (e.g. colonies of seals).

In summary, any decision to proceed with vaccination of free-living wildlife must take many factors into consideration. For many diseases, vaccines are not available and an expensive research and development programme to produce a vaccine for use with wildlife may be very difficult to justify, if there is neither a significant public health risk nor an economic benefit. The most practical approach for development of wildlife vaccines will be to adapt existing and emerging vaccines so that they can be administered orally in bait. It is of paramount importance that they be safety and efficacy tested before use in a range of species. For recombinant poxvirus-vectored vaccines particularly, this involves extensive and expensive government review. If a vaccine is available and licensed for use in wildlife, the risks of vaccination must be weighed against the expected benefits of vaccination. Such assessment must consider not only biological risks but also financial. For most situations, the decision whether to vaccinate free-living wildlife species depends on the answer to three questions:

1. Is the disease (or transmission to man or animals) really a problem?
2. Are there significant benefits to a successful vaccination programme and do these outweigh the associated risks?
3. Is a safe and efficacious vaccine available and usable?

If the answer is an unqualified "yes" to all three questions, the decision to vaccinate simply depends on the availability of money and trained personnel. Rarely, if ever, is this the case!

PART 2   EXISTING AND PROJECTED WILDLIFE VACCINATION PROGRAMMES

In this part of the review, four diseases of wildlife are briefly discussed to reflect the diversity of the need for wildlife vaccines; rabies in foxes and raccoons in Europe and the USA, Aujeszky's Disease (AD) in feral swine in the USA, brucellosis in bison and elk (<u>Cervus elaphus</u>) in North America, and tuberculosis in several species worldwide. When each of these disease problems is analyzed against the parameters established in Part 1, the need for vaccination can be justified. While vaccines are available for use in domestic species with each of these diseases, suitable vaccines are not necessarily available for wildlife.

Rabies

Rabies is a worldwide problem and several wildlife species are involved in the sylvatic perpetuation of the virus. The control of rabies in wildlife populations has been mentioned above as the paradigm for the successful use of vaccines to control a wildlife disease. The early work on the development of vaccines focussed on the use of attenuated strains of rabies virus administered orally to foxes. The success of a programme in Switzerland using vaccine ampules hidden in chicken heads was dramatic (Winkler and Bogel, 1992) and has led to an extensive and continuing programme in several European countries. The attenuated vaccines have been replaced in some areas with a vaccinia virus recombinant expressing the rabies G glycoprotein (Pastoret et al, 1992); the intent is to free Europe of terrestrial rabies within this century. The same vaccinia virus recombinant is currently being used for field trials in the USA to control the disease in raccoons. For additional discussion on the development and use of the vaccinia recombinant, the reader is directed to a paper published in this book by Dr. Kieny and her colleagues.

The excitement over the success of the vaccination programme is tempered somewhat by safety issues and the spectrum of species that can be vaccinated. The safety issues relate to the pathogenicity of both the attenuated vaccine strain of rabies and the vaccinia recombinant for other species, including man (Esposito, 1989). To overcome these problems in the

USA and to address the current epidemic of raccoon rabies in the mid-Atlantic states, Esposito and his colleagues (1988) have developed a raccoonpox recombinant expressing the G glycoprotein and have successfully demonstrated its efficacy in laboratory trials. Raccoonpox virus is believed to be host specific to the raccoon. With regard to the species that can be successfully vaccinated, the above vaccines are less effective in skunks (Mephitis mephitis), an important reservoir of infection in the central states of the USA.

Aujeszky's Diseases in Feral Swine in the USA

Swine (Sus scrofa) were first introduced to the North American continent by Hernando de Soto in 1539. Today, free-living wild swine (feral swine or wild hogs) occur in at least 18 US states with a total population estimated to be between 1 and 2 million. Hog hunting, trapping and related activities such as taxidermy, account for about $8 million/year in Florida alone. The social value of the wild hog also cannot be ignored since hog hunting has long been an integral part of the rural southern lifestyle. In contrast to their economic benefit, wild swine cause extensive ecological damage (root up native flora and pastures, kill fawns and lambs, destroy bird nests). Although many populations remain to be tested, infection with Aujeszky's disease virus (ADV) has been established in free-living wild swine in 11 states and is particularly prevalent in the southeastern states (van der Leek et al, 1993a). Although there is no conclusive evidence of clinical AD in wild swine, limited experimental evidence suggests that adult wild and domestic swine exhibit similar signs. In addition to the presence of ADV antibodies, cases of AD in "hog dogs" and other in-contact species (such as cattle) currently provide the best evidence of ADV circulating within wild swine populations.

Since the inception of a national AD eradication program in the US in 1989 (which has the ultimate goal of eradicating ADV from domestic swine by the year 2000 at an estimated total cost of about $250 million), wild swine have received increased attention. Already incriminated as a source of AD for domesticated swine, the significance of infected wild swine increase as the prevalence of AD in domesticated swine decreases. However, the risk posed by infected wild swine is difficult to quantify.

The depopulation of all wild swine is considered impossible, although small populations in suitable habitats could possibly be eliminated. The only viable long-term option is eradication of ADV from wild swine, since strategies which only address control will ultimately prove incompatible with the national eradication programme goal. While it would be misleading to consider the control of rabies in wildlife and AD in feral swine as similar, the success in controlling rabies justifies exploring this approach for the control of AD in feral swine. The currently available AD vaccines are attenuated or gene deleted, must be given by injection, are thermolabile and therefore unsuitable as wildlife vaccines. In safety studies with the vaccinia rabies recombinant, Brochier et al, (1989) demonstrated that European wild boar seroconverted to rabies after oral administration of the vaccine; this indicates that oral vaccination of swine using a poxvirus vector is feasible. Although AD has been recorded in wildlife species, no wildlife reservoir has been identified. Thus, the long term prospects of controlling AD in feral swine are favorable, if a suitable vaccine can be developed. A recombinant vaccine using ADV genes inserted into swinepox virus (SPV-AD) is under development (van der Leek et al, 1993b) and it is anticipated that it can be used as an oral vaccine. Swinepox virus is believed to be host specific, so the choice of swinepox virus as a vaccine vector should reduce concerns over the safety of using a vaccinia recombinant. This assumes that the SPV-AD vaccine, similar to the existing vaccines used in domestic swine, will significantly reduce the amount of virus excreted by infected pigs, and, when used in combination with methods to decrease population density, could break the transmission cycle. Field trials of sham vaccines in oral baits have established that approximately 90% of swine in study areas can be "vaccinated" with only one distribution of bait (Fletcher et al, 1990).

Brucellosis in Bison and Elk in Greater Yellowstone Ecosystem (GYE)

Whereas the two viral diseases mentioned above have probably always existed in their respective wildlife populations, brucellosis was probably acquired from cattle early in the 20th century (Thorne et al, 1991). Brucellosis is biologically important to wildlife in the GYE, but its significance as a bovine disease gives it great economic and political importance. In a newly infected herd, there is very little difference in the response of cattle, elk and bison to infection; abortion is the most commonly observed problem. It has been calculated that since 1935, the federal government has spent in excess of $1.3 billion to eradicate brucellosis in the USA (this figure does not include support from individual states or losses to the industry). The economic loss through disease in cattle and the zoonotic nature of brucellosis were used to justify the eradication program. Domestic cattle in Wyoming and Montana are now free of brucellosis, so the presence of the disease in bison and elk that trespass outside the confines of the National Park represents not just a threat to local cattle owners (who could lose their entire herd through compulsory slaughter if it became infected), but to the economy of the whole state (which would be penalized by having to follow stringent marketing procedures to prevent the sale of infected cattle). Whereas most people regard the GYE and its associated wildlife as a unique treasure, some, because of its reservoir of brucellosis, see it as a threat to an important international industry and economy. The issue takes on an added dimension of contention when the religious and ceremonial importance of the bison to the Indian tribes of North America is introduced. Despite the fact that the transmission of brucellosis from free-ranging bison and elk has not been confirmed under field conditions, the Wyoming Game and Fish Department initiated studies as early as 1977 to see if the disease could be eradicated from the wildlife by vaccination. The rationale is to achieve a 90% or greater vaccination rate to break the cycle of transmission. Currently, vaccine is administered by "darting" rifle to elk using strain 19 brucella vaccine. In excess of 15,000 elk have been vaccinated. Evaluation of the safety of strain 19 in bison, however, indicates that it is abortogenic in bison. An oral vaccine, perhaps one based on a poxvirus vector or other vector system, would be valuable.

Tuberculosis in Several Species of Wildlife

Similar to brucellosis, tuberculosis is a zoonotic disease of cattle that has become established in wildlife. Also similar to brucellosis, eradication programs in cattle have been highly effective. Unfortunately, it has been recognized that tuberculosis (Mycobacterium bovis) is re-emerging as a disease of wildlife affecting different species in different parts of the world. In Great Britain, badgers (Meles meles) in some areas have a prevalence of up to 50% and in New Zealand the prevalence of tuberculosis in some colonies of brush-tailed possum (Trichosurus vulpecula) is as high as 25%. High on the list of priorities in New Zealand is the development of new vaccines (Jackson, 1991). To further complicate the epidemiology, other species, such as deer in the USA, have recently been found to be infected.

CONCLUSIONS

Ada (1991) summarized the three topics of high priority of the WHO Programme for Vaccine Development in Trans-Diseases Vaccinology as the use of live chimeric vectors, controlled release formulations for vaccine delivery, and oral vaccines. In the same article, Ada delineates the immunological requirements of an ideal vector and concludes that there are advantages to exhaustingly evaluating chimeric live vectors in a veterinary context, before testing them in a medical setting. Since this review on the opportunities and constraints of vaccine development for use in wildlife has established that similar priorities exist for wildlife vaccines, acceptance of Ada's recommendations should be of mutual benefit to human and veterinary medicine.

REFERENCES

Ada, G.L., 1991, Strategies for exploiting the immune system in the design of vaccines, Mol.Immunol., 28:225.

Brochier, B., Blancou, J., Thomas, I., Languet, B., Artois, M., Kieny, M., Lecocq, J.-P., Costy, F., Desmettre, P., Chappuis, G. and Pastoret, P.-P., 1989, Use of recombinant vaccinia-rabies glycoprotein virus for oral vaccination of wildlife against rabies: innocuity to several non-target bait consuming species, J.Wildl.Dis., 25:540.

Carpenter, J.W., Appel, M.J.G., Ericson, R.C. and Novilla, M.N., 1976, Fatal vaccine-induced canine distemper virus infection in black-footed ferrets, J.Amer.Vet.Med.Assoc., 69:691.

Esposito, J.J., 1989, Live poxvirus-vectored vaccines in wildlife immunization programmes: the rabies paradigm, Res.Virol., 140:480.

Esposito, J.J., Knight, J.C., Shaddock, J.H., Novembre, F.J. and Baer, G.M., 1988, Successful oral rabies vaccination of raccoons with raccoon poxvirus recombinants expressing rabies virus glycoprotein, Virology, 165:313.

Fenner, F. and Ratcliffe, F.N., 1965, "Myxomatosis", Cambridge University Press, London and New York.

Fenner, F., Gibbs, E.P.J., Murphy, F.A., Rott, R., Studdert, M.J. and White, D.O., 1992, "Veterinary Virology", 2nd ed., Academic Press, Orlando

Fletcher, W.O., Creekmore, T.E., Smith, M.S. and Nettles, V.F., 1990, A field trial to determine the feasibility of delivering oral vaccines to wild swine, J.Wildl.Dis., 26:502.

Hall, A. and Harwood, J., 1990, "The Intervet guidelines to vaccinating wildlife", Sea Mammal Research Unit, NERC, High Cross, Madingley Road, Cambridge, England.

Jackson, R., ed., 1991, Proceedings Symposium on Tuberculosis. Publication 132, Veterinary Continuing Education Series, Massey University, Palmerston, North Island, New Zealand.

Olmsted, R.A., Langley, R., Roelke, M.E., Goeken, R.M. Adger-Johnson, D., Goff, J.P., Albert, J.P., Packer, C., Laurenson, M.K., Caro, T.M., Scheepers, L., Wildt, D.E., Bush, M., Martenson, J.S. and O'Brien, S.J., 1992, Worldwide prevalence of lentivirus infection in wild feline species: epidemiologic and phylogenetic aspects, J.Virol.,66:6008.

Pastoret, P.-P., Brochier, B., Blancou, J., Artois, M., Aubert, M., Kieny, M.-P. Lecocq, J.-P., Languet, B., Chappuis, G. and Desmettre, P., 1992, Development and deliberate release of a vaccinia-rabies recombinant virus for the oral vaccination of foxes against rabies, in: "Recombinant Poxviruses", M.M. Binns and G.L. Smith, eds., CRC Press, Boca Raton.

Thorne, E.T., Meagher, M. and Hillman, R., 1991, Brucellosis in free-ranging bison: three perspectives, in: "The Greater Yellowstone ecosystem: redefining America's wilderness heritage", R.B. Keiter and M.S. Boyce, eds., Yale University Press, New Haven and London.

Thorne, E.T. and Williams, E.S., 1988, Disease and endangered species: the black-footed ferret as a recent example, Cons.Biol.,2:66.

van der Leek, M.L., Becker, H.N., Pirtle, E.C., Humphrey, P., Adams, C.L., All, B.P., Erickson, G.A., Belden, R.C., Frankenburger, W.B. and Gibbs, E.P.J., 1993a, Seroprevalence of pseudorabies (Aujeszky's Disease) virus in feral swine in Florida, J.Wildl.Dis., 29:80.

van der Leek, M.L., Feller, J.A., Sorensen, G., Isaacson, W., Adams, C.L., Borde, D.J., Pfeiffer, N., Moyer, R.W., Tran, T., Gibbs, E.P.J., 1993b, Evaluation of swinepox virus as a vaccine vector in swine using a pseudorabies (Aujeszky's disease) virus gene insert coding for glycoproteins gp50 and gp63. Vet.Rec. Submitted for publication.

Winkler, W.G. and Bogel, K., 1992, Control of rabies in wildlife, Sci.Amer., 266:86.

World Wildlife Fund, 1992, Anthrax outbreak threatens wildlife in Namibia. Focus, 14:3.

RECOMBINANT PROTEIN ANTIGENS WITH 'BUILT-IN' ADJUVANTICITY

P.Ghiara[1], L.Villa[1], R.Rappuoli[1], S.Gonfloni[2], L.Castagnoli[2] and G.Cesareni[2]

[1]Immunological Research Institute Siena, Siena, Italy and [2]Department of Biology, University of Tor Vergata Rome, Italy

Recombinant DNA technology allows the cloning and expression of an increasing number of proteins from a variety of pathogens which could be considered in the future as possible candidates for human or animal vaccination. However some recombinant proteins are weakly immunogenic and the search for powerful and safe adjuvants for these kinds of antigens is a major task in the field of recombinant vaccine development.

A possible approach to the enhancement of the immune response to such antigens is to utilize immunostimulatory lymphokines, such as IL-1 or TNFalpha as adjuvants (Staruch and Wood 1983, Ghiara et al., 1987, Reed et al., 1989). However their toxic and proinflammatory effects, together with their innate pleiotropicity, strongly limit their practical application in humans. IL-1 for example, is an important modulatory molecule of the immune system and is also induced in vivo after administration of several adjuvant molecules, such as bacterial products, but it cannot be used as such because it is also a potent pyrogen and proinflammatory agent (Dinarello, 1991).

We speculated that, in view of the complexity of its biological properties, discrete domains of the IL-1 molecule could be responsible for single biological activities. Therefore, synthetic stretches of the human IL-1 beta molecule were prepared and tested in different biological assays.

Interestingly, we have found that in the 163-171 sequence of human IL-1 beta (VQGEESNDK) is the minimal structure responsible for the immunostimulatory properties of the entire molecule. Moreover this nonapeptide was shown to be devoid of many in vivo and in vitro undesired proinflammatory activities of IL-1 (Antoni et al., 1986, Boraschi et al., 1988). A summary of the biological properties of the peptide as compared to the entire IL-1 molecule is reported in Table 1.

This peptide has been proposed as adjuvant for poorly immunogenic vaccines, since it could stimulate the immune response to both T-dependent and T-independent antigens (Nencioni et al., 1987). Moreover in the case of HBsAg, the peptide was used both mixed with antigen (Tagliabue et al., 1989) or physically coupled to part of the antigen itself (i.e. the peptide 12-32 of HBsAg) (Rao and Nayak 1990).

New Generation Vaccines, Edited by G. Gregoriadis
et al., Plenum Press, New York, 1993

Table 1. Biological activities of human IL-1beta and of its 163-171 peptide[1]

	IL-1beta	peptide 163-171
IMMUNOLOGICAL EFFECTS		
Adjuvanticity	+++	+++
Immunorestoration	+++	++
Induction IL-2/IL-4	+	++
INFLAMMATION		
Fever	+++	-
Fibroblast PGE	+++	-
Acute phase proteins	+++	-
Plasma cations	+++	-
Corticosteroid induction	+++	-
Glucose homeostasis	+++	-
Hepatic ED induction	+++	-
IL-6 induction	+++	-
ANTITUMORAL ACTIVITY		
Tumor rejection	+	+
NK activation	+/-	+/-
RADIOPROTECTION		
X-ray survival	+++	++
HEMOPOIESIS		
CFU	+++	-
Peripheral leukocytes	+++	+

[1] for more details see Boraschi et al, 1988

Given its small size, a fascinating alternative is to engineer the sequence encoding this peptide within recombinant antigens.

Therefore the peptide VQGEESNDK was genetically grafted into a model protein antigen, i.e. human ferritin H chain, and assessed for its ability to enhance the immune response in vivo against the host protein.

The H chain of human ferritin has been chosen as a test antigen because, due to the high sequence homology with murine ferritin, it is a poor immunogen in the mouse (Miyazaki et al., 1988). Furthermore the three-dimensional structure of the human H ferritin homopolymer is known at high resolution (Lawson et al., 1991) and a detailed mutagenic analysis has revealed tolerant sites that can accept peptide insertions without harming the general fold of the protein (Luzzago and Cesarini, 1989; Jappelli et al., 1992).

Ferritin is an heteropolymer consisting of two types of chains (H and L) that assemble in different proportions in different tissues (Arosio et al., 1987). The monomer consists of a bundle of 5 antiparallel alpha helices (A-E) connected by loops. In this study Ferritin H chain was ef-

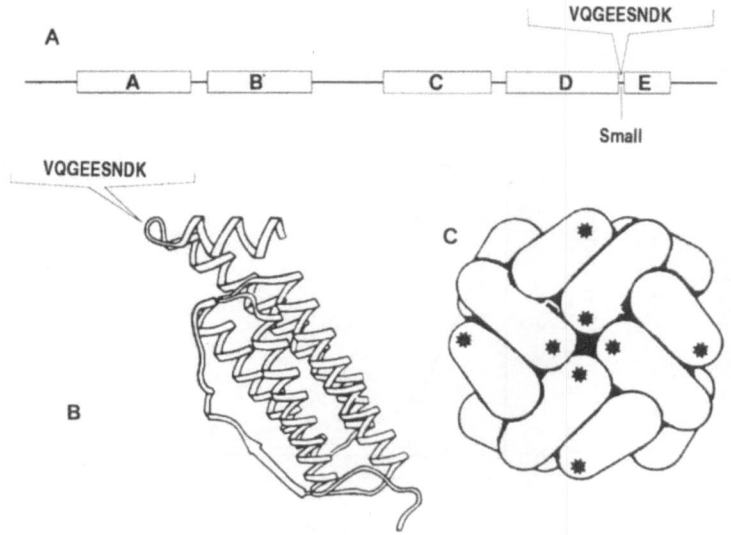

Fig. 1.   a) Schematic representation of human ferritin H chain
           mutant; b) Ribbon-type representation of ferritin alpha
           helices bundle; c) Schematic diagram of ferritin H chain
           polymer; * = insertions points of VQGEESNDK sequence

ficiently expressed in E. coli by using a plasmid vector containing the pL
promoter of lambda bacteriophage. The sequence coding for Ferritin H chain
has been modified by site directed mutagenesis to introduce a unique SmaI
restriction site in the region connecting D and E helices. Two complementary oligonucleotides coding the sequence VQGEESNDK were then introduced
in the unique SmaI site (see Fig 1 for a schematic representation).

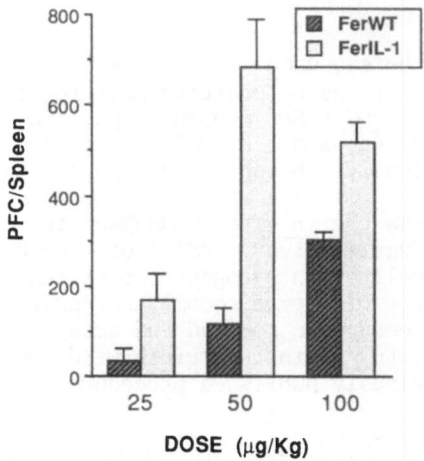

Fig. 2. Specific anti ferritin PFC/spleen. Spleens
        from C3H/HeJ mice immunized with FerWT or
        with FerIL-1 were removed and the number of
        specific anti FerWT PFC/spleen was determined
        using FerWT chemically coupled to SRBC (Sweet
        and Welborn, 1971). Results are expressed as
        mean ± sem of triplicate determinations

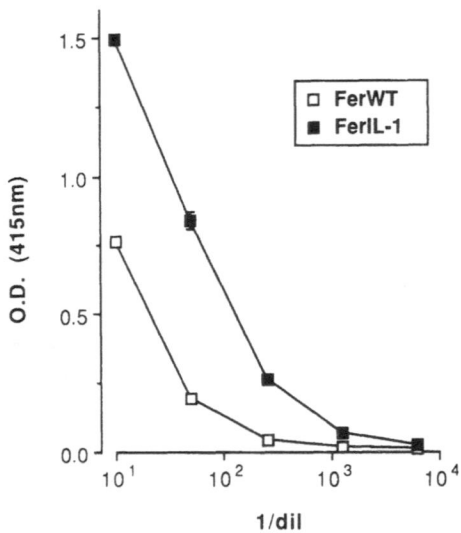

Fig. 3. ELISA anti FerWT. Results are expressed as the
mean ± sd of duplicate determinations. FerWT
(10μg/ml, 0.1ml/well) was absorbed onto wells of
96-well plate (Dynatech) overnight at 4°C. After
saturation with BSA and extensive washings, sera
from immunized animals were added at serial dilutions
and incubated for 2 h at RT. The level of specific
IgG was determined after the addition of alkaline-
phosphatase conjugated goat anti-mouse IgG (Sigma)
for another 2 h at RT. p-Nitrophenyl phosphate was
then added as substrate and the reaction was allowed
to proceed for 30 min in the dark at RT. Absorbance
at 415 nm was determined spectrophotometrically using
a Titertek Multiscan

Correct insertion of the sequence was then verified by sequencing the
mutated expression vector and by positive staining in Western blot of the
purified protein with a VQGEESNDK sequence specific monoclonal antibody
(Boraschi et al., 1989) (data not shown). Wild type or mutated Ferritins
were purified as previously published (Luzzago and Cesareni, 1989).

In order to assess the in vivo immunogenicity of this ferritin mutant
female C3H/HeJ mice (Charles River, Italy) of 4 to 8 weeks of age were
injected intraperitoneally with pyrogen-free saline alone or containing
different amounts of the wild type recombinant protein or mutated contruct.
After 4-5 days the spleens were removed and assayed for the number of
plaque forming cells (PFC) against ferritin wild type (FerWT) coupled to
SRBC according to previously published procedures (Sweet and Welborn, 1971,
Ghiara et al., 1987).

Fig 2 shows the results of an experiment representative of 8 performed. It appears that the Ferritin molecule containing the sequence
VQGEESNDK (FerIL-1) is more immunogenic than the wild type molecule
(FerWT), since it induces a significantly higher number of specific
PFC/spleen at any of the immunization doses used. The enhanced immunogenicity was specifically due to the IL-1 sequence since an unrelated
sequence (GNLLLQTSVV) inserted in the same position was not able to enhance
the immunogenicity of Ferritin (data not shown).

To further assess the enhanced immunogenicity of our Ferritin construct at the level of humoral responses, C3H/HeJ mice were injected i.p. with saline containing 50ug/Kg of FerWT or FerIL-1. Ten days later mice were killed by cervical dislocation and serum was collected and tested by ELISA for the levels of IgG specific for Ferritin wild type. In Fig 3 is reported the result of one experiment representative of the two performed which shows that mice injected with the FerIL-1 had an higher level of specific IgG against wild type Ferritin molecule with respect to animals that received the FerWT as immunizing antigen.

Taken together, the results shown in this report demonstrate that the genetic grafting of the immunostimulatory fragment of IL-1 163-171 (VQGEESNDK) into a recombinant protein, i.e., ferritin H chain, enhances its immunogenicity. The approach of using recombinant proteins with 'built-in' adjuvanticity opens the possibility to design vaccines in which poorly immunogenic proteins are coupled to domains endowed with immunostimulatory properties.

Work is presently in progress in our laboratories to further investigate the immune response to Ferritin with 'built-in' adjuvanticity in vivo, and to extend this approach to other recombinant proteins.

ACKNOWLEDGEMENTS

We thank Dr D Boraschi for helpful discussions and Dr S Abrignani for critical reading of the manuscript. We also wish to thank Mr Corsi for the artwork. The work done in Tor Vergata was supported by the EC BRIDGE program and by the Target project on Biotechnology and Bioinstrumentation of CNR.1.

REFERENCES

Antoni, G., Presentini, R., Perin, F., Tagliabue, A., Ghiara, P., Censini, S., Volpini, G., Villa, L. and Boraschi, D., 1986, A short synthetic peptide fragment of interleukin 1 with immunostimulatory but not inflammatory activity, J.Immunol., 137:3201.

Arosio, P., Adelmann, T.G. and Drysdale, J.W., 1978, On Ferritin deterogenicity. Further evidence for heteropolymers, J.Biol.Chem., 253:4451.

Boraschi, D., Nencioni, L., Villa, L., Ghiara, P., Presentini, R., Perini, F., Frasca, D., Doria, G., Forni, G., Musso, T., Giovarelli, M., Ghezzi, P., Bertini, R., Besedovsky, H.O., del Rey, A., Sipe, J.D., Antoni, G., Silvestri, S. and Tagliabue, A., 1988, In vivo stimulation and restoration of immunoresponse by the non-inflammatory fragment 163-171 of human interleukin 1 beta, J.Exp.Med, 168:675.

Boraschi, D., Volpini, G., Villa, L., Nencioni, L., Scapigliati, G., Nucci, D., Antoni, G., Matteucci, G., Cioli, F. and Tabliabue, A., 1989, A monoclonal antibody to the IL-1 beta peptide 163-171 blocks adjuvanticity but not pyrogenicity of IL-1 beta in vivo, J.Immunol., 143:131.

Dinarello, C.A., 1991, Interleukin-1 and Interleukin-1 antagonism, Blood, 77:1627.

Ghiara, P., Boraschi, D., Nencioni, L., Ghezzi, P. and Tagliabue, A., 1987, Enhancement of in vivo immune response by tumor necrosis factor, J.Immunol., 139:3676.

Jappelli, R., Luzzago, A., Tataseo, P., Pernice, I. and Cesareni, G., 1992, Loop mutations can cause a substantial conformational change in the carboxyl-terminus of the ferritin protein, J.Mol.Biol., in press

Lawson, D.M., Artymiuk, P.J., Yewdall, S.J., Smith, J.M.A., Livingstone, J.
C., Treffey, A., Luzzago, A., Levi, S., Arosio, P., Cesareni, G.,
Thomas, C.D., Shaw, W.V. and Harrison, P.M., 1991, Solving the
structure of human H ferritin by genetically engineering intra-
molecular crystal contacts, Nature, 349:541.

Luzzago, A. and Cesareni, G., 1989, Isolation of point mutation that
affects the folding of H chain of human ferritin in E. coli,
Embo J., 8:569.

Miyazaki, Y., Setoguchi, M., Higuchi, M.Y., Yoshida, S., Akizuki, S. and
Yamamoto, S., 1988, Nucletoide sequence of cDNA encoding the heavy
subunit of mouse macrophage ferritin, Nucl.Acid Res., 16:10373.

Nencioni, L., Villa, L., Tagliabue, A., Antoni, G., Presentini, R., Perin,
F., Silvestri, S. and Boraschi, D., 1987, In vivo immunostimulating
activity of the 163-171 peptide of human IL-1 beta, J.Immunol.,
137:800.

Rao, K.V.S. and Nayak, A.R., 1990, Enhanced immunogenicity of a sequence
derived from hepatitis B virus surface antigen in a composite
peptide that includes the immunostimulatory region from human
interleukin 1, Proc,Natl.Acad.Sci., 87:5519.

Reed, S.G., Phil, D.L. and Grabstein, K.H. 1989, Immune deficiency in
chronic Trypanosoma cruzi infection. Recombinant IL-1 restores Th
function for antibody production, J.Immunol., 142:2067.

Sweet, G.H. and Welborn, F.L., 1971, Use of chromium chloride as the
coupling agent in a modified plaque assay cells producing anti-
protein antibody, J.Immunol., 106:1407.

Staruch, M.J. and Wood, D.D., 1983, The adjuvanticity of interleukin-1
in vivo, J.Immunol., 130:21291.

Tagliabue, A., Antoni, G. and Boraschi, D., 1989, Defining agonist peptides
of human interleukin 1 beta, Lymphokine Res., 8:311.

VACCINATION AGAINST MALARIA; THE ANTI-DISEASE CONCEPT

J.H.L. Playfair, J. Taverne, and Caw Bate

Department of Immunology
UCLMS
London, UK

INTRODUCTION

The prospect of a malaria vaccine seems to be approaching, but more slowly than had been hoped. One problem is the increased evidence for quite rapid and extensive antigenic variation in the blood-stage parasite. Another is the continuing uncertainty as to the exact type of immune response required to eliminate either this stage or the preceding liver stage; at present a range of antibody, cell, and cytokine-mediated responses have been linked to protection in various animal models. And finally, the rather slow development of immunity in those inhabitants of the endemic areas who survive the childhood "danger period" suggest that malaria does not belong with those diseases, like smallpox, measles, polio etc, where elimination of the invading organism is well within the powers of the immune system, provided the initial attack is survived. For all these reasons, we have turned our attention to the symptoms rather than the parasite, asking the question: can the severe complications of malaria be considered as "toxic" manifestations, and if so, are there toxic antigens that could be used as vaccines, in much the same way as tetanus and diphtheria toxins can?

ANTI-TOXIC IMMUNITY

In the early part of this century, malaria was in fact often regarded as toxic disease, and it was recognised that not all patients with the same parasitaemias were equally ill. Thus the simple presence of parasites was not enough to cause symptoms, which must therefore be produced by indirect means. Furthermore it was noted that children could become immune to these toxic effects, and thus suffer less severe symptoms, at an earlier age than they acquired immunity to the parasite (Sinton et al, 1931; Taliaferro, 1949; Hill, 1943). More recent studies in the Gambia confirmed that clinical "tolerance" (defined as "the ability to live asymptomatically with fairly dense parasitaemia") does usually develop by the age of 3-5 despite the persistence of high parasitaemias for up to 5 more years (McGregor et al, 1956). No evidence has been produced that this tolerance was immunological in origin, rather than, for example, a reduced responsiveness to unchanged levels of some stimulating factor, or the presence of some natural blocking factor. However, recent work on the role of cytokines in malaria, reviewed below, does suggest that acquired immunity to symptom-

New Generation Vaccines, Edited by G. Gregoriadis
et al., Plenum Press, New York, 1993

inducing antigens, possibly acting through cytokines such as tumour
necrosis factor (TNF) might be another explanation.

Not all the complications of malaria occur together. Fever, the
hallmark of the disease, is prominent in both P.falciparum and P.vivax
infections, and there is quite good evidence linking it to TNF over-
production (Kwiatkowski et al, 1992; and see below). Anaemia, another
almost universal complication, is particularly seen in the first year of
life, whereas cerebral malaria, the major cause of mortality, tends to
occur from about the second year, and is very seldom seen in P.vivax
infection. Nor is hypoglycaemia, which is frequent in P.falciparum
infection. Thus it is not likely that a single pathogenetic mechanism is
responsible for all the clinical effects of the disease.

TUMOUR NECROSIS FACTOR

However, as Clark was the first to point out (Clark, 1987; Clark et
al, 1989), it can be argued that TNF is involved in almost all the
complications of P.falciparum malaria, since these bear a striking
resemblance to the toxic effects of high-dose TNF administration.
Therefore it is this cytokine that has been most studied so far. There is
a general agreement that raised plasma levels of TNF are associated with
severity and mortality in cerebral malaria (Kern et al, 1989; Grau et al,
1989; Kwiatkowski et al, 1990). In a mouse model resembling cerebral
malaria, anti-TNF antibody protected against cerebral disease but not
against anaemia, but so did antibodies against several other cytokines
(Grau et al, 1988). In a small trial of anti-TNF antibody in human
malaria, a reduction in fever was the only significant effect noted
(Kwiatkowski et al, 1992). Fever and raised TNF levels are also found in
P.vivax infection (Karunaweera et al, 1992), so if TNF is really
responsible for cerebral malaria, some co-factor(s) must be involved too.
These could include differences in the way parasites sequester in deep
tissues, or there may be synergy between TNF and other cytokines or
mediators. Interestingly, additive or synergistic effects are also seen in
the anti-parasitic activity of TNK: killing of gametocytes in vitro appears
to require TNF plus a "complementary factor" (Naotunne et al, 1990). We
therefore chose TNF as a model cytokine and attempted to discover why it
was apparently overproduced in blood-stage malaria.

TNF-INDUCING ("TOXIC") ANTIGENS

In preliminary experiments, using the mouse parasite P.yoelii, we
found that parasitised red cells incubated overnight with peritoneal
macrophages induced levels of TNF comparable to microgram amounts of
bacterial LPS. Lethal and non-lethal strains of the parasite were equally
active. We then found that incubating the parasitised blood overnight
yielded supernatants that themselves induced TNF (Bate et al, 1988). The
human parasite P.falciparum behaved similarly, and either mouse macrophages
or human blood monocytes could be used interchangeably (Taverne et al,
1990). When injected into mice, these supernatants induced TNF sufficient
to be detectable in the serum, and were lethal in mice that had been made
hypersensitive to TNF by pre-treatment with D-galactosamine (Bate et al,
1990). These supernatant factors, while not strictly speaking toxins,
could justifiably be called "toxic".

One obvious possibility was that the TNF induction was due to
contaminating bacterial lipopolysaccharide (LPS), but for reasons to be
summarised later, we believe that this was not the case. However the
active material did resemble LPS in some respects; thus it retained

activity after boiling, digestion with proteases, and deamination by nitrous acid, but was destroyed by lipase treatment, by some types of Phospholipase C, by dephosphorylation with HF, and by deacylation with NaOH. In a two-phase lipid extraction, it appeared in the chloroform/ methanol rather than the water/methanol phase. When injected into normal mice at sub-toxic doses, it induced antibody which blocked its ability to induce TNF both <u>in vitro</u> and <u>in vivo</u>, and was predominantly IgM; we found no evidence for a memory response or a switch to IgG on boosting (Bate et al, 1990). Using this assay, antigens from all the species we have tested so far, including the mouse parasites <u>P.yoelii</u> and <u>P.berghei</u> and the human parasites <u>P.falciparum</u> and <u>P.vivax</u>, show complete cross-blocking - which is in contrast to the species, strain, and variant-specificity of the protein antigens commonly studied (Bate et al, 1992). We concluded that parasitised red cells contain, and can release <u>in vitro</u>, phospholipid molecules that induce TNF and behave as T-independent antigens. We refer to these molecules as "toxic malaria antigens" (Bate et al, 1992). Similar induction of TNF from human macrophages by <u>P.falciparum</u> supernatants has also been reported (Picot et al, 1990).

Further characterisation of the active molecule came from studies in which we used standard phospholipids of the type found in red cell membranes: phosphatidylcholine (PC), phosphatidyl serine (PS), phosphatidyl inositol (PI), and phosphatidyl ethanolamine (PE), as well as phosphatidic acid (PA) and cardiolipin (CL). None of these molecules induced TNF from macrophages, but one of them, PI, completely blocked the ability of exoantigens to induce TNF. Surprisingly, the phosphated monosaccharide inositol monophosphate (IMP) also blocked, and so did preparations of exoantigen treated with either lipase or HF; we refer to these as "detoxified" exoantigens (Bate et al, 1992). At this point we concluded that a phosphate group and an inositol ring were involved in the TNF-triggering part of the exoantigens. This was further supported by the finding that antisera raised by simply immunising mice with PI or IMP, but not the other phospholipids, blocked TNF release. However PC, PS, and PA incorporated into liposomes with or without cholesterol did induce blocking antibody, but this could be absorbed out by PC liposomes, while antibody raised to PI or IMP could not (Bate et al, 1992). From this we concluded that blocking antibodies were of at least two types - one specific for phosphate groups and one also recognising inositol.

RULING OUT ENDOTOXIN

Contamination of laboratory experiments by bacterial LPS is always a possibility, but our toxic antigens appear to differ from LPS in a number of important ways. First, their TNF triggering activity is not blocked by the inclusion of polymyxin B in the cultures, while the blocking antisera, whether against the antigens themselves or against PI or IMP, do not block triggering by LPS. Second, they stimulate macrophages from the LPS-hyporesponsive C3H/HeJ mouse strain. Third, triggering by LPS is not abolished, as that of the exoantigens is, by phospholipase C digestion; contrariwise deamination abolished the activity of LPS but not of the exoantigens. Lastly, neither PI, IMP, or detoxified antigens block triggering by LPS, which may suggest that different macrophage receptors are involved. We therefore feel convinced that LPS alone does not account for our results. However we cannot eliminate the possibility that trace amounts of LPS, insufficient to stimulate on their own, are needed to boost the effects of the toxic antigens, particularly where induction of blocking antibody is concerned (Schuster et al, 1979; Banerji et al, 1990); indeed there would be nothing unphysio-logical if this did occur, since malaria patients probably have increased amounts of circulating LPS (Felton et al, 1980).

Even normal mice, following the injection of toxic antigens, often look ruffled and ill. Investigating this, we found a dramatic drop in blood sugar lasting from 2 until about 8 hours after injection. By comparison with the normal mid-morning blood sugar of 7.6 mmol/L, the 4-hour value averaged 3.4 mmol/L, and values below 2 mmol/L were quite common. Since TNF (and other cytokines) can cause hypoglycaemia, we attempted to inhibit this drop with repeated injections of a monoclonal anti-TNF antibody. Though this antibody protected D-galactosamine pre-treated mice from death, it had no effect on hypoglycaemia. On the other hand, a second and third injection of exoantigen at 2-week intervals induced progressively less hypoglycaemia in normal mice, but this protection was not seen in mice with severe combined immunodeficiency (SCID mice). We concluded that antibody against the exoantigens could protect against hypoglycaemia. In line with the TNF experiments described earlier, we then showed that IMP mixed with the exoantigens blocked their hypo-glycaemic effect, as also did antibody raised against IMP or against a PI-BSA conjugate (Taylor et al, 1992).

The idea that the same molecule(s) induced both TNF and hypoglycaemia is questioned by the fact that we quite often find that exoantigen preparations induce TNF and not hypoglycaemia and vice versa. Some exoantigen preparations do not induce either, and there are still variables in the method of preparation which we have not yet learned to control. Moreover we have occasionally detected low levels of TNF and hypoglycaemia induction with supernatants of normal red cell raising the possibility that the active phospholipids are normal red cell components which are increased in concentration or otherwise modified by the parasite (Hsaio et al, 1991; Vial et al, 1990).

Using rat adipocytes in vitro, we found that malaria exoantigens can act synergistically with small amounts of insulin to promote glucose uptake (manuscript submitted). This tends to support the idea that the hypo-glycaemia the exoantigens induce in vivo is not mediated by TNF.

Anaemia is recognised as one of the most severe complications of malaria in young children, and it is generally felt that it cannot be accounted for purely by red cell parasitisation (Abdalla et al, 1980). In our non-lethal P.yoelii mouse model, the haematocrit falls to 20% of normal in 10 days, despite the fact that the parasite is restricted to reticulo-cytes. The same is true in SCID mice, ruling out a major role for antibody. From about day 8 of infection, 51Cr-labelled normal red cells are eliminated from the circulation unusually rapidly. Neither this effect, nor the anaemia itself, are prevented by prior immunization with toxic antigens, nor by anti-TNF antibody, though TNF has been shown to depress erythropoiesis (Johnson et al, 1989). We are continuing to investigate this important aspect of the infection.

AN IN VITRO ARTEFACT?

Clearly, potentially toxic molecules are released by parasitised red cells in vitro, but we needed to establish whether they were actually released during the infection. We do not detect TNF in the blood of infected mice, at least using the bioassay, nor do infected mice develop hypoglycaemia until about 1 day before they die; this must argue against a significant release of exoantigens in vivo. However serum from infected mice does contain antibodies that block TNF induction by exoantigens, and also antibodies that bind PI and other phospholipids, which suggests that the immune system has been exposed to exoantigens. This discrepancy might be resolved if antibody produced early during infection was able to block both TNF induction and hypoglycaemia, and we are undertaking a detailed study in SCID mice to test this. Another testable explanation is that normal serum contains non-antibody blocking factors that would be capable of binding and neutralising exoantigens until their concentration became overwhelming, analogous to the lipoproteins that have been reported to inactivate LPS (Emancipator et al 1992). Supporting this idea, we have found that normal mouse serum does in fact block TNF induction by exoantigens, but only at dilutions up to about 1/20; by comparison the titres of antibody in immunised mice frequently exceed 1/10000. Acute phase serum induced by oil/turpentine did not have increased blocking activity. One molecule that could fill the role of phospholipid neutralisation might be the normal serum protein β2 glycoprotein 1 (also known as apolipoprotein H), which as well as its effects on coagulation is thought to bind phospholipids and contribute to their antigenicity (McNeil et al, 1990); perhaps it also blocks their ability to trigger macrophages. We plan to test this as well as the other apolipoproteins for their ability to block the induction of TNF.

PROTECTIVE IMMUNITY AND VACCINATION

Returning to the question of vaccination, we can postulate that antibody against the toxic antigens, by blocking TNF, hypoglycaemia, and perhaps other pathological complications, might prevent disease without necessarily affecting the parasite itself. This would be an "anti-disease" vaccine potentially as useful as those against tetanus and diphtheria. (Balb/c x C57BI)F1 hybrid mice have been immunised with exoantigens and challenged with the uniformly lethal P.yoelii 13 days later, and about half recovered after about 3 weeks with very high parasitaemias - in some mice over 80% - but without looking particularly ill. This is, of course, how "anti-disease" immunity would be expected to perform. Similar protection has been obtained in smaller groups of mice immunised with PI and IMP. Neither repeated boosting nor the use of adjuvants improves this level of protection, and though preliminary experiments with exoantigens coupled to protein carriers such as bovine serum albumin and keyhole limpet haemocyanin have yielded substantial increases of blocking antibody, some of which is IgG, the level of actual protection remains the same (Bate et al, in preparation).

We do not understand why the protection, even in genetically homogeneous mice, is only partial. It may simply reflect the fact that our P.yoelii model is not ideal for this type of protection study. For example, as noted above, both vaccinated and non-vaccinated mice become extremely anaemic during infection, and it may be that mice are dying of anaemia though "protected" against other aspects of pathology. This is typical of the difficulties encountered when trying to study immunity in such a multifactorial disease. However a somewhat similar partial protection with exoantigens has also been reported in a P.falciparum model in squirrel monkeys (Ristic and Kreier, 1984), and in cattle with babesiosis (James, 1989).

A HUMAN EQUIVALENT?

We propose that the "anti-toxic" immunity observed in human populations is due to the development of immunity against toxic antigens, probably mediated by antibody which blocks their ability to induce TNF, hypoglycaemia etc. If true, this leads to certain predictions. "Tolerant" children with high parasitaemia but mild or no symptoms ought to have blocking antibody in their blood, and this can clearly be tested. In

immune adults who leave the endemic areas, clinical immunity (IgM?) might
be expected to wane more rapidly than immunity to the parasite itself
(IgG?), which does often seem to be the case. The conserved nature of the
toxic antigens should allow anti-toxic immunity to show a wider degree of
cross-protection than anti-parasite immunity, but this is still
controversial (Jeffery, 1966). Most important, antibody against the
relevant antigens ought to correlate with clinical protection. One such
antigen, a TNF-inducing molecule found in P.falciparum culture supernatants
and originally named by its discoverers Antigen 7, fits the latter
criterion, since antibody levels in Gambian children peak at the age of
about 5, which is when clinical immunity develops; antibody to most other
exoantigens peaks considerably later (Jakobsen et al, 1991). Moreover, T
cell responses to Antigen 7, assayed by IFN$_\gamma$ release, correlate with
clinical severity, which might be important in view of the synergy between
IFN$_\gamma$ and TNF-inducing molecules (38). This kind of analysis is further
complicated by the fact that several of these antibodies also seem to
affect parasitaemia, which by reducing the amounts of toxic antigens, could
obviously obscure the "anti-disease" effect. Careful studies of TNF and
other cytokine responses to a range of exoantigens, in the presence of sera
from both immune and non-immune patients, will be necessary before the
existence or otherwise of true "anti-toxic" immunity in man can be
established.

VACCINATING HUMANS

Whether or not "anti-disease" immunity occurs naturally, the ability
to induce it would be extremely valuable. Compared to the candidate
vaccines already under study, the anti-disease approach has both advantages
and disadvantages. In its favour is the fact that it aims to deal with
what really matters - disease. Also the apparent lack of antigenic
variation noted in the mouse experiments might suggest that a whole series
of species and variant-specific antigens may not be needed. Moreover, by
acting against cytokine-inducing antigens rather than against the cytokines
themselves, as, for example, antibody to TNF or soluble TNF receptors do,
it might exert an effect on other cytokines induced at the same time, such
as IL-1 or IL-6 and thus avoid the need for a complex "cocktail" of
inhibitors. Against it is the lack of T-dependent antibody, with memory,
IgG etc, in the mouse experiments which, if it also applies to the human,
would mean a vaccine injection every few weeks! Another theoretical danger
is that the phospholipid antigens responsible for disease might turn out to
be genuine autoantigens, or at least to cross-react with host molecules;
the vaccine would then induce autoimmunity. Preliminary results (Bate et
al, unpublished) suggest that conjugation to protein carriers such as
bovine serum albumin or keyhole limpet haemocyanin does in fact enable
phospholipids to induce very high titres of IgG, without visible ill-
effects to the mice. At present, then, we visualise an eventual vaccine
consisting of the relevant malaria-derived phospholipid coupled to a
protein - preferably also derived from the parasite. Time will tell
whether this will indeed achieve the goal of disease prevention.

Acknowledgements

The work from our laboratory described in this chapter was supported
by grants from the Medical Research Council of Great Britain and the
European Community.

REFERENCES

Abdalla, S., Weatherall, D.J., Wickramasinghe, S.N. and Hughes, M., 1980,
      The anaemia of P.falciparum malaria. Brit.J.Haematol., 46, 171.

Banerji, B. and Alving, C.R., 1990, Antibodies to liposomal phosphatidylserine and phosphatidic acid, Biochem.Cell Biol., 68, 96.

Bate, C.A.W., Taverne, J. and Playfair, J.H.L., 1988, Malarial parasites induce TNF production by macrophages, Immunology, 64, 227.

Bate, C.A.W., Taverne, J. and Playfair, J.H.L., 1989, Soluble malaria antigens are toxic and induce the production of tumour necrosis factor in vivo, Immunology, 66, 600.

Bate, C.A.W., Taverne, J., Dave, A. and Playfair, J.H.L., 1990, Malaria exoantigens induce T-independent antibody that blocks their ability to incude TNF, Immunology, 70, 315.

Bate, C.A.W., Taverne, J., Karunaweera, N.D., Mendis, K.N., Kwiatkowski, D. and Playfair, J.H.L., 1992, Serological relationship of tumour necrosis factor-inducing exoantigens of Plasmodium falciparum and Plasmodium vivax, Infection and Immunity, 60, 1241.

Bate, C.A.W., Taverne, J., Roman, E., Moreno, C. and Playfair, J.H.L., 1992, Tumour necrosis factor induction by malarial exoantigens depends on phospholipid, Immunology, 75, 129.

Bate, C.A.W., Taverne, J. and Playfair, J.H.L., 1992, Detoxified exoantigens and phosphatidylinositol derivatives inhibit tumor necrosis factor induction by malarial exoantigens, Infection & Immunity, 60, 1894.

Bate, C.A.W., Taverne, J., Bootsma, H.J., Mason, R.C. StH, Skalko, N., Gregoriadis, G. and Playfair, J.H.L., 1992, Antibodies against phosphatidylinositol and inositol monophosphate specifically inhibit tumour necrosis factor induction by malaria exoantigens, Immunology, (in press).

Bate, C.A.W., Taverne, J., Kwiatkowski, D. and Playfair, J.H.L., (in preparation), Malaria exoantigens coupled to carriers induce IgG antibody that blocks their ability to induce TNF.

Clark, I.A., 1987, Cell mediated immunity in protection and pathology of malaria, Parasitol.Today, 3, 300.

Clark, I.A., Chaudri, G. and Cowden, W.B., 1989, Roles of tumour necrosis factor in the illness and pathology of malaria, Trans.Roy,Soc.Trop,Med.Hyg., 83, 436.

Emancipator, K., Csako, G. and Elin, R.J., 1992, In vitro inactivation of bacterial endotoxin by human lipoproteins and apolipoproteins, Infection and Immunity, 60, 596.

Felton, S.C., Prior, R.B, Spagna, V.A. and Kreier, J.P., 1980, Evaluation of Plasmodium berghei for endotoxin by the limulus lysate assay, J.Parasitol., 66, 846.

Grau, G.E., Taylor, T.E., Molyneux, M.E. et al, 1989, Tumor necrosis factor and disease severity in children with falciparum malaria, N.Eng.J.Med., 320, 1586.

Grau, G.E., Fajardo, L.F., Piguet, P.F, Allet, B., Lambert, P.-H. and Vassali, P., 1988, Tumor necrosis factor (cachectin) as an essential mediator in murine cerebral malaria, Science, 237, 1210.

Hill, R.B., 1943, Am.J.Trop.Med.Hyg., 23, 147.

Hsaio, L.L., Howard, R.J., Aikawa, M. and Taraschi, T.F., 1991, Modification of host cell membrane lipid composition by the intra-erythrocytic human malaria parasite Plasmodium falciparum, Biochem.J., 274, 121.

Jakobsen, P.H., Riley, E.M., Allen, S.J., Larsen, S.O., Bennett, S., Jepsen, S. and Greenwood, B.M., 1991, Differential antibody response of Gambian donors to soluble Plasmodium falciparum antigens, Trans.Roy.Soc.Trop.Med.Hyg., 85, 26.

James, M.A., 1989, Application of exoantigens of Babesia and Plasmodium in vaccine development, Trans.Roy.Soc.Trop.Med.Hyg., 83, 67.

Jeffrey, G.M., 1966, Epidemiological significance of repeated infections with homologous and heterologous strains and species of Plasmodium, Bull.Wld.Hlth.Org., 35, 873.

Johnson, R.A., Waddelow, T.A., Caro, J., Oliff, A. and Roodman, G.D., 1989, Chronic exposure to tumour necrosis factor in vivo preferentially inhibits erythropoiesis in nude mice, Blood, 74, 130.

Karunaweera, N.D, Grau, G.E., Gamage, P., Carter, R. and Mendis, K.N., 1992, Dynamics of fever and serum TNF lelvels are closely associated during clinical paroxysms in P.vivax malaria, Proc.Nat.Acad.Sci,USA, in press.

Kern, P.C., Hemmer, J., Van Damme, J., Gruss, H.-J. and Dietrich, M., 1989, Elevated tumor necrosis factor and interleukin 6 serum lavels as markers for complicated Plasmodium falciparum malaria, Amer.J.Med., 57, 139.

Kwiatkowski, D., Molyneux, M.E., Stephens, S., Curtis, N., Klein, Pointaire, P., Smit, M., Allan, R., Brewster, D.R., Grau, G.E., and Greenwood, B.M., 1992, Anti-TNF therapy in the treatment of childhood cerebral malaria, Submitted.

Kwiatkowski, D., Hill, A.V.S., Sambou, I, et al, 1990, TNF levels in fatal cerebral, non-fatal cerebral, and uncomplicated Plasmodium falciparum malaria, Lancet, 336, 1201.

McGregor, I.A., Gilles, H.M., Walters, J.H., Davies, A.H. and Pearson, F.A., 1956, Effects of heavy and repeated malarial infections on Gambian infants and children, Br.Med.J., 686.

McNeil, H.P., Simpson, R.J., Chesterman, C.N. and Krilis, S.A., 1990, Antiphospholipid antibodies are directed against a complex antigen that includes a lipid-binding inhibitor of coagulation: 2 glycoprotein 1 (apolipoprotein H), Proc.Natl.Acad.Sci.USA, 87, 4120.

Naotunne, T.Des, Karunaweera, N.D., Del Giudice, G., Kularatne, M., Grau, G.E., Carter, R. and Mendis, K.N., 1990, Cytokines kill malaria parasites during infection crisis: extracellular complementary factors are essential, J.Exp.Med., 17, 523.

Picot, S., Peyron, F., Vuillez, J.-P., Barbe, G., Marsh, K. and Ambroise-Thomas, P., 1990, Tumor necrosis factor production by human macrophages stimulated in vitro by Plasmodium falciparum, Infection and Immunity, 58, 214.

Riley, E.M., Jakobsen, P.H., Allen, S.J., Wheeler, J.G., Bennett, S., Jepsen, S. and Greenwood, B.M., 1991, Eur.J.Immunol., 21, 1019.

Ristic, M. and Kreier, J.P., 1984, in: Malaria and Babesiosis: Research Findings and Control Measures, Ristic M, Ambroise-Thomas, P. Kreier, J.P. Martinus, Nijhoff, eds, pp 3-33.

Schuster, B.G., Neidig, M., Alving, B.M. and Alving, C.R., 1979, Production of antibodies against phosphocholine, phosphatidyl choline, sphingomyelin, and lipid A by injection of liposomes containing lipid A, J.Immunol., 122, 900.

Sinton, J.A., Harbhagwan, and Singh, J., 1931, The numerical prevalence of parasites in relation to fever in chronic benign tertian malaria, Ind.J.Med.Res., 18, 871.

Taliaferro, W.H., 1949, Immunity to the malaria infections, in: MF Boyd, W.B. Saunders (eds) Malariology, Philadelphia.

Taverne, J., Bate, C.A.W, Sarkar, D.A., Meager, A., Rook, G.A.W. and Playfair, J.H.L., 1990, Human and murine macrophages produce TNF in response to soluble antigens of Plasmodium falciparum, Parasite Immunol., 12, 33.

Taylor, K., Bate, C.A.W, Carr, R.E., Butcher, G.A., Taverne, J. and Playfair, J.H.L., 1992, Phospholipid-containing malaria exoantigens induce hypoglycaemia, Clin.Exp.Immunol., 90, 1-5.

Vial, H.J., Ancelin, M.-L., Philippot, J.R. and Thuet, M.J., 1990, Biosynthesis and dynamics of lipids in Plasmodium-infected mature mammalian erythrocytes, Blood Cells, 16, 531.

PROGRESS IN THE DEVELOPMENT OF EPSTEIN-BARR VIRUS VACCINES

Andrew J Morgan

Department of Pathology & Microbiology
University of Bristol Medical School
Bristol, BS8 1TD, UK

INTRODUCTION

Epstein-Barr virus (EBV) is one of the small number of human herpes-viruses (reviewed by Kieff and Liebowitz, 1990; Miller, 1990). These are complex viruses with large double-stranded DNA genomes, the EBV gene being approximately 170 Kd and coding for at least 70 genes. A prominent characteristic of these viruses in this group is that they are able to persist in a latent infectious state in humans in the face of sustained and comprehensive immune response which include virus-neutralising antibodies in the sera of infected individuals. EBV infects greater than 90% of the human population world-wide, infection usually occurring early in child-hood, but in the Western world infection is frequently delayed until adolescence when there is a 50% chance of infectious mononucleosis occurring. EBV can infect only B lymphocytes and certain classes of epithelial cell by binding to the complement receptor CR2 of B lymphocytes and a similar molecule on epithelial cells (Hutt-Fletcher, 1991). The mode of infection is by horizontal transmission through an oral portal of entry leading to infection of epithelial cells in the oropharynx. It is here that the virus is able to reside in a state of productive infection generating infectious virus in the saliva which can infect other individuals. B lymphocytes in the circulation passing in the region of these infected epithelial cells become infected also but only a very small number ever progress to productive infection. Thus it has been proposed that the virus persists in a dynamic equilibrium where infected B cells are removed from the circulation by specific T cell responses and what infectious virus is produced in the circulation is removed by virus-neutralising antibodies but the supply of newly infected B cells is maintained by virtue of the fact that virus can replicate and infect B cells in the oropharyngeal epithelium.

Where the immune system is impaired as in the case of AIDS patients and organ transplant recipients undergoing immunosuppressive therapy, the normal control of EBV infection is reduced. In these individuals there is a very high incidence of B cell lymphoma which appears to originate from EBV-infected lymphocytes (Nalesnik, 1991). Those tumours that occur in organ transplant recipients frequently regress when the immunosuppressive regime is reduced or removed. A rare and invariably fatal disorder, known as the Duncan syndrome, is an X-linked lymphoproliferative disease arising in boys and is due to a specific defect in the cell-mediated response to

Epstein-Barr virus-infected cells (Purtilo, 1991). It can be argued that EBV vaccine development is justifiable to prevent or reduce the above EBV associated conditions. However in numerical and world-wide terms they are nowhere near as important as undifferentiated nasopharyngeal carcinoma (NPC). NPC is a major world health problem with more than 80,000 new cases per year being reported (Parkin et al, 1984). Indeed NPC is the commonest cancer in men and the second most common in women in S. China. The apparent association of Epstein-Barr virus (EBV) with NPC led to the proposal that a vaccine should be developed to prevent primary EBV infection (Epstein, 1976). Although the distinct geographical distribution of NPC requires that factors other than the virus must be involved, it is thought that vaccination to prevent primary infection could intervene to disrupt the sequence of events leading to the development of this tumour. Cofactors in the causation of NPC have not yet been identified (Chen et al, 1990). There does appear to be a predisposition to the disease within certain major histocompatibility types (Lu et al, 1991). A strong association also exists between EBV infection and endemic Burkitt's lymphoma (BL)(McGrath, 1990), the tumour in which the virus was first discovered (Epstein et al, 1964) and clearly holoendemic malaria is one of the cofactors (Facer and Playfair, 1989).

That there is an association between EBV, BL and NPC is indicated because viral gene products are always found in properly authenticated tumour cells and that there is a characteristic serum antibody profile against EBV antigens at the onset of NPC and prior to and during the course of endemic BL (Ooka et al, 1991). These observations when taken with the knowledge that EBV can immortalise B cells and directly induce lymphoma in certain New World primates give strong evidence of an association but are not proof. As is the case with other tumours in humans associated with virus infection, proof of a causative relationship will only be established when the reduction in the incidence of, or elimination of the tumour in susceptible populations is achieved with an effective vaccine against the virus in question. More recently certain Hodgkins' lymphomas have been found to contain EBV DNA and to express some of the EBV latent proteins (Palleson et al, 1991). Furthermore there is a limited correlation between serum antibody profiles against EBV and the likelihood of Hodgkins' lymphomas developing later in life (Muller et al, 1989). Hodgkin's lymphoma is a major tumour in the Western world and the latter observations should provide a further incentive for EBV vaccine development.

ANIMAL MODEL OF EBV-INDUCED LYMPHOMA

An animal model is an essential pre-requisite in the development of a safe human vaccine. EBV, when given in a sufficiently high dose induces malignant lymphoma in the cottontop tamarin (Saguinus oedipus oedipus) and the owl monkey (Aotus trivirgatus) (Shope et al, 1973; Epstein et al, 1975). Tumours induced in the cottontop tamarin by EBV have been examined closely (Cleary et al, 1985) and resemble the large cell lymphomas which occur with high frequency in organ transplant patients undergoing immuno-suppression. Multiple tumours arise in a number of lymph nodes within 14-21 days, each tumour being mono- or oligoclonal in origin. Unlike NPC and BL the tamarin tumours express all the latent EBV antigens (Young et al, 1988). The expression of lytic phase viral antigens is not observed until tumour cells are grown in tissue culture when a small percentage enter the productive phase of infection. This animal model is not ideal since the primary characteristics of human EBV infection are an oral portal of entry and persistence in the face of a broad-ranging humoral and cell-mediated response. Tamarins do not sustain a persistent infection, exactly as seen in humans, nor have they been infected by the oral route. Furthermore this animal exhibits only a limited major histocompatibility complex class 1 polymorphism (Watkins et al, 1988). The common marmoset (Callithrix jaccus) can develop an ill-defined mononucleosis syndrome (Emini et al, 1989) but no evidence of persistent EBV infection has been obtained. Despite the cottontop tamarin being less than an ideal model for human infection and disease, the induction of lymphoma by a large dose of virus is completely reproducible and could be considered as representing a "worst-case" infection and in this respect is a useful measure of the efficacy of any experimental vaccination.

SUBUNIT VACCINES

That vaccination can prevent herpes virus induced tumours in animals was demonstrated some time ago with Marek's disease of chickens and herpes saimiri in non-human primates using crude cell membrane preparations containing viral glycoproteins (Kaaden and Dietzchold, 1974: Laufs and Steinke, 1975). Analysis of the cell membrane composition of lympho-blastoid cell lines infected with EBV reveals that the major EBV envelope glycoprotein is a molecule of 340 Kd of which 50% is carbohydrate (North et al, 1980; Morgan et al, 1984; Beisel et al, 1985). The gp340 molecule was shown to induce virus-neutralising antibodies and to be the ligand by which the virus attaches to the host cell through the complement receptor CR2 (Tanner et al, 1988). Other viral envelope glycoproteins have been identified but whether it will be necessary to include one or more in a human vaccine remains to be determined. Gp85 in particular has an essential function in the fusion of the viral envelope with the host cell membrane at the point of infection (Haddad and Hutt Fletcher, 1989).

Authentic gp340 purified from infected cells grown in bulk culture (David & Morgan, 1988; Epstein et al, 1985) was found to induce protective immunity in the tamarin against a lymphomagenic dose of EBV when presented in artificial liposomes (Epstein et al, 1985), immunostimulating complexes (Morgan et al, 1988) and with a threonyl muramyl dipeptide (MDP) adjuvant formulation (Morgan et al, 1989). Curiously when a natural gp340 product was isolated using monoclonal antibody affinity chromatography and used to vaccinate tamarins, virus neutralising antibodies were induced to a high titre but the animals were completely unprotected against a lymphomagenic challenge with EBV (Epstein et al, 1986). No satisfactory explanation has yet been put forward for this observation other than that antibody response as a whole may not be involved in protective immunity at all in this model and that it is the induction of cell-mediated immune responses that is important in protective immunity in this case (see below).

The production of gp340 as a natural product is a very expensive and difficult procedure and is wholly unsuitable for the production of material for human trials and beyond. An additional complication is the possible contamination with the potentially oncogenic EBV DNA in the final product. Consequently a number of laboratories have expressed the gp340 gene in bacterial (Beisel et al, 1985; Pither et al, 1992a and b), yeast (Schultz et al, 1987) and mammalian (Emini et al, 1988; Whang et al, 1987; Motz et al, 1987; Madej et al, 1991) expression systems. The mammalian cell products are antigenically very similar to the authentic molecule in that they induce virus-neutralising antibodies and are also recognised by a range of monoclonal antibodies made against the authentic molecule (Madej et al, 1992). It has been necessary to remove the membrane anchor sequence from the gp340 gene to allow the molecule to be secreted in culture because when the protein is expressed in a membrane bound form it appears to be cytotoxic for the cells expressing it. Substantial quantities of gp340 can now be produced in conditions compatible with good manufacturing practice and in the certain absence of any potentially oncogenic EBV DNA. It is hoped that phase I human trials will begin in the near future using such material. It has already been established that this genetically engineered, secreted gp340 made in a bovine papillomavirus expression system induces protective immunity in the tamarin when used in conjunction with a threonyl MDP adjuvant formulation (Finerty et al, 1992).

The question of which adjuvant is to be used in a Phase I human trial remains to be answered because of commercial and regulatory difficulties. Alum is the only adjuvant acceptable for general human use at present and is a weak adjuvant when compared with immuno-stimulating complexes (Iscoms) (Morein et al, 1990) or MDP derivatives (Byars and Allison, 1987; Byars et al, 1991). Similar adjuvants to the latter are undergoing initial human trials with herpes simplex virus vaccines but are not yet generally available. Protection experiments are currently in progress in the cotton-top tamarin to evaluate the use of alum. Cell-mediated immune responses appear to be important in protection against EBV-induced lymphoma in the tamarin model and whether or not alum can induce the appropriate T cell immune responses remains to be seen. Recently, this adjuvant has been shown to produce T cell immune responses against influenza in mice (Dillon et al, 1992). Biodegradable microparticles as controlled antigen delivery systems provide an attractive alternative to the above but have yet to be tested in the tamarin lymphoma animal model (O'Hagan, 1991).

RECOMBINANT VIRUS VECTORS

The disadvantages of using subunit vaccines are that not only will they require an effective adjuvant which is acceptable to regulatory authorities but they may induce only limited immune responses and the particular type of immune responses induced may not be appropriate for the viral infection in question. While the adjuvants so far tested in the tamarin model of EBV lymphoma have been effective it does not necessarily follow that this will be the case in humans. These factors have to be weighed against the important advantage of using a pure and well-defined biologically non-replicative material. In general, the induction of specific antibody responses alone might not be expected to give protection against infection or a reduction in the level of infection except with a few viruses. Furthermore, the relative importance of systemic or mucosal immunity in the prevention of infection by particular viruses may be quite variable. The induction of cell-mediated immune responses is essential to enable good levels of specific antibody to be made and perhaps more importantly for the generation of specific cytotoxic T cells and the laying down of immunological memory.

In order to proceed with human trials of a gp340 subunit vaccine it will be necessary to either demonstrate that gp340 with alum can induce protective immunity in the tamarin model or for Iscoms and MDP formulations to be fully approved by the regulatory authorities. In any event should these adjuvants prove acceptable to the regulatory authorities and effective in human EBV vaccination, it seems unlikely that they will be as effective in the induction of appropriate immune responses in vaccination as live virus vectors. While there are apparent disadvantages in the use of live virus vectors, they offer possible solutions to the major problems in vaccination. Live recombinant virus vectors are likely to induce long-term, broad-ranging immune responses of an appropriate kind and may only need to be given once. They will be inexpensive to produce and could, in the case of vaccinia and adenovirus vectors, be given orally. Furthermore recombinants can be made expressing a range of antigens necessary to protect against a number of different viral infections.

In the tamarin model of EBV-induced lymphoma only the relatively virulent WR strain vaccinia gp340 recombinant was effective and not the recombinant derived from the Wyeth vaccine strain (Morgan et al, 1988), indicating that attenuated recombinants must be constructed in such a way as to retain their immunogenicity. Such vaccinia recombinants are becoming available where attenuation has been achieved by the precise deletion of a number of specific open reading frames from the vaccinia viral genome but where the ability to induce strong immune responses is retained (Tartaglia et al, 1992). Other pox virus vectors have been shown to be effective in inducing immune responses where attenuation is achieved in mammalian cells by virtue of the fact that the pox virus in question is an avian canarypox virus and replicates poorly in human cells (Taylor et al, 1992).

Adenovirus gp340 recombinants have been made and induce complete protective immunity in the tamarin, when inoculated by intramuscular injection, against EBV-induced disease. These recombinants are derived from a non-oncogenic strain of adenovirus (Strain 5) and are replication-defective following the removal of the El region (Ragot et al, 1992a & b). It would appear that replication per se of viral vaccine vectors may not be necessary to induce appropriate and effective immune responses since the replication-defective adenovirus, recombinant avian pox viruses and specifically attenuated vaccinia cannot replicate to any extent in mammalian cells yet induce effective immune responses. Examination of the tissues of animals vaccinated with replication-defective adenovirus reveals the expression of the incorporated foreign viral antigen for at least three months after immunisation (T. Ragot, personal communication). The safety margin in the use of such vectors would appear to be high and may now allow them to be considered for human use. The problem remains, however, that replication-defective adenovirus recombinants must be propagated in transformed cell lines which provide a helper function for the deleted El region of the virus. The possibility of constructing replication-defective virus vectors such that they contain a number of antigens form important disease-causing viruses is attractive. Furthermore it would appear that both replication-defective vaccinia and adenovirus recombinants could be given orally. Indeed vaccinia rabies virus glycoprotein recombinants are being successfully used in the wild in bait for several species. It should be remembered also that very large numbers of people in the US Armed Services received live adenovirus vaccines orally without recorded unfavourable side-effects. Nasal immunisation of rabbits with a replication-defective adenovirus gp340 recombinant generated serum antibodies to gp340 (Ragot at al, 1992a). Significant levels of serum antibody against adenovirus proteins are made in tamarins following multiple nasal inoculation but only low levels of gp340 antibodies are detectable by Western blotting. Whether this route of inoculation will be effective in the tamarin lymphoma model is being tested at the present time.

CELL MEDIATED IMMUNE RESPONSES TO GP340

Studies in humans on the cell-mediated immune response to EBV have been extensive but have almost exclusively concentrated on responses to latent viral antigens (Moss et al, 1991). This is perhaps not surprising since 99% of EBV infected B cells in the human circulation are only latently infected and do not express any viral genes associated with productive infection. This does not mean, however, that cell-mediated immune responses to productive infection antigens are not important since infection must proceed via the productive phase and the cells supporting this essential phase of the EBV life cycle could be key targets for cytotoxic T cells. Specific T cell responses to gp340 have been detected in normal seropositives and gp340 itself can stimulate the production of T cells which are capable of inhibiting virus-induced transformation in vitro (Bejarano et al, 1991). Furthermore, it has been possible to map several T cell epitopes in the gp340 molecule with the aid of synthetic peptides and gp340-specific T cell clones prepared from normal seropositives (Ulaeto et al, 1988, Wallace et al, 1991).

Whatever the role of T cell immune responses in normal human seropositives and in those individuals who succumb to EBV-related disease, it seems likely that T cell immune responses generated by gp340 following vaccination will be important in affording adequate protection against infection. Although there is evidence in the case of the tamarin lymphoma model that immune protection is cell-mediated it does not automatically follow that this will be the case in humans since, as has been stated above, the tamarin model of EBV infection differs in two important respects from infection in humans. Humans sustain a lifelong persistent infection and are infected by oral, horizontal transmission of the virus. It will be important to investigate and understand T cell immune responses in humans following any EBV vaccination with either subunit or live recombinant viral vectors. The relative importance of antibody or cell-mediated immune responses following vaccination of course cannot be determined until trials take place in humans themselves. Studies on cell-mediated immune responses in the tamarin model are extremely limited primarily for the practical reason that only very small volumes of blood can be obtained and that so few animals are available for experiments. The issue is further complicated by the fact that the tamarin has a limited major histocompatibility complex Class 1 polymorphism and human immunological T cell marker reagents do not crossreact with the tamarin analogues.

In the case of the EBV lymphoma model in the cottontop tamarin it appears that protective immunity against the lymphoma is afforded by T cells and that antibodies, whether they be virus-neutralising or not, may make little or no contribution to protection. The evidence to support this view comes from several experiments using live recombinant virus vectors expressing the EBV envelope glycoprotein gp340. Gp340 vaccinia recombinants were made both from the virulent WR strain and the attenuated vaccinia strain Wyeth. Firstly, only the virulent strain induced protection in tamarins against EBV-induced lymphoma but no detectable antibodies were generated against gp340 itself. High levels of antibodies were generated against the components of the vaccine vectors themselves, the level of which reflected the virulence of the strain from which the recombinant had been made (Morgan et al, 1988). Secondly, when tamarins were immunised with subunit gp340 purified by monoclonal antibody affinity chromatography, high levels of virus-neutralising antibody were generated but the animals were completely unprotected against EBV challenge (Epstein et al, 1986). Thirdly, when adenovirus recombinants expressing gp340 have been used in the tamarin model, complete protection against EBV-induced lymphoma was achieved and specific gp340 antibodies were generated but these were not virus-neutralising in vitro in contrast to antibodies obtained in rabbits immunised with the same adenovirus gp340 vector (Ragot et al, 1992a & b). It may therefore be that when Iscoms and MDP formulations have been used with gp340 to give protective immunity in the tamarin that this protection was the result of the induction of cell-mediated immune responses and was not dependant on the high levels of antibody against gp340 that were induced.

ARTIFICIAL VACCINES

Efforts have been made to map B and T cell peptides on the gp340 molecule with the long term aim of identifying peptides which are important in inducing protective immune responses in humans. The aim will be to produce a synthetic peptide vaccine which would contain peptides which are functionally important in a protective immune response and these may include both T and B cell epitopes. The gp340 molecule is large and at least 50% of its mass is carbohydrate (Morgan et al, 1984). The contribution to the antigenic profile of this molecule by the carbohydrate portion has not been investigated but may well have some significance. When the gp340 gene is expressed in yeast the carbohydrate moiety of the glycoprotein is radically different from that of the authentic molecule expressed in mammalian cells to the extent that immunisation of rabbits with this material generates specific antibodies which do not neutralise EBV in vitro and show limited crossreactivity with antibodies against the authentic molecule (Emini et al, 1988). Yeast or bacterial products may well be reconsidered should specific antibody induction not be important in protective immunity and the appropriate T cell responses can be obtained with an effective adjuvant.

The gp340 gene has been expressed as fragments in β-galactosidase fusion proteins in bacteria where of course no carbohydrate moieties are attached to the gene product. These fragments were used in Western blot analysis to detect linear epitopes recognised by antibodies in normal seropositive individuals and individuals suffering from EBV-induced disease (Pither et al, 1992a). A number of immunodominant linear epitopes were identified and were recognised equally by serum antibodies from healthy individuals and patients suffering from BL and NPC. None of these linear epitopes bound antibodies which can neutralise EBV in vitro with the possible exception of epitopes located at the N-terminal end of the molecule (Pither et al, 1992b). This observation is consistent with the fact that the virus receptor binding region is located in the extreme amino terminal end of the gp340 molecule. It would appear that the bulk of the epitopes which induce virus neutralising antibodies are confined to discontinuous and other conformation-dependent regions of the gp340 molecule.

Bacterial fusion proteins containing overlapping fragments covering the whole length of the gp340 molecule have also been used to locate regions containing T cell epitopes based on their capacity to induce proliferative responses in gp340-specific T cell clones. Only limited studies have been carried out here using the few available gp340 specific clones isolated from normal seropositive donors. The bacterial fusion protein approach itself offers a quicker method for mapping T cell epitopes which are capable of inducing a proliferative response in gp340 specific T cells. A great deal more work needs to be carried out to map functionally important epitopes in the gp340 molecule and the apparent importance of T cell responses in the tamarin lymphoma model and the likelihood that T cell responses will be very important in protecting humans against infection underscore this objective. A synthetic peptide-based artificial EBV vaccine is therefore not likely to become a reality for a number of years.

CONCLUSION

Several strong candidates for subunit EBV vaccines based on the major envelope glycoprotein gp340 have been developed and can be produced from recombinant sources in mammalian cells. It remains, however to determine which adjuvant should or can be used for Phase I human trials of this subunit vaccine. Although the tamarin lymphoma model for EBV lymphoma has been very useful, its shortcomings make it very difficult to speculate on the outcome of human vaccination with gp340 products. The relevant importance of cell-mediated and humoral immune responses and whether systematic or mucosal immunity will be important will only be determined in human vaccine trials, as will the route of inoculation, frequency and size of dosage. The desirability of inducing mucosal immune responses in the form of IgA antibodies in the oropharynx has been brought into question as it has been found that EBV can enter and infect normally refractory epithelial cells via the secretory component-mediated IgA transport system when bound to polymeric but not monomeric IgA (Sixbey and Yao, 1992). The development of live virus vaccine vectors that are replication defective or deficient offers yet more interesting possibilities but again the most important questions will be answered not in the tamarin EBV lymphoma model but in properly controlled human trials.

Acknowledgements

The author is grateful to Susan Finerty for reading the manuscript and to Glenise Maytham for secretarial assistance and acknowledges financial support from the Cancer Research Campaign of the United Kingdom.

REFERENCES

Beisel, C., Tanner, J., Matsuo, T., Thorley-Lawson, D., Kedy, F. and Kieff, E., 1985, Two major outer envelope glycoproteins of Epstein-Barr virus are encoded by the same gene, J. Virol., 54:665.

Bejarano, M. T., Masucci, M. G., Morgan, A., Morein, B., Klein, G. and Klein, E., 1990, Epstein-Barr virus (EBV) antigens processed and presented by B cells, B blasts and macrophages trigger T-cell-mediated inhibition of EBV-induced B-cell transformation, J. Virol., 64:1398.

Byars, N. E. and Allison, A. C., 1987, Adjuvant formulation for use in vaccines to elicit both cell-mediated and humoral immunity, Vaccine, 5:223.

Byars, N. E., Nakaon, G., Welch, M., Lehman, D. and Allison, A. C., 1991, Improvement of hepatitis B vaccine by the use of a new adjuvant, Vaccine, 9:309.

Chen, C. J., Liang, K. Y., Chang, Y. S., Wang, Y. F., Hsieh, T., Hsu, M. M., Chen, J. Y. and Liu, M. Y., 1990, Multiple risk factors of nasopharyngeal carcinoma: Epstein-Barr virus, malarial infection, cigarette smoking and familial tendency, Anticancer Res., 10:547.

Cleary, M. L., Epstein, M. A., Finerty, S., Dorfman, R. F., Bornkamm, G. W., Kirkwood, J. K, Morgan, A. J. and Sklar, J., 1985, Individual tumors of multifocal EBV-induced malignant lymphomas in tamarins arise from different B-cell clones, Science, 228:722.

David, E. M. and Morgan, A. J., 1988, Efficient purification of Epstein-Barr virus membrane antigen gp340 by fast protein liquid chromatography, J. Immunol. Methods, 108:231.

Dillon, S. B., Demuth, S. G., Schneider, M. A. Weston, C. B., Jones, C. S., Young, J. F., Scott, M., Bhatnaghar, P. K., LoCastro, S. and Hanna, N., 1992, Induction of protective class I MHC-restricted CTL in mice by a recombinant influenza vaccine in aluminium hydroxide adjuvant, Vaccine, 10:309.

Emini, E. A., Schleif, W. A., Armstrong, M. E., Silberklang, M., Schultz, L. D., Lehman, D., Maigetter, R. Z., Qualtiere, L. F., Pearson, G. R. and Ellis, R. W., 1988, Antigenic analysis of the Epstein-Barr virus major membrane antigen (gp350/220) expressed in yeast and mammalian cells: implications for the development of a subunit vaccine, Virology, 166:387.

Emini, E. A., Schleif, W. A., Silbeklang, M., Lehmanm, D. and Ellis, R. W., 1989, Vero cell-expressed Epstein-Barr virus (EBV) gp350/220 protects marmosets from EBV challenge, J. Med. Virol, 27:120.

Epstein, M. A., 1976, Epstein-Barr virus - is it time to develop a vaccine program? Journal of the National Cancer Institute, 56:697.

Epstein, M. A. Achong, B. G. and Barr, Y. M., 1964, Virus particles in cultured lymphoblasts from Burkitt's lymphoma, Lancet, i:703.

Epstein, M. A., Morgan, A. J., Finerty, S., Randle, B. J. and Kirkwood, J. K., 1985, Protection of cottontop tamarins against Epstein-Barr virus-induced malignant lymphoma by a prototype subunit vaccine, Nature, 318:287.

Epstein, M. A., North, J. R. and Morgan, A. J., 1983, Purification and properties of the gp340 component of Epstein-Barr virus membrane antigen in an immunogenic form, J. Gen. Virol., 64:455.

Epstein, M. A., Randle, B. J., Finerty, S. and Kirkwood, J. K., 1986, Not all potently neutralizing, vaccine-induced antibodies to Epstein-Barr virus ensure protection of susceptible experimental animals, Clin. Exp. Immunol., 63:485.

Epstein, M. A., zur Hausen, H., Ball, G. and Rabin, H., 1975, Pilot experiments with EBV in owl monkeys (Aotus trivirgatus). III: Serological and biochemical findings in an animal with reticuloproliferative disease, Int. J. Cancer, 15:17.

Facer, C. A. and Playfair, J. H., 1989, Malaria, Epstein-Barr virus, and the genesis of lymphomas, Adv. Cancer Res., 53:33.

Finerty, S., Tarlton, J., Mackett, M., Conway, M., Arrand, J. R., Watkins, P. E. and Morgan, A. J., 1992, Protective immunization against Epstein-Barr virus-induced disease in cottontop tamarins using the virus envelope glycoprotein gp340 produced from a bovine papillomavirus expression vector, J. Gen. Virol, 743:449.

Haddad, R. S. and Hutt-Fletcher, L. M., 1989, Depletion of glycoprotein gp85 from virosomes made with Epstein-Barr virus proteins abolishes their ability to fuse with virus receptor-bearing cells, J. Virol., 63:4998.

Hutt-Fletcher, L., 1991, Epstein-Barr virus tissue tropism: a major determinant of immunopathogenesis, Springer Seminars in Immunopathology, 13:117.

Kaaden, O. R. and Dietzschold, B. 1974, Alterations of the immunological specificity of plasma membranes of cells infected with Marek's disease and turkey herpesviruses, J. Gen. Virol., 25:1.

Kieff, E. and Liebowitz, D., 1990, Epstein-Barr virus and tis replication, in: "Virology," B. N. Fields, D. M. Knipe, R. M. Chanock, M. S. Hirsch, J. O. Melnick, T. P. Monath and B. Roizman, eds., Raven Press, New York.

Laufs, R. and Steinke, H., 1975, Vaccination of non-human primates against malignant lymphoma, Nature, 253:71.

Lu, S., Day, N. E., Degos, L., Lepage, V., Wang, P-C., Chan, S-H., Simons, M., McKnight, B., Easton, D., Zeng, Yi and de-The, G., 1991, Linkage of a nasopharyngeal carcinoma susceptibility locus to the HLA region, Nature, 346:470.

Madej, M., Conway, M. J., Morgan, A. J., Sweet, J., Wallace, L., Arrand, J. and Mackett, M., 1992, Purification and characterisation of Epstein-Barr virus gp340/220 produced by a bovine papilloma virus vector system, Vaccine, 10:777.

Magrath, I., 1990, The pathogenesis of Burkitt's lymphoma, Adv. Cancer Res., 55:133-270.

Miller, G., 1990, Epstein-Barr virus, in: "Virology," B. N. Fields, D. M. Knipe, R. M. Chanock, M. S. Hirsch, J. O. Melnick, T. P. Monath and B. Roizman, eds., Raven Press, New York.

Morein, B., Fossum, C., Lovgren, K. and Hoglund, S., 1990, The Iscom - a modern approach to vaccines, Seminars in Virology, 1:49.

Morgan, A. J., Allison, A. C., Finerty, S., Scullion, F. T., Byars, N. E. and Epstein, M. A., 1989, Validation of a first-generation Epstein-Barr virus vaccine preparation suitable for human use, J. Med. Virol., 29:74.

Morgan, A. J., Finerty, S., Lovgren, K., Scullion, F. T. and Morein, B., 1988, Prevention of Epstein-Barr (EB) virus-induced lymphoma in cottontop tamarins by vaccination with the EBV envelope glycoprotein gp340 incorporated into immune-stimulating complexes, J. Gen. Virol., 69:2093.

Morgan, A. J., Mackett, M., Finerty, S., Arrand, J. R., Scullion, F. T. and

Epstein, M. A., 1988, Recombinant vaccinia virus expressing Epstein-Barr virus glycoprotein gp340 protects cottontop tamarins aginst EBV-induced malignant lymphomas, J. Med. Virol., 25:189.

Morgan, A. J., North, J. R. and Epstein, M. A., 1983, Purification and properties of the gp340 component of Epstein-Barr virus membrane antigen in an immunogenic form, J. Gen. Virol., 64:455.

Morgan, A. J., Smith, A. R., Barker, R. N. and Epstein, M. A., 1984, A structural investigation of the Epstein-Barr (EB) virus membrane antigen glycoprotein, gp340, J. Gen. Virol, 65:397.

Moss, D. J., Misko, I. S., Sculley, T. B., Apolloni, A., Khanna, R. and Burrows, S. R., 1991, Immune regulation of Epstein-Barr virus (EBV): EBV nuclear antigen as a target for EBV-specific T cell lysis, Springer Seminars in Immunopathology, 13:147.

Motz, M. Deby, G. and Wolf, H., 1987, Truncated versions of the two major Epstein-Barr viral glycoproteins (gp250/350) are secreted by recombinant Chinese hamster ovary cells, Gene, 58:149.

Mueller, N., Evans, A., Harris, N. L., Comstock, G. W., Jellum, E., Magnus, K., Orentriech, N., Polk, B. F. and Vogelman, J., 1989, Hodgkin's disease and Epstein-Barr virus. altered antibody pattern before diagnosis, N. Engl. J. Med., 320:689.

Nalesnik, M. A., 1991, Lymphoproliferative disease in organ transplant recipients, Springer Seminars in Immunopatholsogy, 13:181.

O'Hagan, D. T. Rahman, D., McGee, J. P., Jeffry, H., Davies, M. C., Williams, P., Davis, S. S. and Challacombe, S. J., 1991, Biodegradable microparticles as controlled release antigen delivery systems, Immunology, 73:239.

Ooka, T., de Turenne-Tessier, M. and Stolzenberg, M. C., 1991, Relationship between antibody production to Epstein-Barr virus (EBV) early antigens and various EBV-related diseases, Springer Seminars in Immunopathology, 13:233.

Pallesen, G., Hamilton Dutoit, S. J., Rowe, M. and Young, L. S., 1991, Expression of Epstein-Barr virus latent gene products in tumour cells of Hodgkin's disease, Lancet, 337:320.

Parkin, D. M., Stjemsward, J. and Muri, C. S., 1984, Estimates for the worldwide frequency of twelve major cancers , Bulletin of the World Health Organisation, 62:163.

Pither, R. J., Zhang, C. X., Shiels, C. Tarlton, J. Finerty, S. and Morgan, A. J., 1992a, Mapping of B-cell epitopes on the polypeptide chain of the Epstein-Barr virus major envelope glycoprotein and candidate vaccine molecule gp340, J. Virol., 66:1246.

Pither, R. J., Nolan, L., Tarlton, J., Walford, J. and Morgan, A. J., 1992b, Distribution of epitopes within the amino acid sequence of the Epstein-Barr virus major envelope glycoprotein, gp340, recognized by hyperimmune rabbit sera, J. Gen. Virol., 73:1409.

Purtilo, D. T., 1991, X-linked lymphoproliferative disease (XLP) as a model of Epstein-Barr virus-induced immunopathology, Springer Seminars in Immunopathology, 13:181.

Ragot, T., Tosoni-Pittoni, E., Finerty, S., Morgan, A. J. and Perricaudet, M., 1991, Recombinant adenoviruses which express the Epstein-Barr virus membrane antigen gp340/220, gp220 only or a secreted form of gp340 induce persistent virus-neutralizing antibodies in rabbits in: "Epstein-Barr Virus and Human Disease," D. V. Ablahsi, ed., Humana Press, Clifton, New Jersey.

Ragot, T., Tosoni-Pittoni, E., de Mazancourt, S., Finerty, S., Morgan, A. J. and Perricaudent, M., 1992a, Recombinant adenoviruses which express the Epstein-Barr virus membrane antigen gp340/220 and used as a live vaccine induce persistent virus-neutralizing antibodies in rabbits, J. Gen. Virol., (submitted for publication).

Ragot, T., Finerty, S., Watkins, P., Perricaudet, M. and Morgan, A. J., 1992b, Protective immunity against EBV lymphoma in the cottontop tamarin is induced by adenovirus recombinants expressing the EBV envelope glycoprotein gp340, J. Gen. Virol., 74:501.

Schultz, L. D., Tanner, J., Hofmann, K. J., Emini, E. A., Condra, J. H., Jones, R. E., Kieff, E. and Ellis, R. W., 1987, Expression and secretion in yeast of a 400-kDa envleope glycoprotein derived from Epstein-Barr virus, Gene, 54:113.

Shope, T., Dechairo, D. and Miller, G., 1973, Malignant lymphoma in cotton-top marmosets after inoculation with Epstein-Barr virus, Proc. Natl. Acad. Sci. USA, 70:2487.

Sixbey, J. W. and Yao, Q-Y., 1992, Immunoglobulin A-induced shift of Epstein-Barr virus tissue tropism, Science, 255:1578.

Tanner, J., Whang, Y., Sample, J., Sears, A. and Kieff, E., 1988, Soluble gp350/220 and deletion mutant glycoproteins block Epstein-Barr virus adsorption to lymphocytes, J. Virol., 62:4452.

Tartaglia, J., Perkus, M. E., Taylor, J., Norton, E. K., Audonnet, J-C., Cox, W. I., Davis, S. W., Vanderhoeven, J., Meignier, B., Riviere, M., Languet, B. and Paoletti, E., 1992, NYVAC: A highly attenuated strain of vaccinia virus, Virology, 88:217.

Taylor, J., Weinberg, R., Tartaglia, J., Richardson, C., Alkhatib, G., Briedis, D., Appel, M., Norton, E. and Papoletti, E., 1992, Nonreplicating viral vectors as potential vaccines: recombinant canarypox virus expressing measles virus fusion (F) and hemagglutinin (HA) glycoproteins, Virology, 187:321.

Ulaeto, D., Wallace, L., Morgan, A. J., Morein, B. and Rickinson, A. B., 1988, In vitro T cell responses to a candidate Epstein-Barr virus vaccine: human CD4+ T-cell clones specific for the major envelope glycoprotein, gp340, Eur. J. Immunol., 18:1689.

Wallace, L. E., Wright, J., Ulaeto, D. O., Morgan, A. J. and Rickinson, A. B., 1991, Identification of CD4+ T cell epitopes of the candidate Epstein-Barr virus vaccine glycoprotein gp340, I. Virol., 65:3821.

Watkins, D. I., Hodi, F. S. and Letvin, N. L., 1988, A primate species with a limited major histocompatability complex class I polymorphism, Proc. Natl. Acad. Sci. USA, 85:7714.

Whang, Y., Silberklang, M., Morgan, A., Munshi, S., Lenny, A. B., Ellis, R. W. and Kieff, E., 1987, Expression of the Epstein-Barr virus gp350/220 gene in rodent and primate cells, J. Virol., 61:1796.

Young, L. S., Finerty, S., Rickinson, A. B. and Morgan, A. J., 1988, Epstein-Barr virus gene expression in lymphomas induced by the virus in the cottontop tamarin, J. Virol., 63:1967.

REQUIREMENTS FOR INDUCTION OF SEMLIKI FOREST VIRUS NEUTRALIZING ANTIBODIES

BY A NON-INTERNAL IMAGE MONOCLONAL ANTIBODY

Cornelis A. Kraaijeveld, Tom A. M. Oosterlaken and Harm
Snippe

Eijkman-Winkler Laboratory for Medical Microbiology
University of Utrecht, Utrecht, The Netherlands

INTRODUCTION

Recently we identified a noninternal image monoclonal anti-idiotypic antibody (ab2β MAb), designated 1.13A321 (IgG1), that, cross-linked by glutaraldehyde to keyhole limpet haemocyanin (KLH) and combined with the adjuvant Quil A, was able to evoke Semliki Forest Virus (SFV) neutralizing anti-anti-idiotypic (ab3) antibodies in BALB/c mice (Oosterlaken et al, 1991; Oosterlaken et al, 1992). These antibodies protected mice against an otherwise lethal infection with SFV. In this study conditions for induction of SFV-neutralizing ab3 antibodies were investigated. It is shown that cross-linking of 1.13A321 to itself by glutaraldehyde is sufficient to evoke SFV-neutralising ab3 antibodies. Furthermore a recombinantly expressed protein consisting of cro-β-galactosidase at the N-terminus and aminoacid residues 115 to 151 of the E2 membrane protein of SFV at the C-terminus was used as carrier molecule (Snijders et al, 1989). The SFV fragment contains two T-helper cell epitopes which might potentially provide an SFV specific T-cell memory upon immunization (Snijders et al, 1992). Furthermore the potential of ab2 MAb 1.13A321 as carrier for a linear B-cell epitope of SFV was investigated.

MATERIALS AND METHODS

Virus

The avirulent prototype strain of SFV (Garoff et al, 1980) was obtained from H. Garoff, The Karolinska Institute, Huddinge University Hospital, Huddinge, Sweden. It was used for determination of SFV-neutralizing antibodies. The virulent strain of SFV, SF/LS 10 Cl/A (Bradish, Allner and Maber, 1971), was received from C. J. Bradish, The Porton Down Microbiological Research Establishment, Salisbury, UK. The 50% percent lethal dose (LD$_{50}$) for 10-14 week old BALB/c mice proved to be 1 to 2 plaque forming units (PFU) when intraperitoneally (i.p.,) injected in 0.5 ml phosphate buffered saline (PBS) of pH 7.2.

Cells and Media

L cells, a continuous line of mouse fibroblasts, were maintained in Dulbecco's modified Eagle's medium (DMEM), buffered with 0.01 M N-2-

hydroxyethylpiperazine-N'-ethane sulfonic acid, supplemented with 5% calf serum, 0.2% tryptose and antibiotics.

SFV-specific MAb

SFV-neutralizing MAb UM5.1 (IgG2a), conjugated to horseradish peroxidase (HRPO) by the periodate method (Nakane and Kawaoi, 1974) was used for detection of SFV in cell culture (Van Tiel et al, 1986).

Anti-idiotypic MAb

Anti-idiotypic (ab2) MAb 1.13A321 (IgG1) was induced against SFV-neutralizing MAb UM 1.13 as described previously (Oosterlaken et al, 1991).

Carrier Molecules for ab2 MAb 1.13A321

The conventional carrier keyhole limpet haemocyanin (KLH) was purchased from Calbiochem, La Jolla, CA, USA.

A recombinantly expressed protein consisting of cro-β-galactosidase at the N-terminus and amino acid residues 115 to 151 of the E2 membrane protein of SFV at the C-terminus was prepared and purified as previously described (Snijders et al, 1989). The SFV fragment contains two T-helper cell epitopes located at positions 115-129 and 137-151. (Snijders et al, 1991). The latter epitope is immunodominant and able to induce and elicit delayed type hypersensitivity to SFV (Snijders et al, 1989; Snijders et al, 1992).

SFV Specific Synthetic Peptide

A peptide with an amino acid sequence corresponding to region 240-255 [with three additional amino acids (CGG) at the NH_2 terminus: CGGPFVPRADEPARKGKVH (COOH] of the membrane protein E2 was synthesized by E. Freund (Hubrecht Laboratory, Utrecht, The Netherlands). The peptide, designated 240-255, contains a B-cell epitope of SFV. High serum levels (titers > 3000) of non-neutralizing but SFV-reactive antibodies evoked to this linear B-cell epitope protected BALB/c and DBA/2 mice against lethal challenge with virulent SFV (Snijders et al, 1991; Snijders et al, 1992).

Coupling of ab2 MAb to KLH, recombinant protein or synthetic peptide

Ab2 MAb 1.13A321 was purified by protein G sepharose affinity chromatography (Akerstrom et al, 1985) and subsequently coupled to either KLH, recombinantly expressed protein or synthetic peptide. Antibody was covalently conjugated to either molecule with glutaraldehyde as described previously (Oosterlaken et al, 1988). In brief: 0.8 mg purified 1.13A321 in 0.2 ml PBS was mixed with 1 mg of either substance (in 0.2 ml distilled water) and then covalently coupled to each other by the addition of 0.06 ml of 2.5% glutaraldehyde. After 20 min incubation at room temperature the reaction was stopped with 0.06 ml of 0.2 M glycine. After addition of 0.48 ml distilled water, the mixture was dialysed overnight at + 4°C against distilled water to remove excess glycine. Thereafter the conjugate (1.0 ml) was used for anti-idiotypic immunization of mice.

In one experiment synthetic peptide was coupled to ab2 MAb 1.13A321, according to Lee et al. (1980), using N-gamma-maleimidobutyryloxysuccinimide (GMBS, Calbiochem, Hoechst, San Diego, USA) as the coupling agent.

Mice

BALB/c mice, DAB/2 mice and nude (+/+) BALB/c mice were obtained from the National Institute of Health and Environmental Protection (RIVM), Bilthoven, The Netherlands. The mice were kept in the animal house of the University of Utrecht until use at an age of 10 to 14 weeks.

Anti-anti-idiotypic immunization

Mice were intracutaneously immunized once or twice (5 weeks apart) with conjugated ab2 MAb 1.13A321 (doses equivalent to 40 µg ab2 MAb per animal) combined with the adjuvant Quil A (50 µl per animal). Quil A, a purified saponin (Morein et al, 1984; Kensil et al, 1991), was obtained from Superfos Biosector, Denmark. The mixture (0.1 ml) was injected at two sites (2 x 0.05 ml) in the vicinity of the draining lymph nodes in the groins. Blood for determination of SFV-neutralizing antibodies in serum was obtained from ether-anaesthetized mice by retro-orbital puncture.

Determination of SFV-Neutralizing Antibodies

SFV-neutralizing antibodies were determined by neutralization enzyme immunoassay (N-EIA) as described earlier with slight modifications (Van Tiel et al, 1986). In brief: serum samples obtained from individual mice were serially diluted in DMEM, supplemented with 5% calf serum, in wells of 96-well plates. Each serum dilution (0.05 ml) was mixed with a standard infectious dose of 1000 PFU of SFV (0.05 ml) and incubated for 1h at 37°C. Subsequently 20,000 L cells (0.1 ml) were seeded into each well to form monolayers. Non-neutralized SFV was allowed to multiply for 18h at 37°C. Then the L cells monolayers were fixed by addition of 0.05% glutaraldehyde for 10 min at room temperature. After washing with tapwater and rinsing with PBS direct EIA of SFV was performed with HRPO-labelled MAb UM 5.1. Preincubation of the virus inoculum with SFV neutralizing serum reduces

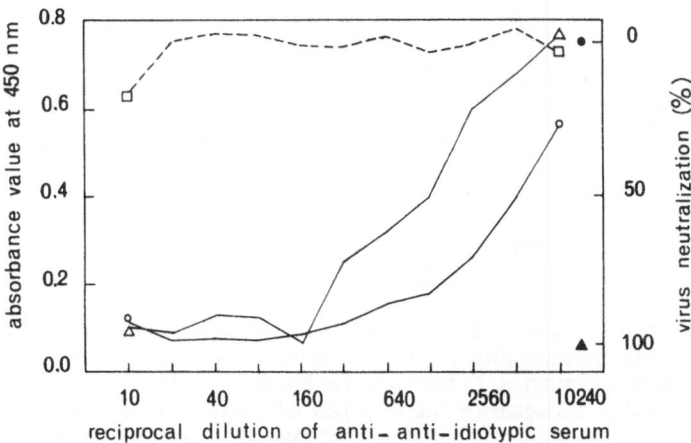

Fig. 1. Titration of SFV-neutralizing mouse anti-anti-idiotypic sera. Tested were two immune sera (o, △) and normal mouse serum (◻). Female BALB/c mice received two subcutaneous immunizations, 5 weeks apart, with 1.13A321 coupled to KLH and mixed with Quil A. Serum samples for titration by N-EIA were obtained on day 84. The single symbols to the right represent the mean (n=10)A_{450} of the virus control (●) and the cell control (▲). The interconnected points represent single A_{450} values

Table 1. Induction of SFV-neutralizing anti-anti-idiotypic antibodies in normal BALB/c mice but not in nude (+/+) BALB/c mice

Mouse strain[a]	MEAN (± SD) $\log_{10}$ Neutralization Titre[b]	Survival Ratio[c]
BALB/c	1.3 ± 1.3	5/5
nude BALB/c	< 0	0/6

[a] Mice received two subcutaneous immunizations, 5 weeks apart, with 1.13A321 coupled to KLH and mixed with Quil A.

[b] Serum samples were obtained on day 49 for determination of SFV neutralizing antibodies by N-EIA.

[c] Mice were intraperitoneally challenged with 400 PFU of SFV on day 62.

virus multiplication and thereby the appearance of the spike proteins on the surface of L cells as indicated by inhibition of absorbance in the EIA. Inhibition of virus multiplication by immune serum could be calculated as a percentage of control: % inhibition ("virus neutralization") = 100% − $[(A_{450}$ serum − A_{450} cell control)/(A_{450} virus control − A_{450} cell control)] x 100%. The titer of immune serum can be arbitrarily defined as that dilution causing 50% inhibition. Absorbance values were measured at 450 nm with a Titertek multiskan photometer (Flow Laboratories, Irvine, UK).

Protection of Mice

Immunized mice and control mice were intraperitoneally injected with 400 PFU (equivalent to 250 LD_{50} units) of the virulent strain of SFV in 0.5 ml PBS of pH 7.2. To quantitate protection mice were observed for 21 days although nonprotected mice died within eight days after challenge.

Results

No induction of SFV-neutralizing anti-anti-idiotypic antibodies in nude BALB/c mice by ab2 MAb 1.13A321.

Normal BALB/c mice immunized with KLH cross-linked ab2β MAb 1.13A321 develop SFV-neutralizing antibodies in serum which could be titrated by N-EIA. An example of such a titration is shown in Fig. 1. Nude BALB/c mice are unable to make detectable quantities of neutralizing antibodies upon booster immunization and consequently those mice are not protected against challenge with virulent SFV (Table 1).

Induction of SFV neutralizing antibodies by ab2 MAB 1.13A321 cross-linked to either itself or to a B-cell epitope of SFV.

Cross-linking of ab2 MAb to itself by glutaraldehyde is a sufficient condition for induction of SFV neutralizing ab3 antibodies as shown in the next experiment.

Both DBA/2 and BALB/c mice were intracutaneously immunized with

1.13A321 cross-linked to either itself and to a B-cell epitope of SFV. Cross-linking of the B-cell epitope to 1.13A321 was established by either glutaraldehyde or maleimide. Control mice received only the adjuvant Quil A. After 35 days the mice were boosted with the same antigens by the same route. At day 63 blood was obtained for determination of SFV neutralizing ab3 antibodies in serum. Seven days later on day 70 all mice were challenged with 400 PFU of virulent SFV. As shown in Table 3 BALB/c mice did develop SFV neutralizing antibodies after glutaraldehyde cross-linking and not after maleimide coupling of peptide to ab2βMAb. Due to genetic restriction DBA/2 did not develop SFV neutralizing antibodies (Oosterlaken et al, 1992). However, in both mouse strains antibodies appeared against B-cell epitope 240-255 indicating that cross-linked 1.13A321 provided very likely T-cell help to the B-cell epitope. The antibody titers to the B-cell epitopes (240-255) were, however, too low to protect the DBA/2 mice against lethal challenge.

Rapid development of protective immunity after immunization with ab2 MAb 1.13A321 cross-linked to recombinant protein.

A state of protective immunity in BALB/c mice to virulent SFV evolves quickly upon immunization with ab2βMAb 1.13A321 cross-linked to recombinant protein as shown in the next experiment.

Thirty female BALB/c mice were intracutaneously immunized with ab2 MAbβ1.13A321 coupled to recombinant protein and combined with the adjuvant Quil A. Ten mice were immunized similarly with free ab2 MAb1.13A321 and Quil A. Separate groups (n=5) of mice were intraperitoneally challenged with 400 PFU of SFV respectively 3, 5, 7, 14, 21, and 28 days after primary immunization. Just before infection blood samples were taken from ether-anaesthetized mice for determination of SFV-neutralizing anti-anti-idiotypic antibodies in individual sera. As shown in Fig. 2 partial protective immunity was attained at day 7 and full protective immunity at day 14 concomitantly with the appearance of neutralizing antibodies. The mice that were immunized with free ab2βMAb 1.13A321 developed neither SFV-neutralizing antibodies nor protective immunity in the 4 weeks after primary immunization.

Discussion

In this study we show that ab2αMAb 1.13A321, cross-linked to either KLH or recombinant protein, induced rapidly protective immunity to SFV in BALB/c mice due to the appearance of SFV-neutralizing antibodies. Such antibodies are not inducible in nude (+/+) BALB/c mice at least not after booster immunization. Nude BALB/c mice are unable to mount a secondary IgG response and therefore the T-helper epitopes located on the anti-idiotypic vaccine are of no use in these animals.

In BALB/c mice use of a carrier molecule is not an essential requirement for induction of SFV neutralizing ab3 antibodies. Mere cross-linking of 1.13A321 by glutaraldehyde proved to be a sufficient condition for induction of SFV-neutralizing ab3 antibodies (Table 2). This result suggests that on glutaraldehyde aggregated 1.13A321 T-helper epitopes(s) are located. The concept of T-helper epitope(s) on 1.13A321 is sustained by the observation that this ab2 MAb could serve as carrier molecule for a linear B-cell epitope of SFV (Table 2). High levels of non-neutralizing antibodies to the SFV-specific peptide could protect DBA/2 and BALB/c mice against lethal SFV infection (Snijders et al, 1992). In the present study the anti-peptide antibody levels were obviously too low to protect the DBA/2 mice.

An ab2α MAb binds with its antigen combining site (paratope) to

Table 2. Genetically restricted induction of SFV neutralizing ab3 antibodies. Ab2α MAB 1.13A321 serving as carrier molecule for a B-cell epitope of SFV

Anti-anti-idiotypic immunization of mice with ab2 MAb 1.13A321 [a]		Mean (± SD) ¹⁰log serum antibody titer [b] against peptide	Mean (± SD) ¹⁰log serum titer of SFV neutralizing antibodies [c]	Survival ratio
mouse strain	cross-linking to ab2 MAb or peptide 240-255			
DBA/2	control	< 2.0	< 0.5	0/5
BALB/c	control	< 2.0	< 0.5	0/4
DBA/2	1.13A321 (GA)	< 2.0	< 0.5	0/5
BALB/c	1.13A321 (GA)	< 2.0	1.1 ± 0.6	5/5
DBA/2	240-255 (GA)	2.5 ± 0.4	< 0.5	0/5
BALB/c	240-255 (GA)	3.3 ± 0.4	1.5 ± 1.1	4/4
DBA/2	240-255 (MAL)	< 2.0	< 0.5	1/4
BALB/c	240-255 (MAL)	2.2 ± 0.2	< 0.5	1/4

a Mice were intracutaneously immunized with ab2 MAbs (40 µg per animal) cross-linked by glutaraldehyde (GA) or maleimide (MAL) to peptide (40 µg per animal) combined with Quil A (50 µl per animal). The mice were given similar boosters at day 35. Mice were infected intraperitoneally at day 70 of immunization with 400 PFU of SFV in 0.5 ml PBS of pH 7.2.

b Determined by solid phase EIA using −¹⁰log dilutions (2.0, 2.5, 3.0, 3.5, 4.0, 4.5 and 5.0). Titres were for sera from individual mice 7 days before challenge.

c Detected by N-EIA using −¹⁰log dilutions (0.5, 1.0, 1.5, 2.0, 2.5, and 3.0) of the same sera used for determination of anti-peptide antibodies.

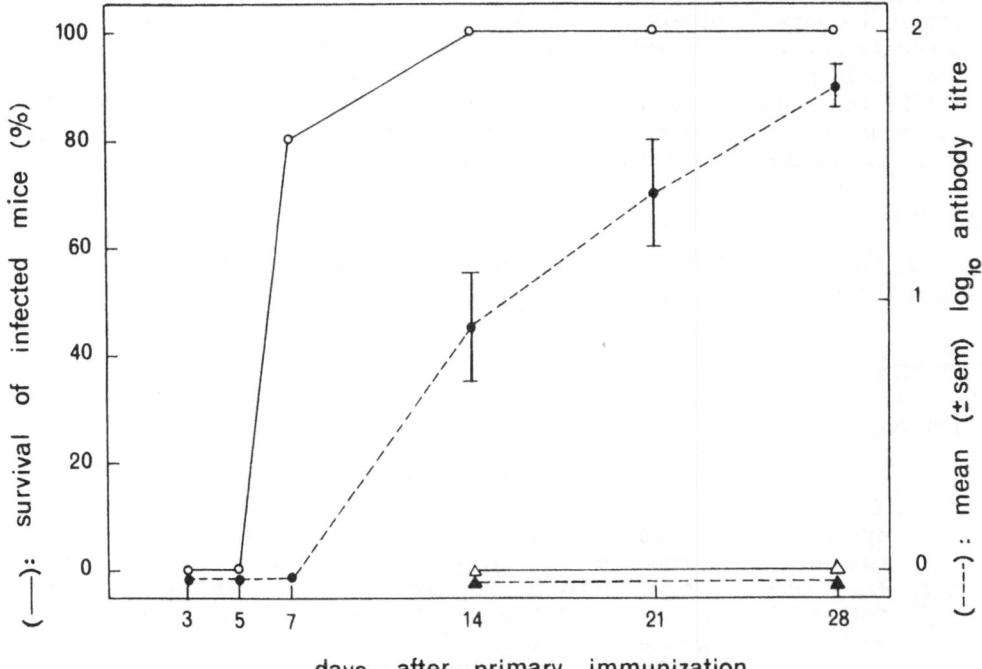

Fig. 2. Rapid development of protective immunity after immunization with noninternal image ab2 MAb 1.13A321 cross-linked to a recombinantly expressed protein. Groups (n=5) of mice were subcutaneously immunized with either cross-linked or free ab2 MAb. In either case an equivalent of 40 μg of ab2 MAb per animal was combined with 50 μl of Quil A per animal. Mice that were immunized with free ab2 MAb were challenged on either day 14 or day 28 with 400 pFU of SFV. The mice that were immunized with cross-linked ab2 MAb were challenged after the indicated time intervals with the same dose (400 pFU). Just before challenge blood was obtained from etheranaesthetized mice for determination of SFV-neutralizing ab3 antibodies in the individual sera

recurring idiotopes located on antigen receptors of B-cells destined to produce SFV neutralizing antibodies. Such binding of ab2 MAb to B-cells might be the initial trigger for proliferation and maturation to antibody producing plasma cells (Hiernaux, 1988; Jerne et al., 1982; Kohler et al., 1989; Rimmelzwaan et al, 1989; Roitt et al, 1985). Presumably the binding of aggregated ab2 MAb to B-cells is advantageous for the triggering of antibody production. A second function of glutaraldehyde mediated aggregation might be the masking and destruction of the Fc parts of ab2 MAb preventing thereby their binding to the Fc receptors on B-cells. That binding of ab2 MA b to Fc$_\gamma$ receptors would provoke a negative feedback mechanism on B-cell activation (Wofsy and Goldstein, 1990). Besides aggregation and functional obliteration of Fc parts induction of ab3 antibody production might be enhanced by provision of T-helper cell epitopes on carrier molecules like KLH and recombinant protein.

Furthermore we demonstrate in the present study that a recombinantly expressed protein could be used as carrier molecule for ab2 MAb 1.13A321. Hybrid protein consists of cro-β-galactosidase and aminoacid residues 115-151 of E2. The SFV specific part contains two T-helper cell epitopes which potentially might provide SFV specific T-helper cell memory to anti-idiotypic immunized animals (Snijders et al, 1992). However, conclusive evidence is not yet obtained that the SFV specific T-helper cell epitope(s) in the hybrid protein is a functional one when cross-linked to ab2MAb 1.,13A321 (our unpublished results). Nevertheless the described approach might be useful to enhance the longterm effectiveness of anti-idiotypic vaccines by induction of virus specific T-helper cell memory.

Acknowledgement

We thank Alies Snijders, Barry Benaissa-Trouw, Theo Harmsen and Geert Ekstijn for their cooperation.

REFERENCES

Akerstrom, B., Broding, T., Reis, K. and Bjorck, L., 1985, Protein G: a powerful tool for binding and detection of monoclonal and polyclonal antibodies, J. Immunol., 135:2589.

Bradish, C. J., Allner, K. and Maber, H. B., 1971, The virulence of original and derived strains of Semliki Forest Virus for mice, guinea-pig and rabbits, J. Gen. Virol., 12:141.

Garoff, H., Frischauf, A. M., Simons, K., Lehrach, H. and Delius, H., 1980, Nucleotide sequence of cDNA coding for Semliki Forest Virus membrane glycoproteins, Nature, 288:236.

Hiernaux, J. R., 1988, Idiotypic vaccines and infectious diseases, Inf. and Imm., 56:1407.

Jerne, N. K., Roland, J. and Cazenave, P. A., 1982, Recurrent idiotopes and internal images, EMBO J., 243.

Kensil, C. R., Patel, U. P., Lennick, M. and Marciani, D., 1991, Separation and characterization of saponins with adjuvant activity from Quillaja saponaria Molina Cortex, J. Immunol., 146:431.

Kohler, H., Kaveri, S., Kieber-Emmons, T., Morrow, W.J.W., Muller, S. and Raychaudhuri, S., 1989, Idiotypic networks and nature of molecular mimicry: an overview, Meth. Enzymol., 178:3.

Lee, A. C. J., Powell, J. E., Tregear, G. W., Niall, H. D., and Stevens, V. C., 1980, A method for preparing -HCG COOH peptide-carrier conjugates of predictable composition, Molecular Immunology, 17:749.

Morein, B., Sundquist, B., Hoglund, S., Dalsgaard, K. and Osterhaus, A., 1984, ISCOM: a novel structure for antigenic presentation of membrane proteins from enveloped viruses, Nature, 308:457.

Nakane, P. K. and Kawaoi, A., Peroxidase-labelled antibody: a new method of conjugation, 1974, J. Histochem. Cytochem., 22:1084.

Oosterlaken, T. A. M., Harmsen, M., Tangerman, C., Schielen, P., Kraaijeveld, C. A. and Snippe, H., 1988, A neutralization-inhibiton enzyme immunoassay for anti-idiotypic antibodies that block monoclonal antibodies neutralizing Semliki Forest Virus, J. Immunol. Methods, 115:255.

Oosterlaken, T. A. M., Harmsen, M., Jhagjhoor-Singh, S. S., Kraaijeveld, C. A., and Snippe, H., 1991, A protective monoclonal anti-idiotypic vaccine to lethal Semliki Forest Virus infection in BALB/c mice, J. Virology, 65:98.

Oosterlaken, T. A. M., Harmsen, M. Ekstijn, G. L., Kraaijeveld, C. A. and Snippe, H., 1992, IgVh determined genetic restriction of a non-internal image monoclonal anti-idiotypic vaccine against Semliki Forest virus, Immunology, 75:224.

Rimmelzwaan, G. G., Bunschoten, E. J., UytdeHaag, F. G. C. M. and Osterhaus, A. D. M. E., 1989, Monoclonal anti-idiotypic antibody vaccines against poliovirus, canine parvovirus, and rabies virus, Meth. Enzymol., 178:275.

Roitt, I. M., Thanavala, Y. M., Male, D. K. and Hay, F. C., 1985, Anti-idiotypes as surrogate antigens: structural consideration, Immunol. Today, 6:265.

Snijders, A., Benaissa-Trouw, B. J., Oosting, J. D., Snippe, H. and Kraaijeveld, C. A., 1989, Identification of a DTH-inducing T-cell epitope on the E2 membrane protein of Semliki Forest Virus, Cellular Immunology, 123:23.

Snijders, A., Benaissa-Trouw, B. J., Oosterlaken, T. A. M., Puijk, W. C., Posthumus, W. P. A., Meloen, R. H., Boere, W. A. M., Oosting, J. D., Kraaijeveld, C. A. and Snippe, H., 1991, Identification of linear epitopes on Semliki Forest Virus E2 membrane protein and their effectiveness as a synthetic peptide vaccine, J. Gen. Virology, 72:557.

Snijders, A., Benaissa-Trouw, B. J., Visser-Vernooij, H. G., Fernandez, I., Snippe, H. and Kraaijeveld, C. A., 1992, A DTH-inducing T-cell epitope of Semliki Forest virus mediates effective T-helper activity for antibody production, Immunology, 77:322

Van Tiel, F. H., Boere, W. A. M., Harmsen, M., Benaissa-Trouw, B. J., Kraaijeveld, C. A. and Snippe, H., 1986, Rapid determination of neutralizing antibodies to Semliki Forest Virus in serum by enzyme immunoassay in cell culture using virus specific monoclonal antibodies, J. Clin. Microbiol., 24:665.

Wofsy, C. and Goldstein, B., Cross-linking of Fc receptors and surface antibodies, 1990, J. Immunol., 145:1814.

POLYSACCHARIDE VACCINES

M.R. Lifely

Department of Cell Biology, The Wellcome Research
Laboratories, Langley Court, Beckenham, Kent, BR3 3BS, UK.

INTRODUCTION

Many of the bacteria that cause life-threatening diseases in humans
have polysaccharide capsules. Neisseria meningitidis, Haemophilus
influenzae and Streptococcus pneumoniae are principle causes of meningitis
and septicaemia, particularly in children under 2 years of age, in both the
developed and the developing world. Prophylactic use of vaccines, rather
than antibiotic treatment, is appropriate and justifiable for two reasons:
firstly, the course of disease is rapid with 5% mortality and 10% serious
sequelae, even with the most sophisticated treatment; and, secondly, the
emergence of antibiotic-resistant strains is a global problem due to the
widespread use of antibiotics. The focus of attention on capsular
polysaccharides as attractive vaccine candidates stems from their surface
location on the organism, resulting in their direct interaction with the
immune system, and the discovery that the presence of anti-capsular
antibodies correlates with protection from disease. The fact that >90% of
disease caused by N. meningitidis is associated with only three
structurally and serologically distinct capsular polysaccharides,
designated group A, B and C, and that virtually all H. influenzae disease
is associated with a single capsular polysaccharide (type b) has encouraged
the belief that effective vaccines containing one or a few components would
be forthcoming. This review will discuss progress towards this goal.

H. INFLUENZAE TYPE b (Hib) POLYSACCHARIDE VACCINE

Virtually all H. influenzae disease is caused by encapsulated type b
strains and is confined almost exclusively to children under 5 years old.
This led to the evaluation of the purified type b polysaccharide,
polyribosylribitolphosphate (PRP), in immunogenicity and efficacy trials,
with controversial results. In short, a prospective trial in Finland of
almost 100,000 children between 3 and 71 months of age demonstrated
protective efficacy of 90% in children over 18 months of age (Peltola et
al, 1984). As expected, the PRP vaccine demonstrated no protective
efficacy in younger children. Following this trial, in 1985 the vaccine
was licensed for use in the USA in all children over 2 years of age.
Subsequently, efficacy of the vaccine was evaluated in several case-control
studies, with highly uncertain results. Vaccine efficacy in different
geographical locations ranged from 88% (95% confidence interval of 74 to

New Generation Vaccines, Edited by G. Gregoriadis
et al., Plenum Press, New York, 1993

- Covalent or Non-Covalent Binding

- Carrier Protein

- Oligosaccharide or Polysaccharide

- Method of Coupling; 'Spacer' Group

- Characterisation and Reproducibility

- Dose; Route of Immunisation; Interval between Doses

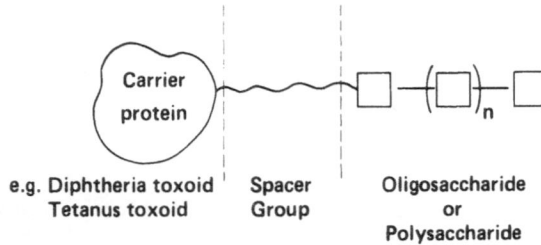

Fig. 1. Schematic diagram of the construction of polysaccharide-protein conjugates, and variables that
may influence immunogenicity

96%) in Connecticut/Dallas/Pittsburgh to -55% (95% confidence interval of -238 to 29%) in Minnesota (Shapiro and Berg, 1990). A more careful examination of the results of the Finnish study showed that although the PRP vaccine had a protective efficacy of 80% in infants in the age range of 24 to 35 months, the 95% confidence interval was 7 to 95%, reflecting the uncertainties that were subsequently encountered. Although these studies led to far-reaching conclusions about the need to conduct and interpret carefully-designed efficacy trials, it was clear that the PRP vaccine was far from optimal in protection from H. influenzae type b (Hib) disease. The incentive to design more efficacious vaccines, particularly in the youngest age groups, was clear.

Hib CONJUGATE VACCINES

Much scientific endeavour has gone into the study of a new generation of Hib vaccines that are composed of covalent conjugates of PRP and carrier proteins. Four such conjugate vaccines are at different stages of investigation in immunogenicity and efficacy trials, and three have already been licensed for use in children of different age ranges. The four conjugate vaccines differ markedly in their composition, and, although they are clearly superior to the PRP vaccine alone, they differ also in their immunogenicity. This serves to highlight the large number of variables that can be associated with the construction of conjugate vaccines (Figure 1), and, importantly, this allows us the opportunity to investigate those factors that are necessary for good immunogenicity and protective immunity.

PRP-Diphtheria Toxoid (PRP-D)

The first of the Hib conjugate vaccines was manufactured by Connaught Laboratories (Gordon, 1984), and consisted of the heat-sized polymer, PRP, coupled to diphtheria toxoid via a 6-carbon spacer molecule (PRP-D). The conjugate was found to be superior in immunogenicity to the PRP vaccine in children older than 18 months (Ward, 1991), which led to its licensure in the USA in 1987 for this age group. However, PRP-D had only limited immunogenicity in infants immunized at 2,4 and 6 months of age. Nevertheless, a large prospective efficacy trial in Finland was conducted on infants immunized at 3,4,6 and 14 months of age (Eskola et al., 1987; Eskola et al, 1991), and, despite the low geometric mean antibody titer at seven months (0.42 ug/mL), protective efficacy was 94% (95% confidence interval of 83 to 98%). This unexpectedly high efficacy was attributed to the priming effect of PRP-D, which would allow a rapid anti-PRP response to be mounted on encountering invasive H. influenzae type b organisms (Eskola et al, 1987).

As with the PRP vaccine, however, a second efficacy study in a different geographical location gave very different results. In Alaskan infants, a high risk group, who were immunized at 2,4 and 6 months of age with PRP-D, vaccine efficacy was estimated to be only 35% (95% confidence interval of -57 to 73%) (Ward et al, 1990). This was consistent with the low anti-PRP response observed in these infants after receiving three doses of PRP-D vaccine.

PRP Oligosaccharide-Diptheria Toxoid Mutant (HbOC)

Prototype vaccines were initially developed by Anderson and colleagues, and consisted of short oligosaccharides covalently coupled by reductive amination to diphtheria toxoid or to a non-toxic mutant of diphtheria toxoid (CRM_{197}) (Anderson, 1982; Anderson et al, 1985)). Several immunogenicity studies on both adults and infants were performed with these vaccines in order to investigate structure/immunogenicity relationships. In one study (Anderson et al, 1985), 12-16 month old infants were immunized with a vaccine of PRP oligosaccharides of 3-10 repeating units coupled either to diphtheria toxoid or to the cross-reactive mutant, CRM_{197}. Only in the latter case was a significant increase in the anti-PRP response seen, suggesting the importance of the carrier protein.

In a second study (Anderson et al, 1986), the effect of the oligo-saccharide chain length on immunogenicity was investigated. Adults and 1 year old infants were immunized with oligosaccharides of 8 or 20 repeating units, coupled to diphtheria toxoid (Dpo8 and Dpo20, respectively). In adults after one immunization both conjugates gave a good anti-PRP response, whereas in infants after two immunizations a significantly higher anti-PRP response was obtained with Dpo20. Although this suggested that oligosaccharide chain length may affect immunogenicity in some age groups, interpretation of the results was complicated by the possibility of differential cross linking of the carrier protein made with biterminally activated oligosaccharides, as was the case here. The results also suggested that responses in adults were not predictive of the responses in infants.

Lastly (Anderson et al, 1989), the effects of chain length, exposure of terminal groups and hapten loading were investigated in a series of experiments with the carrier CRM_{197}. Three vaccines, C-4, C-6 and C-12 were synthesised from oligosaccharides with mean repeating units of 4,6 and 12. Only the reducing end of the (uniterminally activated) oligosaccharide fractions were able to couple to the carrier, thereby eliminating the potential problem of cross linking of the biterminally activated Dpo8 and Dpo20 oligosaccharides. One year old infants, immunized with C-4, C-6 or C-12, showed an increase in the geometric mean anti-PRP response over preimmune levels after both primary and secondary immunization (Table I). The fact that there were no significant differences between the means suggested that, for these conjugates, chain length was not important. In

Table 1. PRP oligosaccharides (OS) conjugated to CRM_{197}: Effect of saccharide chain length and hapten loading

Vaccine	Ribose/protein ($\mu g/\mu g$)	OS chain length	Anti-PRP ($\mu g/mL$) response in 1-year-old infants		
			Pre-	Post-1°	Post-2°
C-4	0.03	4	0.23	0.51	4.8
C-6	0.02	6	0.06	0.48	1.8
C-12	0.03	12	0,33	0.65	3.9
C-7	0.10	7	0.20	2.4	58
HbOC	0.20	20	0.20	5.9	27

contrast, the vaccine, C-7, made from an oligosaccharide with mean repeating unit of 7, and activated by introducing an aldehyde group into the reducing ribitol moiety, had a 3-5 fold higher hapten loading than C-4, C-6 and C-12, and a correspondingly higher anti-PRP response in 1 year old infants. A similar vaccine to C-7 with a different exposed terminal group was not significantly different in immunogenicity (results not shown). Overall, it was concluded that chain length and exposed terminal groups were less important than hapten loading in governing immunogenicity. One of these prototypes, HbOC, synthesised by biterminal activation of a PRP oligosacharide with mean repeating unit of 20, and coupling by reductive amination to the carrier, CRM_{197}, was selected for manufacture by Praxis Biologics.

This vaccine has demonstrated good anti-PRP responses, after two or three doses, in several immunogenicity trials in infants under 6 months of age (Kayhty et al, 1991; Decker et al, 1992), and, importantly, to be significantly more immunogenic than PRP or PRP-D. A comparative efficacy trial was established in Finland in 1988 in which children were immunized at 4,6 and 14-18 months of age with HbOC or PRP-D in a randomized trial. Although the follow-up is still continuing, early results (Ward, 1991; Eskola et al, 1991) indicated that, in accord with significantly higher anti-PRP responses, HbOC seemed to have a higher protective efficacy than PRP-D (2 vaccine failures in the HbOC group, 1 after the first and 1 after the second immunization, versus 11 vaccine failures in the PRP-D group, 6 after one dose and 5 after two doses). In a second efficacy trial in California, preliminary results (Ward, 1991) indicated that HbOC is highly efficacious in infants immunized at 2,4 and 6 months of age. The vaccine has now bee licensed in the USA for routine use in infants with recommended immunizations at 2,4,6 and 18 months of age.

PRP-Outer Membrane Protein Conjugate (PRP-OMP)

This conjugate was prepared by Merck Sharp & Dohme by covalent coupling of native PRP to partially purified outer membrane proteins (OMP) from N. Meningitidis via a bigeneric thioester (Marburg et al, 1986). This vaccine has also been studied in several immunogenicity trials (Decker et al, 1992; Einhorn et al, 1986; Weinberg and Granoff, 1988). The kinetics of the response appeared to be different from other Hib conjugate vaccines in that a single dose was immunogenic in infants as young as 2 months of age (Einhorn et al, 1986), although no clear booster response was seen after a second immuniztion (Decker et al, 1992; Einhorn et al, 1986; Weinberg and Granoff, 1988). An efficacy study recently conducted on Navajo Indians immunized at 2 and 4 months of age has indicated (Ward, 1991) high protective efficacy with no cases of Hib disease in the vaccinated cohort but 21 cases in the control group. This vaccine has recently been licensed in the USA for routine administration at 2,4 and 12 months of age.

PRP-Tetanus Toxoid Conjugate (PRP-T)

Pioneering work by Schneerson, Robbins and colleagues in the early 1980s (Schneerson et al, 1980; Schneerson et al, 1986) resulted in the development of several polysaccharide-protein conjugates coupled through 6-carbon spacer molecules. In conjunction with Institut Merieux native PRP has been coupled in this manner to tetanus toxoid (Schneerson et al, 1986). This vaccine is immunogenic in infants as young as 2 months of age, although, unlike PRP-OMP, one or two booster doses are required for high immunogenicity (Kayhty et al, 1991; Decker et al, 1992). Efficacy trials in infants are underway but results are not yet available.

Comparative Immunogenicity of PRP-D, HbOC, PRP-OMP and PRP-T

It is accepted that interassay and interlaboratory variations in the determination of anti-PRP levels can be considerable (Ward et al, 1988). For that reason, comparison of the vaccines based on the published literature is difficult. Several studies, as mentioned above, have already compared PRP-D with the other conjugates, leading to the clear conclusion that PRP-D is significantly less immunogenic in infants below the age of 6 months. One recently published study (Decker et al, 1992) in which anti-PRP titers have been determined in a single laboratory for the comparative assessment of all four conjugate vaccines will be discussed in detail here. Infants were immunized at 2,4 and 6 months of age, and anti-PRP antibody titers measured at 6 months (2 months after second immunization) and at 7 months (1 month after third immunization).

Comparative immunogenicity of PRP-D, HbOC, PRP-OMP and PRP-T

A sub-population of infants in the study was also assessed for antibody levels at 4 months (2 months after primary immunization). Mean anti-PRP levels after three doses of HbOC and PRP-T (Table II) were 3.08 µg/mL and 3.64 µg/mL, respectively, significantly higher than after PRP-OMP (1.14 µg/mL) or after PRP-D (0.28 µg/mL). As observed in previous studies, however, only PRP-OMP gave a significant response in the subset of infants evaluated after primary immunization. Other comparative immunogenicity studies are in progress.

Results of the immunogenicity trials have led to some discussion of the relative merits of these conjugate vaccines in a vaccination programme. On the basis of the low immunogenicity and questionable efficacy, it is difficult to support the use of PRP-D in infants. PRP-OMP appears to have both advantages (in raising a clinically useful response after one immunization) and disadvantages (lower response after two or three immunizations) compared with HbOC and PRP-T. Many issues related to the differences in immunogenicity of these conjugate vaccines have been raised by these studies, and it would be unsurprising if such differences were to occur with other polysaccharide conjugate vaccines.

Table 2. Anti-PRP responses to four Hib conjugate vaccines

Vaccine	Anti-PRP response (µg/mL) [a]			
	Pre-	Post-1° [b]	Post-2°	Post-3°
PRP-D (N=62)	0.07	0.06	0.08	0.28
HbOC (N=61)	0.07	0.09	0.13	3.08
PRP-OMP (N=64)	0.11	0.83	0.84	1.14
PRP-T (N=63)	0.10	0.05	0.30	3.64

[a] Infants vaccinated at 2,4 and 6 months

[c] Subset of infants (N=13 or 14)

Issues Raised by Hib Conjugate Vaccines

Although not covered in detail in this review, there are many issues related to the Hib conjugates which deserve some mention, not least because they are the same issues which will need to be addressed in evaluation of conjugate vaccines against other diseases, notably N. meningitidis Group B.

Clearly, optimisation of the immune response is dependent on parameters which define the chemical composition of the conjugates, as highlighted in Figure 1. Because of the complexity of the products, which often contain covalent three-dimensional cross-linkages between PRP and carrier protein, their characterization may not, at present, be attainable. This leads directly to concerns of lot-to lot variations in a vaccine.

The large variability in efficacy trial results (Peltola et al, 1984; Shapiro and Berg, 1990; Eskola et al, 1987; Eskola et al, 1991; Ward et al, 1990) conducted with PRP and PRP-D has been ascribed to geographical location, target population, environmental factors, chance due to the low incidence of disease and many other factors. One lesson to be learnt is that results with vaccines which give less-than-optimal protective efficacy in one study cannot be extrapolated to a different situation.

It has been noted in a number of efficacy studies (Daum et al, 1991) that vaccine failures frequently occurred within 14 days of administration of either PRP or PRP-conjugate vaccines. One simple hypothesis was that injection of the vaccine may result in a transient decline in serum antibody concentration by complex formation between PRP antigen and pre-existing anti-PRP antibody. It was subsequently shown (Sood and Daum, 1990) that immunization of adults and children (24 and 18 months, respectively) with PRP or PRP-D did indeed significantly lower the mean anti-PRP antibody concentration on or before the fourth day after immunization. However, adults or infants at 2 months and 4 months immunized with PRP-OMP did not show a significant decline in mean anti-PRP titers. The reason that PRP and PRP-D but not PRP-OMP vaccination induced transient antibody decline is not clear, but may relate to the small sample size, differences in the chemistry of the vaccines, or age of the subjects.

One suggestion is that the protein carrier for PRP-OMP being mitogenic for B-cells may result in the rapid increase in anti-capsular antibody, which would offset a transient drop in adsorbed anti-PRP antibody.

A recent publication (Clemins et al, 1992) has drawn attention to the possibility that PRP-conjugate vaccines may interfere with the response to other vaccines, notably Diphtheria-Pertussis-Tetanus (DPT) vaccine. Groups of infants immunized at 2,4 and 6 months of age with PRP-T given with DPT mixed in the same syringe or with DPT given at a different site resulted in significantly lower anti-pertussis responses than a group of infants given DPT alone. Anti-diphtheria toxin and anti-tetanus toxin responses were depressed, but not significantly. Further clarification is required to determine whether the diminished anti-pertussis responses are clinically significant, but this result certainly raises important issues concerning immunization policy.

Estimates of the minimum concentration of anti-PRP antibody required for protection against Hib disease have been extensively quoted as 0.15 $\mu g/mL$ for short term protection and 1.0 $\mu g/mL$ for long term protection and many immunogenicity study results are interpreted with regard to the percentage of responders with antibody levels higher or lower than these concentrations (Kayhty et al, 1991; Decker et al, 1992; Einhorn et al, 1986). Granoff and others (Granoff and Holmes, 1991; Amir et al, 1990; Schlesinger et al, 1992) have recently questioned the relationship between anti-PRP levels and functional activity in bactericidal and opsonic assays. Several qualitative differences in anti-PRP antibodies from individuals have been revealed, including avidity, isotype distribution, idiotype expression and fine antigenic specificity. Amir et al, (1990) found a clear correlation between antibody avidity and both bactericidal and opsonic activity. No other correlating factor was found after extensive study. This result was given practical importance in a recent study (Schlesinger et al, 1992) to determine the avidity and bactericidal activity of anti-PRP antibody following immunization with the Hib conjugates, HbOC, PRP-OMP and PRP-T. In a comparative assessment, HbOC elicited antibody of significantly higher avidity than that from PRP-T, which was of significantly higher avidity than that from PRP-OMP. The correlation between avidity and bactericidal activity was also apparent. These results may have implications for vaccine selection in a situation where only low serum antibody levels are attained. It is also suggested that antibody avidity may be an important parameter to consider in immunogenicity studies of the future.

It has been shown in several studies (Granoff and Holmes, 1991; Anderson et al, 1987; Weinberg et al, 1987) that immunization with HbOC or PRP-OMP results in immunological priming. That is, following an initial course of conjugate vaccine, individuals have been shown to respond to a subsequent booster dose of unconjugated PRP, even in infants too young to respond normally to unconjugated PRP. The booster response was no higher if restimulated with the conjugate vaccine. The results suggested that a population of memory B-cells had been induced after immunization with the PRP-conjugates. Of course one implication of these findings, although unproven, is that exposure to Hib bacteria in primed individuals would lead to an anti-PRP memory response and protection from disease. Interestingly, a group of children recovered from Hib disease in infancy did not respond to the unconjugated PRP (Weinberg et al, 1987), suggesting a lack of immunological priming. Although clearly of benefit here, it is not known whether the induction of memory B-cells is a general phenomenon that may occur with other conjugate vaccines. Additional means of inducing cell mediated immunity might be considered. For example, although commonly used protein carriers, such as diphtheria and tetanus toxoids, have the advantage of proven efficacy, they are limited because helper T-cells induced upon vaccination are of no significance during infection. The use of protein carriers from the pathogenic organism (homologous carriers) may be of additional benefit in conjugate vaccines. To be effective, the choice of homologous carrier should be based on the presence of conserved or cross-reactive T-cell epitopes within the molecule from different strains. Thus, immunization in the target population would result initially in protective immunity through the production of anti-capsular antibody. As antibody levels wane, immunological priming through the induction of T-cells reactive against the protein carrier should become important. In this model of protection, exposure to the invading bacteria results in the triggering of memory T-cells with correspondingly rapid anti-capsular antibody production. It is likely that the presence of both memory B- and T-cells would be of even greater benefit in protection from disease.

Several instances have been reported (Granoff and Holmes, 1991; Kaplan et al, 1992) in which Hib conjugate vaccines may have important advantages in individuals with impaired antibody responses to unconjugated PRP. Those conditions associated with poor antibody responses to PRP vaccines include sickle cell disease, malignancies and immunodeficiencies. Whatever the cause of this defect, immunological unresponsiveness can be overcome by immunizing with PRP-conjugate vaccines.

H. influenzae type b is transmitted by airborne droplet spread followed by oropharyngeal colonization. Carriage of the organism is quite common and only in rare cases does it lead to invasive disease. Although immunization with capsular polysaccharide vaccines, including PRP, has not resulted in a decrease in oropharyngeal carriage, a recent report (Takala et al, 1991) has indicated that immunization of children with PRP-D eliminated carriage of Hib. None of 327 3-year-old children who received the vaccine carried Hib, whereas 14 (3.5%) of 398 children who did not receive the vaccine were carriers. Carriage rates of non-type b H. influenzae were the same regardless of the vaccination status of the children. The results suggest that immunization with PRP-conjugates could reduce the prevalence of Hib in the general population and induce herd immunity.

MEMINGOCOCCAL B CONJUGATES

N. meningitidis is a major pathogen worldwide, occurring in both endemic and epidemic forms. The most common forms of meningococcal disease are meningitis (invasion of the meninges) and septicaemia, with a high mortality and morbidity rate. Children under 2 years of age are commonly associated with the disease, although it may be found in older age groups. The organism is subdivided into structurally and serologically distinct serogroups, defined by the capsular polysaccharide. Serogroups A, B and C account for more than 90% of disease, with group B alone responsible for 50-80% of disease in many countries.

The first purified polysaccharide vaccines against meningococcal disease became available in the late 1960s. The A and C polysaccharides proved to be immunogenic in adults and efficacious in field trials, and anti-capsular antibodies were strongly associated with protection (Gotschlich, 1984). However, immunity was not long-lasting, and the vaccines were ineffective in children below the age of 2 years. The group B polysaccharide was found to be non immunogenic and afforded no protection from group B meningococcal disease. The B polysaccharide (Figure 2) is a (2- 8)- -linked homopolymer of N-acetylneuraminic acid (NeuNAc). This is identical to the capsular polysaccharide of E. coli K1, the most common cause of neonatal meningitis, and suggests a common mechanism by which these pathogens evade the host defence systems.

8)—α—D—NeuNAc—(2→   *N. meningitidis* **B**

9)—α—D—NeuNAc—(2→   *N. meningitidis* **C**
   ⋮ 7,8
   OAc

8)—α—D—NeuNAc—(2→9)—α—D—NeuNAc—(2→   *E. coli* **K92**

Fig. 2. Structure of the polysaccharides from Neisseria
         meningitidis group B and C, and Escherichia coli K92

    In our initial studies (Lifely et al, 1981) designed to understand
the poor immunogenicity of B polysaccharide we found that at pH<6 the
polymer underwent an internal lactonisation (Figure 3A) involving the
carboxyl group of one residue and the OH-9 group of an adjacent residue.
This was consistent with the molecular conformation of the polymer
determined by nuclear magnetic resonance and theoretical molecular
mechanics calculations (Figure 3B) (Lifely et al, 1987). Lactonisation of
<10% of residues resulted in dramatic reduction of the antigenicity of the
polysaccharide. We reasoned that stabilisation of the polymer in its
native configuration may increase its immunogenicity. We found that
aluminium ions bound strongly to the carboxyl groups of the B
polysaccharide, markedly reducing lactonisation at low pH. Although this
by itself did not increase the immunogenicity of the polymer in animal
studies, addition of aluminium ions to B polysaccharide-conjugates, as
described below, markedly enhanced the immune response to the
polysaccharide component.

    Non-covalent conjugates of B polysaccharide with outer membrane
proteins (OMP) from N. meningitidis have been prepared in several
laboratories (Zollinger et al, 1979; Moreno et al, 1985). We developed a
methodology for the preparation of naturally-formed conjugates which were
high molecular weight (>20 million daltons) complexes held together by
hydrophobic linkages. Mice immunized with this vaccine gave a transient

A   B

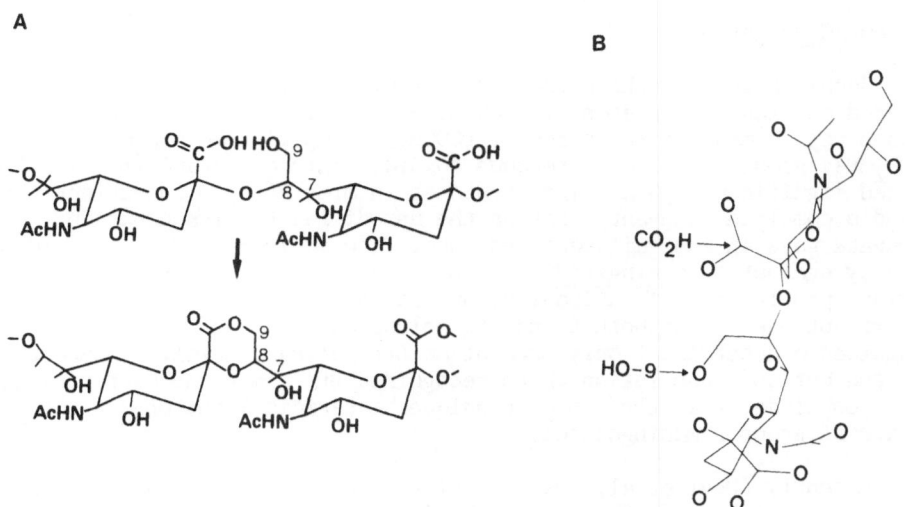

Fig. 3. Neisseria meningitidis group B polysaccharide. A: Lactonisation
        at low pH; and B: Calculated conformation of a disaccharide unit

rise in anti-B antibody which was mainly of the IgM class, typical of anti-B responses seen after infection. Mice were however protected from challenge with N. meningitidis group B strains. Upon addition of aluminium ions to this vaccine to stabilise the B polysaccharide structure, significantly higher antibody responses were observed in mice, which could be substantially boosted by a second immunization (Lifely et al, 1987).

We recently (Lifely et al, 1991) conducted an immunogenicity trial using this vaccine in 25 male adult volunteers, divided into three dose groups and immunized three times at four-weekly intervals. The mean anti-B response was consistently lower in the low (50 µg) dose group throughout the vaccination schedule, suggesting this is a sub optimal dose in adults. The anti-B response in the two higher dose groups (100 µg and 150 µg) was highly significant, rising from 2.84 µg/mL before vaccination to 13.50 µg/mL 1 week after third vaccination. The mean antibody levels attained after three doses of the vaccine were maintained without marked decline 2 months later. Thereafter only a gradual decline in the anti-B response was apparent, which after 1 year was still significantly higher than prevaccination levels (Figure 4). The response was predominantly of the IgM isotype. Biological efficacy of pooled pre- and postvaccination sera was assessed in a murine passive protection model. Mice passively immunized with 11 uL pooled postvaccination sera were protected from lethal challenge with N. meningitidis group B organisms, whereas equivalent protection only occurred in mice injected with 100 uL prevaccination sera. Protection was abrogated when sera had been previously adsorbed with B polysaccharide to remove anti-B antibodies.

In conclusion, this study showed that the vaccine was immunogenic in adults, with two doses given four weeks apart being sufficient to raise anti-B levels 6-fold. A third dose appeared to have a minimal effect. Despite previous findings that anti-B antibodies declined rapidly, mean anti-B levels in this study remained elevated for at least one year, suggesting that long-lived or memory B-cells with specificity for B polysaccharide may have been generated. Persistence of the response may have reflected the high degree of binding of B polysaccharide to OMP in this study and the presence of aluminium ions to stabilize the B polysaccharide. In addition, the selection of the carrier OMP (N. meningitidis serotype 6) may be important since this gave an anti-B response in mice higher than with other B polysacchride-OMP conjugates tested.

Covalent Conjugates

Jennings and coworkers (Jennings et al, 1986; Jennings et al, 1989) prepared covalent conjugates of a chemically-modified B polysaccharide, through replacement of the N-acetyl ($COCH_3$) group with an N-propionyl ($COCH_2CH_3$) group, coupled to tetanus toxoid. This conjugate (BPr-TT) induced significant IgG and IgM responses in mice after 3 injections in Freund's complete adjuvant, whereas the unmodified B polysaccharide conjugate gave an insignificant response. The antisera had bactericidal activity against N. meningitidis group B organisms, and contained two distinct populations of antibodies; one recognised a cross-reactive determinant present on both B and BPr polysaccharides and the other recognised a determinant only present on BPr polysaccharide. Interestingly only the antibody population which recognised BPr alone was bactericidal. It was concluded that BPr mimics a unique bactericidal epitope on the surface of group B meningococci.

Recently (Devi et al, 1991), an interesting approach was adopted by Devi et al. who coupled the polysaccharides from group B or C meningococci, or E. coli K92 (see Figure 2) to tetanus toxoid via a 6-carbon spacer.

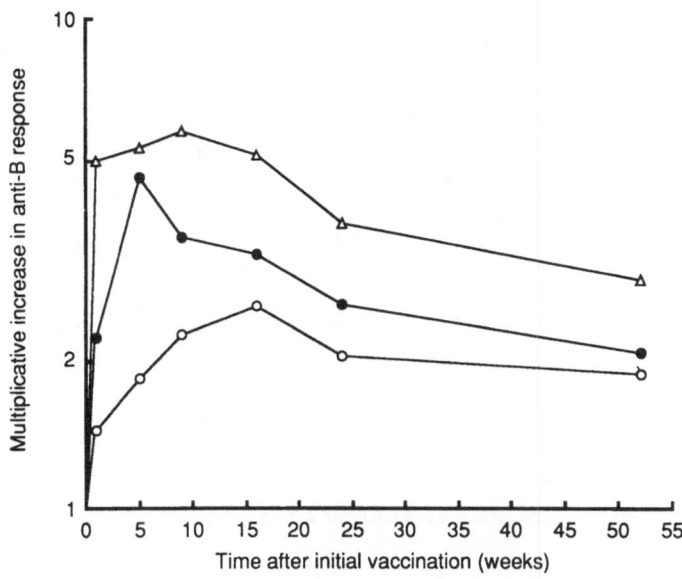

Fig. 4. Persistence of the anti-B response in male adult
volunteers after three immunizations with a meningococcal
B polysaccharide-outer membrane protein (OMP) vaccine.
0, 50 μg dose group; Δ, 100 μg dose group, ●, 150 μg
dose group

Antibody responses measured in mice after the third immunization with these
conjugates are summarized in Table III. Both IgM and IgG anti-B antibodies
are raised following immunization with B-TT but not with C-TT. Conversely,
anti-C antibodies were raised in mice immunized with C-TT. Interestingly,
both anti-B and anti-C antibodies were elicited by immunization with K92-
TT. It is possible that this conjugate may provide a novel route to
vaccinate against N.meningitidis group B and group C, and E. coli K1
disease.

NAGO: A NOVEL ADJUVANT

The inductive interaction of antigen presenting cells (APC) with T-
cells is the primary specific event in most immune responses. In addition,
at the chemical level, Rhodes and coworkers (Rhodes, 1989; Zheng et al,
1992) have shown that reversible interactions between carbonyl (C=O) and
amino (NH_2) functions (Schiff base formation) appear to be essential for
antigen-specific T-cell activation. This suggests a mechanism to enhance
responses by increasing the expression of reactive groups on these cells.
This can be achieved using galactose oxidase (GO) which generates aldehyde
groups on carbon-6 of terminal galactose residues. Since many galactose
residues are only terminally exposed after removal of N-acetylneuraminic
acid (NeuAc) by neuraminidase (NA) treatment, the combination of NA+GO
should optimally generate aldehyde groups on APC and T-cells.

The potency of NAGO as an adjuvant was investigated (Zheng et al,
1992) using the non-covalent conjugate of meningococcal B polysaccharide
coupled to OMP, as described above. Mice were primed with the conjugate
given with or without adjuvant, and boosted 28 days later with the
conjugate in alum. The results (Figure 5) clearly indicate that NAGO given
with or 1 hour before the conjugate is a superior adjuvant to alum in

Table 3. Antibody responses to polysaccharides from group B or C
meningococci, or E.coli K92 conjugated to tetanus toxoid

Immunogen	Anti B response[b]		Anti-C response[b] (μg/mL)	
	IgM(μg/mL)	IgG(U)[c]	IgM	IgG
B-TT	1.50	1.81	ND	ND
C-TT	0.05	0.05	0.53	107.5
K92-TT	1.20	17.2	0.68	21.4

[b] Antibody responses measured by ELISA after third immunization.

[c] Results are expressed as a percentage of a high titered human serum standard.

priming for a secondary anti-B response. This raises the prospect for
enhancement of immune responses in humans to such conjugates using the
mechanism of galactose oxidation.

CONCLUSIONS

A new generation of polysaccharide vaccines is upon us. The concept
of conjugating T-independent polysaccharides to protein carriers to
generate T-dependent responses is not new. Much work over the last decade
with H. influenzae type b conjugates has shown that this concept can be

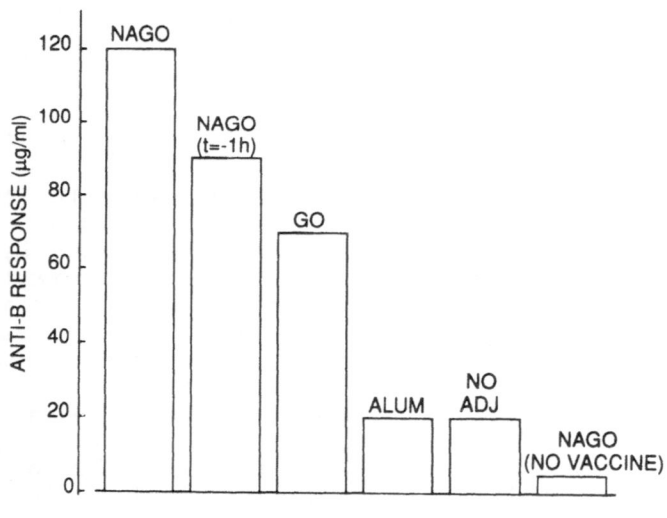

Fig. 5. Immunization with B polysaccharide - OMP (MB6) conjugate.
Anti-B responses in mice primed with a menigococcal B polysaccharide-outer membrane protein (OMP) vaccine with or
without adjuvant. All mice were boosted 28 days later with
the vaccine given with alum.

transformed into reality. Three of these Hib conjugates have now been
licensed for use in infants; they have and undoubtedly will continue to
save lives and prevent serious sequelae in a large number of children.
There are as yet many unanswered questions concerning these conjugates, but
it seems that the success already seen with Hib conjugates will stimulate
further research in this area against other diseases. N. meningitidis
group B disease has stubbornly resisted progress for some time. Recent
initiatives in this area, however, provide hope that effective vaccines may
also soon become available.

ACKNOWLEDGEMENTS

Many of my colleagues at the Wellcome Research Laboratories have made
significant contributions to the work on a meningococcal group B vaccine.
In particular, I would like to thank Carlos Moreno, John Lindon, Jane
Esdaile, Zhen Wang, Susan Roberts and Wilma Shepherd. The work on the NAGO
adjuvant was masterminded by John Rhodes, with support from Biao Zheng,
Sara Brett, John Tite and Tom Brodie.

REFERENCES

Amir, J., Liang, X. and Granoff, D. M., 1990, Variability in the functional
        activity of vaccine-induced antibody to Haemophilus influenzae type
        b, Pediatr. Res., 27:358.
Anderson, P., 1982, Antibody responses to Haemophilus influenzae type b and
        diphtheria toxin induced by conjugates of oligosaccharides of the
        type b capsule with the nontoxic protein CRM$_{197}$,
        Infect. Immun., 39:233.
Anderson, P., Pichichero, M., Edwards, K., Porch, C. R. and Insel, R.,
        1987, Priming and induction of Haemophilus influenzae type b
        capsular antibodies in early infancy by Dpo20, an oligosaccharide-
        protein conjugate vaccine, J. Pediatr., 111:644.
Anderson, P., Pichichero, M. E. and Insel, R. A., 1985, Immunogens
        consisting of oligosaccharides from the capsule of Haemophilus
        influenzae type b coupled to diphtheria toxoid or the toxin protein
        CRM 197, J. Clin. Invest., 76:52.
Anderson, P., Pichichero, M. E., Insel, R. A., Betts, R., Eby, R. and
        Smith, D. H., 1986, Vaccines consisting of periodate-cleaved
        liposaccharides from the capsule of Haemophilus influenzae type be
        coupled to a protein carrier: structural and temporal requirements
        for priming in the human infant, J. Immunol., 137:1181.
Anderson, P., Pichichero, M. E., Stein, E. C., et al., 1989, Effect of
        oligosaccharide chain length, exposed terminal group, and hapten
        loading on the antibody response of human adults and infants to
        vaccines consisting of Haemophilus influenzae type b capsular
        antigen uniterminally coupled to the diphtheria protein CRM197,
        J. Immunol., 142:2464.
Clemens, J. D., Ferreccio, C., Levine, M. M., et al., 1992, Impact of
        Haemophilus influenzae type b polysaccharide-tetanus protein
        conjugate vaccine on responses to concurrently administered
        diphtheria-tetanus-pertussis vaccine, J. Amer. Med. Assoc.,
        267:673.
Daum, R.S., Siber, G.R., Ballanco, G.A. and Sood, S.K., 1991, Serum anti-
        capsular-antibody response in the first week after immunization of
        adults and infants with the Haemoplilus influenzae type b -
        Neisseria meningitidis outer membrane protein complex conjugate
        vaccine, J.Infect.Dis., 164-1154.
Decker, M. D., Edwards, K. M., Bradley, R. and Palmer, P., 1992,
        Comparative trial in infants of four conjugate Haemophilus

influenzae types b vaccines, J. Pediatr., 120:184.

Devi, S. J. N., Robbins, J. B. and Schneerson, R., 1991, Antibodies to poly [(2→8)-⍺-N-acetylneuraminic acid] and poly[(2→9)-⍺-N-acetylneuraminic acid] are elicited by immunization of mice with Escherichia coli K92 conjugates: potential vaccines for groups B and C meningococci and E. coli K1, Proc. Natl. Acad. Sci. USA, 88:7175.

Einhorn, M. S., Weinberg, G. A., Anderson, E. L., Granoff, P. D. and Granoff, D. M., 1986, Immunogenicity in infants of Haemophilus influenzae type b polysaccharide in a conjugate vaccine with Neisseria meningitidis outer-membrane protein, Lancet, ii:229.

Eskola, J., Peltola, H., Takala, A. K., et al., 1987, Efficacy of Haemophiolus influenzae type b polysaccharide-diphtheria toxoid conjugate vaccine in infancy, N. Engl. J. Med., 317:717.

Eskola, J., Takala, A., Kayhty, H., Peltola, H. and Makela, P. H., 1991, Experience in Finland with Haemophilus influenzae type b vaccines, Vaccine, 9:S15.

Gordon, L. K., 1984, Characterization of a hapten-carrier conjugate vaccine: Haemophilus influenzae-diphtheria conjugate vaccine in: "Modern Approaches to Vaccine," R. M. Chanock and R. A. Lerner, eds., Cold Spring Harbor Press, Cold Spring Harbor, New York.

Gotschlich, E. C., 1984, Meningococcal meningitis in: "Bacterial Vaccines," R. Germanier, ed., Academic Press Inc., Orlando.

Granoff, D. M. and Holmes, S. J., 1991, Comparative immunogenicity of Haemophilus influenzae type b polysaccharide-protein conjugate vaccines, Vaccine, 9:S30.

Jennings, H. J., Gamian, A., Michon, F. and Ashton, F. E., 1989, Unique intermolecular bactericidal epitope involving the homosialopolysaccharide capsule of the cell surface of group B Neisseria meningitidis and Escherichia coli K1, J. Immunol., 142:3585.

Jennings, H. J., Roy, R. and Gamian, A., 1986, Induction of meningococcal group B polysaccharide-specific IgG antibodies in mice using an N-propionylated B polysaccharide-tetanus toxoid conjugate vaccine, J. Immunol., 137:1708.

Kaplan, S. L., Duckett, T., Mahoney, D. H. Jr., et al., 1992, Immuno-genicity of Haemophilus influenzae type b polysaccharide-tetanus protein conjugate vaccine in children with sickle hemoglobinopathy or malignancies, and after systemic Haemophilus influenzae type b infection, J. Pediatr., 120:367.

Kauhty, H., Eskola, J., Peltola, H., et al., 1991, Antibody responses to four Haemophilus influenzae types b conjugate vaccines, Amer. J. Dis. Child., 145:223.

Lifely, M. R., Gilbert, A. S. and Moreno, C., 1981, Sialic acid poly-saccharide antigens of Neisseria meningitidis and Escherichia coli: esterification between adjacent residues, Carbohydr. Res., 94:193.

Lifely, M. R., Moreno, C. and Lindon, J. C., 1987, An integrated molecular and immunological approach towards a meningococcal group B vaccine, Vaccine, 5:11.

Lifely, M. R., Roberts, S. C., Shepherd, W. M., et al., 1991, Immuno-genicity in adult males of a Neisseria meningitidis group B vaccine composed of polysaccharide complexed with outer membrane proteins, Vaccine, 9:60.

Marburg, S., Jorn, D., Tolman, R. L., et al., 1986, Bimolecular chemistry of macromolecules: synthesis of bacterial polysaccharide conjugates with Neisseria meningitidis membrane protein, J. Am. Chem. Soc., 108:5282.

Moreno, C., Lifely, M. R. and Esdaile, J., 1985, Immunity and protection of mice against Neisseria meningitidis group B by vaccination, using polysaccharide complexed with outer membrane proteins: a comparison with purified B polysaccharide, Infect. Immun., 47:527.

Peltola, H., Kayhty, H., Virtanen, M. and Makela, P. H., 1984, Prevention of Haemophilus influenzae type b bacteremic infections with the capsular polysaccharide vaccine, N. Engl. J. Med., 310:1561.

Rhodes, J., 1989, Evidence for an intercellular covalent reaction essential in antigen-specific T cell activiation, J. Immunol., 143:1482.

Schlesinger, Y., Granoff, D. M. and the vaccine study group, 1992, Avidity and bactericidal activity of antibody elicited by different Haemophilus influenzae type b conjugate vaccines, J. Amer. Med. Assoc., 267:1489.

Schneerson, R., Barrera, O., Sutton, A. and Robbins, J. B., 1980, Preparation, characterization, and immunogenicity of Haemophilus influenzae type b polysaccharide-protein conjugates, J. Exp. Med., 152:361.

Schneerson, R., Robbins, J. B., Parke, J. C., et al., 1986, Quantitative and qualitative analyses of serum antibodies elicited in adults by Haemophilus influenzae type b and pneumococcus type 6A capsular polysaccharide-tetanus toxoid conjugates, Infect. Immun., 52:519.

Shapiro, E. D. and Berg, A. T., 1990, Protective efficacy of Haemophilus influenzae type b polysaccharide vaccine, Pediatrics, 85:643.

Sood, S. K. and Daum, R. S., 1990, Disease caused by Haemophilus influenzae type b in the immediate period after homologous immunization: immunologic investigation, Pediatrics, 85:698.

Takala, A. K., Eskola, J., Leinonen, M., et al., 1991, Reduction of oropharyngeal carriage of Haemophilus influenzae type b (Hib) in children immunized with an Hib conjugate vaccine, J. Infect. Dis., 164:982.

Ward, J., 1991, Prevention of invasive Haemophilus influenzae type b disease: lessons from vaccine efficacy trials, Vaccine, 9:S17.

Ward, J. I., Brenneman, G., Letson, G., et al., 1990, Limited protective efficacy of an H. influenzae type b conjugate vaccine (PRP-D) in native Alaskan infants, N. Engl. J. Med., 323:1393.

Ward, J. I., Greenberg, D. P., Anderson, P. W., et al, 1988, Variable quantitation of Haemophilius influenzae type b anticapsular antibody by radioantigen binding assay, J.Clin.Microbiol., 26:72.

Weinberg, G. A., Einhorn, M. S., Lenoir, A. A., Granoff, P. D. and Granoff, D. M., 1987, Immunologic priming to capsular polysaccharide in infants immunized with Haemophilus influenzae type b polysaccharide-Neisseria meningitidis outer membrane protein conjugate vaccine, J. Pediatr., 111:22.

Weisnberg, G. A. and Granoff, D. M., 1988, Polysaccharide-protein conjugate vaccines forthe prevention of Haemophilus influenzae type b diseases, J. Pediatr., 113:621.

Zheng, B., Brett, S. J., Tite, J. P., Lifely, M. R., Brodie, T. and Rhodes, J., 1992, Galactose oxidation in the design of immunogenic vaccines, Science, 256:1560.

Zollinger, W. D., Mandrell, R. E., Grifiss, J. M., Altieri, P. and Berman, S., 1979, Complex of meningococcal group B polysaccharide and type 2 outer membrane protein immunogenic in man, J. Clin. Invest., 63:836.

LIPOSOMAL DELIVERY OF CYTOKINES: A MEANS TO IMPROVE THEIR THERAPEUTIC

PERFORMANCE

Yechezkel Barenholz[1], Orna Palgi[1,2], Galina Golod[1,2], Noam Emanuel[2], Yaron Rutkowski[2], Efrat Braun[2] and Eli Kedar[2]

[1]Department of Biochemistry and [2]Lautenberg Center of Immunology
Hebrew University – Hadassah Medical School
PO Box 1172, Jerusalem 91010, Israel

INTRODUCTION

Animal studies indicate that cytokines have the potential of being efficacious therapeutic agents in a broad spectrum of diseases (Meager, 1990; De Vita et al, 1991). All cytokines are peptides, most of them in the range of 20 kDa molecular mass. Their specificity is related to the fact that they operate through a receptor-mediated process at very low concentrations (10^{-10}-10^{-12} M) and at a short range in autocrine or paracrine fashion. The diversity and complexity of their activity is described in Table 1.

This complexity explains (i) why the use of combinations of cytokines can produce additive-synergistic or even antagonistic effects; (ii) why the effects of cytokines are influenced by the timing of delivery with respect to the clinical status and other treatments of the subject.

Therefore, the net effect of cytokine treatment may be beneficial, indifferent, or even damaging (when their toxic effects exceed their benefits). Most of the clinical efforts so far in therapy by cytokines were directed towards cancer therapy (De Vita et al, 1991; Rosenberg, 1991; Kedar and Klein, 1992).

CYTOKINES AND CANCER THERAPY

The focussing on various aspects of cancer therapy is based on extensive tissue culture and animal studies which demonstrate that under suitable conditions cytokines have a broad spectrum of activities as immunotherapeutic agents. These effects are related to the type of cytokine, to the cancer model, and to the exact timing and mode of cytokine delivery. The activities are summarized in Table 2.

The marked therapeutic potential on the one hand and, on the other hand, the excellent feasibility through biotechnology of sufficient availability of the desired recombinant cytokine at the quality required of a pharmaceutical agent, permit a broad range of clinical trials with various cytokines, such as GM-CSF (Lieschke and Burgess, 1992), G-CSF

New Generation Vaccines, Edited by G. Gregoriadis
et al., Plenum Press, New York, 1993

Table 1. Profile of cytokine biological activities.

1. Diversity: One cell has the potential to produce many cytokines.

2. Pleitropy: one cytokine has many biological activities.

3. Cascade: One cytokine may trigger production of other cytokines and cytokine receptors.

4. Redundancy: Several cytokines may have the same activity on the same target cells (Meager, 1990).

Table 2. Potential activities of cytokines in cancer immunotherapy.

1. Direct effects on the tumor: cytolysis, cytostasis, vasculature damage, terminal differentiation (e.g., *TNFs, IFNs).

2. Enhancing the expression of TAA, MHC antigens, and adhesion molecules (e.g., IL-2, TNFs, IFNs, CSFs, IL-4, IL-6).

3. Recruiting, expanding, and stimulating endogenous effector cells (T-cells, NK/LAK cells, macrophages, e.g., IL-2, IL-4, IL-6, CSFs, IFNs).

4. Maintaining and enlarging adoptively transferred lymphocyte populations (mainly IL-2).

5. Protecting against chemo/radiotherapy-induced myelosuppression (CSFs, IL-1, IL-6).

(For review, see Kedar and Klein, 1992)

*see List of Abbreviations

(Metcalf and Morstyn, 1991), IL-2 (Lotze and Rosenberg, 1991), TNF (Alexander, 1991), and erythropoietin (for general reviews see, De Vita et al, 1991, Oettgen, 1991; Kedar and Klein, 1992). Results of several randomized controlled studies in which the cytokine treatment was compared with a placebo are disappointing, however (Bronchud, 1990; Groopman, 1991; Lancet Editorial, 1991).

STRATEGIES TO OVERCOME PROBLEMS IN THE CLINICAL USE OF CYTOKINES

One explanation for the disappointing results in clinical trials with cytokines is related to their very short half-life ($t\frac{1}{2}$) in the circulation and the nonspecific biodistribution, namely unfavourable pharmacokinetics (De Vita et al, 1991; Kedar and Klein, 1992). These disadvantages are more pronounced in humans than in mice (where most preclinical trials have been performed).

Two different approaches may be applied to modify the pharmacokinetics of cytokines in order to improve their therapeutic index:

(i) Chemical modification: Designed to prolong circulation half-life ($t\frac{1}{2}$) and possibly modify interactions with various cell types.

(ii) Encapsulation in drug delivery system, such as liposomes, in order to obtain slow release, extend t½, modify biodistribution, reduce toxicity, and possibly improve targeting.

In this paper we evaluate the use of liposomes as cytokine carriers for three cytokines: recombinant human TNF (Cetus, USA), recombinant mouse GM-CSF (Immunex, USA), and recombinant human IL-2 (Cetus, USA), which differ in their spectrum of biological activities as well as physical and chemical properties (Meager, 1990). For IL-2 we will compare a chemically modified IL-2 with liposomal IL-2 (Barenholz et al, 1991; Barenholz et al, in preparation; Kedar et al, 1991, 1992). These three cytokines were selected because the outcome of their clinical use in free, soluble form is questionable (Bronchud, 1990; Groopman, 1991; Lotze, 1991; Lotze and Rosenberg, 1991; Kedar and Klein, 1992).

WHY LIPOSOMES?

The reasons for selection of liposomes as a suitable cytokine controlled delivery system are related to liposome biocompatibility, the great versatility in liposome types as well as in liposome methods of preparation, and last but not least, the pharmaceutical feasibility of producing liposomal dosage forms as required by the pharmaceutical industry in industrial settings (for review see Barenholz and Crommelin, 1993; Barenholz, 1992). Finally, animal studies with liposomal IL-2 formulations show encouraging results (reviewed in Kedar and Klein, 1992).

THE DESIGN OF A LIPOSOMAL CYTOKINE DELIVERY SYSTEM

The design of a liposomal delivery system requires working in two parallel lines:

(A) **Biological aspect:** Having the appropriate animal model systems required to assess the pharmacokinetics, toxicity and therapeutic efficacy of the cytokine. These are described in Table 3.

(B) **Engineering of liposomal cytokine formulation:** This aspect is rather complex and cannot be dealt with in detail here (for review see Barenholz and Amselem, 1993; Barenholz and Crommelin, 1993; Barenholz et al, 1993).

The selection of lipids for the liposomes is dependent on the requirements for the liposome performance. For the purpose of this study we classified the liposomes into two groups:

1. Classical large liposomes (>200 nm) which are composed of fluid lipids, namely, those with gel-to-liquid crystalline phase transition temperature (Tm) lower than 37°C. These liposomes have a short circulation time and fast uptake by the reticuloendothelial system (RES); therefore they are referred to as RES directed liposomes (Senior, 1987; Barenholz and Crommelin, 1993).

2. Stealth™ - small liposomes (<100 nm) which due to their low rate of uptake by the RES have prolonged circulation time, and due to their small size have the ability to extravasate with good localization in tumors at various sites (Gabizon, 1992; Lasic et al, 1991; for review see Barenholz and Crommelin, 1993 and references therein).

Recently we demonstrated the differences in pharmacokinetics and biodistribution of encapsulated doxorubicin in cancer patients (Gabizon

Table 3. Quantification and efficacy models of liposomal cytokine formulations.

	In vitro assay	In vivo model in mice	Dependence of biological activity in vitro on cytokine release
TNFα	Cytotoxicity using TNF-sensitive cells (BALB/c clone 7 fibrosarcoma)	Antitumor effect using sarcoma (MCA 105) and carcinoma (M109)	YES
IL-2	Cell proliferation of IL-2 dependent T-cell line (CTLL-2)	as above	NO
GM-CSF	Cell proliferation using GM-CSF dependent cell line (32Dcl) and colony formation assay (GM CFU)	Hematopoietic recovery following chemotherapy or bone marrow transplantation, (in vivo/in vitro), CFU-S (in vivo in spleen), resistance to microorganisms	NO

et al, 1991, 1992). It was found that while the RES directed liposomes have a t½ of 15-20 min and are distributed mainly to the RES, the small Stealth liposomes have a t½ of ca. 42h, with greater drug levels detected in the patient's malignant effusions and cells in these effusions.

The design of the delivery system is dependent on the physico-chemical properties of the cytokines, which affect the approach for liposome preparation and also determine what size of liposomes has a therapeutic feasibility. For example, small liposomes cannot be used for delivery of cytokines with low encapsulation efficiency. The physico-chemical characterization of the three cytokines used in our study is summarized in Table 4, which also describes encapsulation efficiencies relative to those of a marker for aqueous trapped volume (^{3}H-inulin, Lichtenberg and Barenholz, 1988).

Finally, for the selection of liposome preparation method, the stability of the cytokines has to be assessed under conditions used in various methods of liposome preparation. Table 5 summarizes the stability data.

SUMMARY OF THE DESIGN OF LIPOSOMAL-CYTOKINE FORMULATIONS

Various assays have been used to quantify the activity of TNF, GM-CSF, and IL-2 and to assess their therapeutic efficacy in vivo. The stability profile, and, especially, the almost complete loss of activity of TNF (but not of GM-CSF and IL-2) in contact with organic solvents and detergents drastically limit the means available for preparing liposomal TNF. For liposomal GM-CSF and liposomal IL-2 formulations, almost all approaches available for liposome preparation can be applied (Lichtenberg and Barenholz, 1988; Barenholz and Crommelin, 1993). For liposomal TNF, only methods which do not require organic solvents or detergents can be

---

Stealth^{TM} is a trade mark of Liposome Technology Inc., referring to liposomes having long circulation time due to the presence of special lipids.

Table 4. Physico-chemical characterization of cytokines used.

(Source)	State of aggregation	Molecular mass (kDa)	Glycosylation	Hydrophobicity	3D structure	Encapsulation efficiency (cytokine/aqueous trapped volume)
rh TNFα (Cetus)	Homotrimer	17 x 3' (157 x 3aa)	-	Low	High resolution 3 antiparallel β-sandwich which forms "jelly roll" structural motif	~ 1.0
rh-IL-2 (Cetus)	Monomer	14 (149aa)	-	Very hydrophobic	High resolution 5-6 helical stretches	>> 1.0
rm-GM-CSF (Immunex)	Monomer	22	+	Contain 4 hydrophobic stretches	Low resolution containing protein segments that form 4 helices	> 1.0

Table 5. Cytokine stability under conditions used in various methods of liposome preparation and characterization.

Treatment	TNF	IL-2	GM-CSF
Freezing and thawing (10 times)	+	+	+
Ultrasonic irradiation 30 min (bath)	+	+	+
Diethyl ether	-	?	+
Tertiary butanol	-	+	+
*SDS (7-35 mM)	-	+	+
Heating 40°C; 50°C; 60°C (15 min)	???	+; +-; -	+; +; ?

+, stable; -, not stable; ?, under investigation.
* The detergent has to be nontoxic to the cells used in the bioassay.

used. The high sensitivity of TNF is related to the importance of its quaternary structure to its activity. The other major limitation of the liposomal TNF system and its clinical application is the low encapsulation of TNF (Table 4) in small-size (<100 nm) Stealth extravasating liposomes.

The very high percentage of encapsulation (>85%) of IL-2 and GM-CSF (Barenholz et al, 1991) in the liposome, which is higher than the percentage encapsulation of ^{3}H-inulin (liposome-aqueous phase marker), suggests that at least part of the cytokine is associated with the liposome membrane. This may be related to the presence of hydrophobic domains of these two peptides (Table 4). This is also supported by the fact that GM-CSF and IL-2 can interact with their target cells in culture without first being released from the liposomes (Table 3).

BIOLOGICAL ACTIVITY AND THERAPEUTIC PERFORMANCE OF LIPOSOMAL CYTOKINE FORMATIONS

Liposomal GM-CSF and liposomal IL-2 exhibited an improved therapeutic efficacy as compared with the soluble cytokine. We showed that: (i) liposomal GM-CSF (RES directed) was more efficacious than soluble GM-CSF for boosting extramedullary hematopoiesis in mice; (ii) lower doses and less frequent administrations were required (compared with soluble GM-CSF) to stimulate hematopoiesis; (iii) liposomal GM-CSF kept its activity for at least 5 months at 4°C (for more details see Barenholz et al, 1991).

Liposomal IL-2 was compared with unmodified IL-2 and IL-2 that was chemically modified by covalently attaching polyethylene glycol (PEG-IL-2) in order to extend its circulation time (Knauf et al, 1988; Kedar and Klein, 1992; Francis, 1992), for antitumor activity in mice, in combination with chemotherapy (Kedar et al, 1991, 1992). In order to achieve extravasation of liposomes in the tumors we selected small unilamellar Stealth liposomes (Gabizon, 1992; Gabizon et al, 1992) as a delivery system for

Table 6. Soluble IL-2, PEG-IL-2, and Stealth liposomal IL-2: Pharmacokinetics and immunomodulation in mice.

Parameter	IL-2	PEG-IL-2	Lip-IL-2
$t_{1/2\alpha}$ plasma	1.5 min	50 min	20 min
Duration of plasma level of ≥ 100 U/ml	40 min	12-24h	8h
% increase in WBC	0-20	0-100	0-40
% increase in spleen cell number	0-100	200-400	100-350
% increase in peritoneal exudate cell number	0-80	100-600	20-400
Fold increase in LAK cell activity (vs. IL-2)		20-100	10-30

Pharmacokinetics: 5×10^4 Cetus units (CU) injected iv.

Immunomodulation: IL-2 (4×10^4 CU/dose) administered iv once daily on days 1-5; PEG-IL-2 or S-Lip-IL-2 (1×10^5 CU/dose) iv, days 1,4; tests performed on days 6-8.

unmodified IL-2. Hereafter in this paper, liposomes refers to Stealth liposomes. Lipid composition selected was: egg phosphatidylcholine (PC), N-carbamylpolyethylene glycol methyl ether 1,2-distearoyl-sn-glycero-3-phosphoethanolamine triethylammonium salt, and cholesterol (55:5:40 mole ratio). The fast loss of activity of IL-2 at 60°C (Table 5) and the need to prepare the liposome above the gel-to-liquid crystalline temperature (Tm) forced us to use egg PC and not the more saturated hydrogenated soy PC.

Table 6 summarizes the comparison of pharmacokinetics and immuno-modulation between soluble IL-2, PEG-IL-2 and liposomal IL-2. Indeed the encapsulation of IL-2 in liposomes and, more so, "Pegylation" of IL-2, extends plasma circulation time. Of special interest is the time in which IL-2 level in plasma is above 100 Cetus units/ml (CU/ml), which was extended from 40 min for soluble IL-2 to 8 h and 12-24 h for liposomal IL-2 and PEG-IL-2, respectively. The relatively short plasma t½ of IL-2 delivered via Stealth liposomes when compared with the much longer t½ of the Stealth liposomes with or without drugs (>10 h) (Gabizon, 1992; Gabizon et al, 1992; Braun, 1992) is related to the rapid in vivo release of IL-2 from the liposomes and its partition into cells and other plasma components which serve as a sink for the highly amphipathic IL-2 (Barenholz and Amselem, 1993; Barenholz and Crommelin, 1993).

Comparing the immunomodulating activity of IL-2 (Table 6), it is clear that the improvement is related to the degree of prolongation of IL-2 circulation time. The level of effect is in the order of PEG-IL-2 > liposomal IL-2 >> IL-2.

Liposomal IL-2 was found more stable than soluble IL-2 in solution at 4°C and 37°C. It retained full activity (>95%) at 4°C and 37°C for at least 6 weeks, while at the same storage time soluble IL-2 lost most of its activity. The encouraging results on the improved immunomodulation of liposomal IL-2 and PEG-IL-2 led us to perform therapy experiments (one experiment is illustrated in Fig. 1), in which a combination of cyclophos-phamide (CY) chemotherapy and IL-2 immunotherapy was used in mice carrying

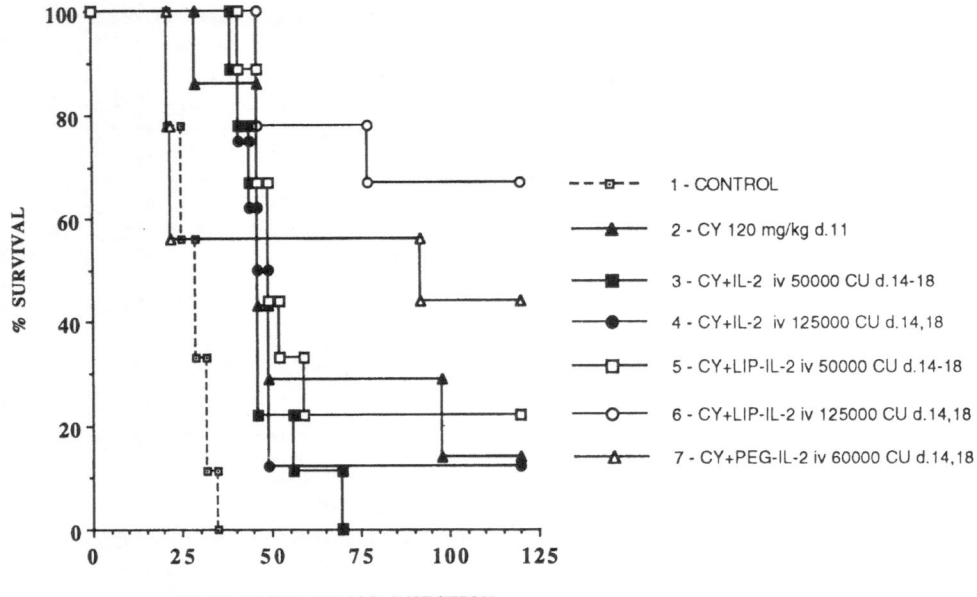

Fig. 1. Chemo-immunotherapy of Balb/C mice carrying M109
(lung metastases) carcinoma.

5 x 10^5 M109 tumor cells were injected i.v. on
day 0 to Balb/c mice: CY (cyclophosphamide) (120 mg/kg)
was administered i.p. on day 11; IL-2 and LIP-IL-2
were given i.v. (50,000 CU daily) either for 5 days
starting on day 14, or twice (125,000 CU/ day) on
days 14 and 18; PEG-IL-2 (60,000 CU/daily) was given
i.v. on days 14 and 18. In group no. 7, 4/9 mice
died of toxicity following the second injection of
PEG-IL-2. The liposomes used were small (<100 nm)
long-circulating liposomes composed of egg phosphatidylcholine, PEG-DSPE (polyethylene glycol of
1900 daltons molecular mass attached to the amino
group of distearoyl phosphatidylethanolamine) and
cholesterol (50:5:45 mole ratio).

M109 carcinoma pulmonary metastases. Figure 1 demonstrates that the best
combination treatment was achieved when IL-2 was delivered via Stealth
liposome in 2 doses of 125,000 CU each (70% of the mice were long-term
survivors). A lower dose of IL-2 in Stealth liposomes given daily was less
efficacious. PEG-IL-2 showed good therapeutic efficacy although it was
toxic already at two treatments of 60,000 CU. During our study we used
several batches of PEG-IL-2 which varied in their toxicity. In some
batches toxicity started to show up at 100,000 CU/dose, and in others at
60,000 CU/dose, given twice at a 4-day interval. Liposomal IL-2 did not
show any toxicity at these doses.

In conclusion, it seems that IL-2 delivered via Stealth liposomes is
superior to soluble, free IL-2 as an immunostimulatory and therapeutic
agent. Experiments using other murine tumor systems and optimization of
the combined chemo-immunotherapy protocol are now underway in order to
evaluate the advantages of Stealth liposome IL-2 formulation.

List of Abbreviations: IL - Interleukin; IFN - Interferon; TNF - tumor necrosis factor; CSF - colony stimulating factor; TAA - tumor association antigens; MHC - major histocompatibility complex; NK - natural killer; LAK - lymphokine activated killer; Lip - liposomal; S - Stealth; RES - reticuloendothelial system; PEG - polyethylene glycol; CY - cyclophosamide.

Acknowledgement

This work was supported in part by grants from the Israel Cancer Association and the Israel Cancer Research Fund. The authors would like to thank Liposome Technology Inc. (Menlo Park, California) for generously supplying the PEG-DSPE, and Cetus, USA and Immunex, USA for providing the cytokines; Sigmund Geller for help in editing the manuscript and Beryl Levene for typing it.

REFERENCES

Alexander, R.B., 1991, in: "Biological Therapy of Cancer", V.T. De Vita, S. Hellman and S.A. Rosenberg, eds., Lippincott, Philadelphia.

Amselem, S., Gabizon, A. and Barenholz, Y., 1993, in: "Liposome Technology", 2nd edition, Vol. 1, G. Gregoriadis, ed., CRC Press, Boca Raton, FL.

Barenholz, Y., 1992, in: "Liposome Dermatics", O. Braun-Falco, H.C. Korting and H.I. Maibach, eds., Springer-Verlag, Berlin.

Barenholz, Y. and Amselem, S., 1993, in: "Liposome Technology", 2nd edition, Vol. 1, G. Gregoriadis, ed., CRC Press, Boca Raton, FL.

Barenholz, Y. and Crommelin, D.J.A., 1993, Liposomes as pharmaceutical dosage forms, in: "Encyclopedia of Pharmaceutical Technology", J. Swarbrick and J.C. Boylan, eds., Marcel Dekker, New York, in press.

Barenholz, Y., Palgi, O., Golod, G. and Kedar, E., 1991, Cytokine, 3:487.

Barenholz, Y., Amselem, S., Goren, D., Cohen, R., Gelvan, D., Samuni, A., Golden, E.B. and Gabizon, A., 1993, Medicinal Res.Rev., in press.

Braun, E., 1992, M.Sc. Thesis, Hebrew University, Jerusalem, Israel,

Bronchud, M.H., 1990, Eur.J.Cancer, 26:928.

De Vita, V.T., Hellman, S. and Rosenberg, S.A., eds., 1991 "Biological Therapy of Cancer", Lippincott, Philadelphia.

Francis, G.E., 1992, Focus on Growth Factors, 3:4.

Gabizon, A., 1992, Cancer Res., 52:891.

Gabizon, A., Chisin, R., Amselem, S., Druckman, S., Cohen, R., Goren, D., Fromer, I., Peretz, T., Sulkes, A. and Barenholz, Y., 1991, Br.J.Cancer, 64:1125.

Gabizon, A., Catane, R., Uziley, B., Kaufman, B., Safree, T., Barenholz, T. and Huang, A., 1992, Proc.Am.Soc.Clin.Oncol., 11, 124.

Groopman, J.E., 1991, J.Natl.Inst.Health Res., 3:75.

Kedar, E. and Klein, E., 1992, Adv. Cancer Res., 59:245.

Kedar, E., Rutkowski, Y., Braun, E., Emanuel, N. and Barenholz, Y., 1991, Cytokine, 3, 490.

Kedar, E., Braun, E., Rutkowski, Y., Emanuel, N. and Barenholz, Y., 1992, Proc.8th Int.Congress of Immunology, Budapest.

Knauf, M.J., Bell, D.P., Hirtzer, P., Luo, Z.-P., Young, J.D. and Katre, N.V., 1988, J.Biol.Chem., 263:15064.

Lancet Editorial 1991, Lancet, 338, 217.

Lasic, D.D., Martin, F.J., Gabizon, A., Huang, S.K. and Papahadjopoulos, D., 1991, Biochim.Biophys.Acta, 1070:187.

Lichtenberg, D. and Barenholz, Y., 1988, in: "Methods in Biochemical Analysis", Vol. 33, D. Glick, ed., Wiley, New York.

Lieschke, G.J. and Burgess, A.W., 1992, N,Engl.J.Med., 327:28 and 99.

Lotze, M.T., 1991, in: "Biological Therapy of Cancer", V.T. De Vita, S. Hellman and S.A. Rosenberg, eds., Lippincott, Philadelphia.

Lotze, M.T. and Rosenberg, S.A., 1991, in: "Biological Therapy of Cancer",
     V.T. De Vita, S. Hellman, and S.A. Rosenberg, eds., Lippincott,
     Philadelphia.
Meager, A., 1990, "Cytokines", Open University Press, Buckingham, UK.
Metcalf, D. and Morstyn, G., 1991, in: "Biological Therapy of Cancer",
     V.T. De Vita, S. Hellman and S.A. Rosenberg, eds., Lippincott,
     Philadelphia.
Mule, J.J., 1991, in: "Biological Therapy of Cancer", V.T. De Vita, S.
     Hellman and S.A. Rosenberg, eds., Lippincott, Philadelphia.
Oettgen, H.F., 1991, Current Opinion in Immunology, 3:699.
Rosenberg, S.A., 1991, Cancer Res., 51:5074.
Senior, J.H., 1987, Crit.Rev.Therap.Drug Carrier Systems, 3:123.

Participants of the NATO Advanced Studies Institute "New-Generation Vaccines: The Role of Basic Immunology" held at Cape Sounion Beach, Greece during 24 June – 5 July, 1992. The Organizing Committee included A.C. Allison (ASI Co-Director), K. Dalsgaard, G. Gregoriadis (ASI Director and Chairman), G. Poste (ASI Co-Director) and H. Snippe.

CONTRIBUTORS

Akerblom, L., Swedish University of Agricultural Sciences and the National Veterinary Institute, Department of Virology, Biomedical Centre, Box 585, S-751 23 Uppsala, Sweden

Austyn, J.M., Nuffield Department of Surgery, University of Oxford, John Radcliffe Hospital, Headington, Oxford OX3 9DU, UK

Barber, B.H., Department of Immunology, Medical Sciences Building, University of Toronto, Toronto, Canada M5S 1A8

Barenholz, Y., Department of Biochemistry, The Hebrew University - Hadassah Medical School, P.O. Box 1172, Jerusalem 91010, Israel

Bate, C., Department of Immunology, UCLMS, London, UK

Braun, E., Lautenberg Center of Immunology, The Hebrew University - Hadassah Medical School, P.O. Box 1172, Jerusalem 91010, Israel

Brochier, B., Service de Virologie-Immunologie, Faculte de Medecine Veterinaire, Universite de Liege, 45 Rue des Veterinaires, B-1070 Bruxelles, Belgique

Castagnoli, L., Department of Biology, University of Tor Vergata, Rome, Italy

Cesareni, G., Department of Biology, University of Tor Vergata, Rome, Italy

Diminsky, D., Department of Biochemistry, The Hebrew University - Hadassah Medical School, P.O. Box 1172, Jerusalem 91010, Israel

Emanuel, N., Lautenberg Center of Immunology, The Hebrew University - Hadassah Medical School, P.O. Box 1172, Jerusalem 91010, Israel

Even-Chen, Z., Biotechnology General Ltd, Rehovot 76326, Israel

Fernandez, I.M., Eijkman-Winkler Laboratory for Medical Microbiology, University of Utrecht, Heidelberglaan 100, The Netherlands

Francis, M.J., Department of Virology and Process Development, Pitman-Moore Ltd, Breakspear Road South, Harefield, Uxbridge, Middlesex, UB9 6LS, UK

Ghiara, P., Immunological Research Institute Siena, Siena, Italy

Gibbs, E.P.J., College of Veterinary Medicine, University of Florida, Gainesville, Florida 32610, USA

Golod, G., Department of Biochemistry and Lautenberg Center of Immunology, The Hebrew University - Hadassah Medical School, P.O. Box 1172, Jerusalem 91010, Israel

Gonfloni, S., Department of Biology, University of Tor Vergata, Rome, Italy

Gregoriadis, G., Centre for Drug Delivery Research, School of Pharmacy, University of London, 29-39 Brunswick Square, London WC1N 1AX, UK

Groves, M.J., Institut for Tuberculosis Research, College of Pharmacy, University of Illinois and Chicago, 840 West Taylor Street, Chicago, IL 60607-7019, USA

Kedar, E., Lautenberg Center of Immunology, The Hebrew University - Hadassah Medical School, P.O. Box 1172, Jerusalem 91010, Israel

Kieny, M.P., Transgene S.A., 11 Rue de Molsheim, 67082 Strasbourg Cedex, France

Kit, S., Baylor College of Medicine and NovaGene, Inc, Houston, TX, USA

Klergerman, M.E., Institut for Tuberculosis Research, College of Pharmacy, University of Illinois and Chicago, 840 West Taylor Street, Chicago, IL 60607-7019, USA

Kraaijeveld, C.A., Eijkman-Winkler Laboratory for Medical Microbiology, University of Utrecht, Utrecht, The Netherlands

Kreuter, J., Institut fur Pharmazeutische Technologie, J.W. Goethe-Universitat, D-6000 Frankfurt/Main, Germany

Lifely, M.R., Department of Cell Biology, The Wellcome Research Laboratories, Langley Court, Beckenham, Kent, BR3 3BS, UK

Lovgren, K., Swedish University of Agricultural Sciences and the National Veterinary Institute, Department of Virology, Biomedical Centre, Box 585, S-751 23 Uppsala, Sweden

Lou, Y., Institut for Tuberculosis Research, College of Pharmacy, University of Illinois and Chicago, 840 West Taylor Street, Chicago, IL 60607-7019, USA

Morein, B., Swedish University of Agricultural Sciences and the National Veterinary Institute, Department of Virology, Biomedical Centre, Box 585, S-751 23 Uppsala, Sweden

Morgan, A.J., Department of Pathology & Microbiology, University of Bristol Medical School, Bristol BS8 1TD, UK

Oosterlaken, T.A.M., Eijkman-Winkler Laboratory for Medical Microbiology, University of Utrecht, Utrecht, The Netherlands

Palgi, O., Department of Biochemistry and Lautenberg Center of Immunology, The Hebrew University - Hadassah Medical School, P.O. Box 1172, Jerusalem 91010, Israel

Pastoret, P.-P., Service de Virologie-Immunologie, Faculte de Medecine Veterinaire, Universite de Liege, 45 Rue des Veterinaires, B-1070 Bruxelles, Belgique

Playfair, J.H.L., Department of Immunology, UCLMS, London, UK

Rappuoli, R., Immunological Research Institute Siena, Siena, Italy

van Rooijen, N., Department of Cell Biology, Faculty of Medicine, Vrije Universiteit, van der Boechorstraat 7, 1081 BT Amsterdam, The Netherlands

Rutkowski, Y., Lautenberg Center of Immunology, The Hebrew University – Hadassah Medical School, P.O. Box 1172, Jerusalem 91010, Israel

Skea, D.L., Department of Immunology, Medical Sciences Building, University of Toronto, Toronto, Canada M5S 1A8

Snidjers, A., Eijkman-Winkler Laboratory for Medical Microbiology, University of Utrecht, Heidelberglaan 100, The Netherlands

Snippe, H., Eijkman-Winkler Laboratory for Medical Microbiology, University of Utrecht, Utrecht, The Netherlands

Taverne, J., Department of Immunology, UCLMS, London, UK

Villa, L., Immunological Research Institute Siena, Siena, Italy

Villacres-Eriksoon, M., Swedish University of Agricultural Sciences and the National Veterinary Institute, Department of Virology, Biomedical Centre, Box 585, S-751 23 Uppsala, Sweden

Wang, Z., Centre for Drug Delivery Research, School of Pharmacy, University of London, 29-39 Brunswick Square, London WC1N 1AX, UK

INDEX

Abortion, 127
  cell-associated virus in, 127
Accessory cells, 12
  cellular interactions, 12
Adenovirus recombinants, 167, 168
  expression of gp340, 168
  as vaccines, 167
Adjuvants, 9, 10, 53, 61, 62, 73, 83, 149, 166,
  175, 195
  alum, 53, 166
  aluminium hydroxide, 62
  interleukin-1, 149
  ISCOM, 61, 166
  lymphokines, 149
  multiple accessory cell concept, 10
  nanoparticles, 73
  NAGO, 195
  optimization, 83
  properties of, 9
  Quil A, 175
  tumour necrosis factor alpha, 149
Alum, 53, 101, 166
  as adjuvant, 53
  effectiveness of, 101
  potency, 166
Anthrax, 120
  attenuation of, 120
Antibodies, 35, 175
  anti-idiotypic, 175
  antipeptide, 35
  presentation, 15
Antigen-antibody conjugate, 102
  immunopotentiating mechanism, 102
Antigen presenting cells, 7, 23
  B-cells, 23
  dendritic cells, 7
Antigens, 8, 13, 14, 16, 17, 84
  conjugates, 13
  humoral immune response, 16
  liposome-entrapped, 14
  liposome-mediated targeting, 14
  non-particulate, 14
  presentation, 15
  proteins, 14

Antigens (cont'd)
  Semliki Forest Virus, 84
  targeting to dendritic cells, 8
  targeting of, 17
  thymus dependant, 13
  thymus independant, 13
  TNP-KLH, 14
  TNP-sheep red blood cells, 14
Anti-idiotypic antibodies, 175, 176
  against SFV, 176
  KLH, cross linked to, 175
  monoclonal, 175
Aujeszky's Diseases, 145
  eradication programs, 145
  in feral swine, 145
  in free-living wild swine, 145
  in "hog dogs", 145
  occurence, 145
  recombinant vaccine, 145
  vaccines against, 145
  wildlife reservoir, 145
Avian canarypox virus, 167
  replication, 167
Avian pox viruses, 167
  recombinant, 167
Avidin, 102
  as an antigen, 102

Bacillus Calmette-Guerin, *see also* BCG, 111
  as an immunostimulant, 111
  targeting to tumours, 111
Bacteria, 185
  polysaccharide capsules, 185
Bacterial antigens, 24
BCG, 112, 113
  alternatives to, 113
  antitumour response, 112
  clinical use, 113
  Copenhagen substrain, 112
  digestion of, 112
  fibronectin receptors, 112
  heat-killed, 113
  local response, 112

BCG (cont'd)
    mimicking of, 113
    non-phagocytic cells, uptake by, 112
    phagocytosis of, 112
    proteins excreted by, 112
    response induced, 112
    side effects, 113
    tumour cells, killing of, 112
    uptake by cells, 112
B-cell epitopes, 23, 88, 101
    linear, 88
    synthesis, 101
    in synthetic peptides, 23
B cells, 7, 15, 16, 17, 106
    antigen presentation, 15
    antigen specific, 106
    cooperation with T-cells, 15
    generation, 17
    generation of memory, 16
    memory, 7, 17
    proliferation of, 17
    secondary responses, 7
B-cell peptides, 43
    genetic restriction, 43
B and Th cell peptides, 24
    co-polymers of, 24
B-galactosidase, 37
    FMDV VP1 peptide, fused to, 37
B-Polysaccharide antigens, 193, 194
    conjugates, 193
    immunogenicity, 193, 194
B-Polysaccharide vaccines, 194
    covalent conjugates, 194
    human trials, 194
    immune responses, 194
Brucellosis, 144
    in bison, 144
    in elk, 144
    eradication programs, 144
    vaccination against, 144
    in wildlife, 144

Carrier proteins, *see also* Carriers, 33
    in B-cell antibody production, 33
    forms, 33
    function, 33
    roles, 33
    synthetic peptides, linkage to, 33
    for synthetic peptides, 33
    T-cell help, 33
Carriers, 34, 35, 37, 83
    albumin, 35
    keyhole limpet haemocyanin, 35
    for natural peptides, 34
    optimization, 83
    recombinant, 37
    sperm whale myoglobin, 35
    for synthetic peptides, 34

Carriers (cont'd)
    whale myoglobin, 35
Cell mediated immunity, 65
    by ISCOM, 65
Class II gene products, 102
    expression on B-cells, 102
Class I MHC-restricted immunity, 16
    macrophages, role of, 16
Class II MHC, 102
Conjugate vaccines, 188, 189, 190, 191
    against Hib disease, 191
    efficacy, 190
    Hib, 190
    immune responses, 190
    immunogenicity, 190
    in infants, 190
    interference with other vaccines, 191
    with tetanus toxoid, 189
Core fusion particles, 38, 39
    immunogenicity, 38, 39
Coupling methods, 35, 36
    coupling to carriers, 35
    problems, 36
Coupling reagents, 35
    gluteraldehyde, 35
    heterobifunctional cross-linkers, 35
    problems, 36
Cytokines, 201, 202
    in cancer therapy, 201
    chemical modification, 202
    circulation time, 202
    clinical trials, 202
    clinical use, 202
    in liposomes, 201
    properties, 201
    side effects, 202
Cytotoxic T-cells, 67
    ISCOM, induction by, 67

Dendritic cells, 1-6, 8
    as adjuvants, 1-10
    antigen internalization by, 5
    antigen presentation, 6
    antigen processing by, 5
    cytokines effect on, 6
    cytokines, action on, 3
    distribution, 1
    functions, 1
    immature, 2
    and immunization, 5
    immune responses, initiation of, 5
    non-lymphoid, 6
    markers, 2
    mature, 4
    maturation, 3
    migratory, 4
    migratory forms, 1
    migratory routes, 6

Dendritic cells (cont'd)
    origin, 2
    precursors, 2
    roles, 2
    subsets, 4
    targeting of, 8
    vaccination with, 8
    veiled cells, 2
Dichloromethylene diphosphonate, 13
    in liposomes, 13
    in macrophage removal, 13

EBV, 168
    infectivity, 168
    lymphoma model, 168
EBV vaccine, 166
    cell-mediated immunity, 166
Effector cells,11
Epstein-Barr virus, 165
    gp340, 165
    subunit vaccine, 165
    vaccine, 165
Epstein-Barr virus, see also EBV, 163
    in AIDS patients, 163
    gene structure, 163
    infected cells, 163
    infectivity, 163
    in organ transplant recipients, 163
    virus neutralising antibodies, 163

Feline immunodeficiency virus, see also FIV, 142
Ferritin-IL-1 peptide mutant, 150
    humoural response, 152, 153
    immunogenicity, 152
    immunization with, 152
FIV, 142
    clinical disease, 142
    in domestic cats, 142
    in Florida panther, 142
    in wild Felidae, 142
FMDV peptides, 26
    antibody response to, 26
    epitopes, 26
FMDV peptides, 27
    T-cell epitopes, 27
Follicular dendritic cells, 2, 7, 16, 17, 18
    antigen targeting to, 17
    distribution, 2
    localization, 7
    memory B-cells, generation of, 16
    in the spleen,18
Foot-and-mouth disease, see also FMDV, 25
    B-cell epitopes, 25
    peptides, 25
    T-cell epitope, 25
Freund's complete adjuvant, 101
    toxicity, 101
Fusion, 126

Fusion (cont'd)
    herpesvirus-infected cells, 126
Fusion proteins, 37, 169
    bacterial, 169
    B-galactosidase, 37
    B surface antigen, 37
    bacterial, 37
    hepatitis B core antigen, 37
    particulate, 37
    in peptide presentation, 37
    peptides fused to, 37
    TrpLE, 37

Gelatin microspheres, 113
    targeting to tumour cells, 113
Gp340, 165, 166, 167, 168, 169
    bacteria, expressed in, 169
    carbohydrate content, 169
    cell-mediated immunity to, 167
    cell-mediated protection, 168
    epitopes in, 167
    gene, 165, 169
    genetically engineered, 166
    immune responses, 168
    with ISCOMS, 168
    isolation, 165
    mammalian cells, expressed in, 169
    with muramyl dipeptide, 168
    production of, 165
    protective immunity, 165
    recombinants, 167
    T cell epitopes, 169
    T cell responses to, 167
    vaccine, 166
    yeast, expressed in, 169

H. influenzae type b, see also Hib, 186
Haemophilus influenza, 185
    in children, 185
    transmission type b, 192
    composition, 51
    lipids in, 53-55
    purified, 52
Helper T cells, 7
Hepatitis B, 28
    immune response, 28
Hepatitis B core antigens, 37
    as fusion proteins, 37
Hepatitis B surface antigen, 37, 44, 55
    composition, 51
    as fusion protein, 37
    lipids in, 53-55
    pre-S peptide, 44
    purified, 52
    S-peptide, 44
    structure of, 55
Hepatitis B vaccines, 51
    characterization, 51

Hepatitis B vaccines (cont'd)
  composition, 51
  efficacy, 51
  HBsAg particles, 51
  liposomal, 51
  preparation, 51
Herpesviridae, 117
  bovine rhinotracheitis virus, 117
  swine pseudorabies, 117
  vaccines for, 117
  virus species, 117
Herpesvirus genomes, 117
  encoded proteins, 117
  location, 117
Herpesvirus vectors, 123, 124
  design of, 124
  FMDV VP1 epitope, insertion of, 124
  foreign genes, insertion of, 123, 124
  recombinant, 123
Herpesviruses, 117, 119, 120
  foreign genes, vector for, 119
  genes, 117
  genomes of, 117
  virulence factor, 120
Hib vaccines, 186
  immunogenicity, 186
Hib-conjugated vaccines, 186
  in children, 187
  composition, 186
  immunogenicity, 186
HRV peptide, 39
  immunogenicity, 39
  neutralising activity, 39
Human herpes simplex virus, 117
Human herpes viruses, 163
  characteristics, 163
  Epstein-Barr virus, 163
  structure, 163
Human rhinovirus, 28
  immune responses, 28
  peptides, 28
  T-cell epitopes, 28
Humoural immune response, 16
  generation of, 16

IBRV marker vaccine, 122
  attenuation of, 122
  efficacy, 122
  gene-deleted, 122
  protection of calves, 122
  safety, 122
IBRV vaccine, 127
  construction of, 127
IBRV/FMDV vaccine, 125
  cost, 125
  delivery, 125
  efficacy, 125
  safety, 125

IFA-S-peptide, 46
  immune response, against, 46
IFA-pre-S_1 peptide, 46
  immune response, against, 46
IL-1 peptide, 150
  ferritin H chain, grafted into, 150-152
IL-2, 207, 208
  in cancer treatment, 208
  chemoimmunotherapy, 208
  immunomodulating activity, 207
  stability, 207
Immunization, 8, 177, see also Vaccination
  anti-anti-idiotypic, 177
  by the intraperitoneal route, 8
  by the intravenous route, 8
Immunoadjuvant activity, 11
  antibody response, induction of, 11
Immunogenicity, 23
  synthetic peptides, 23
Immunological adjuvants, see also Adjuvants
  alum, 101
  block copolymers, 43
  immunostimulating complexes, 43, 61
  liposomes, 43
  nanoparticles, 43
  new-generation, 43
Immunopotentiation, 14, 101
  adjuvant independent, 101
  mechanism of, 14
Immunostimulating complexes, 61, see also ISCOM
  as adjuvants, 61
  composition, 61-62
  immune response, 61
  structure, 61-62
Immunotargeting, 101-106
  antibody response, 103
  to Class II MHC, 102
  IgG subclass response, 104
  immunopotentiation, rationale for, 104, 105
  of influenza, 106
  Langerhans cells, 106
  mechanism, 104
  memory response, 103
  monoclonal antibodies used, 105
  primary response, 103
  variables involved, 105
Incomplete Freund's adjuvant, 46, see also IFA
  pre-S_1-peptide in, 46
  S-peptide in, 46
Influenza hemagglutinin, 106, 107
  antibody responses to, 107
  immunotargeting in mice, 107
  in non-murine species, 107
Intergitating cells, 2
Interleukin-1, 64, see also IL-1
  as an adjuvant, 149
  domain of, 149
  ISCOM, induced by, 64

Interleukin-1 (cont'd)
　peptides from, 149
Interleukin-1 beta, 150
　adjuvanticity, 150
　antitumoral activity, 150
　biological activity, 150
　effect on hemopoiesis, 150
　immunological effects, 150
　immunorestoration, 150
　induction of IL-2/IL-4, 150
　inflammation, 150
　peptides from, 150
　radioprotection, 150
　side effects, 150
Interleukin-2, *see also* IL-2
　in liposomes, 206
　PEG, conjugated to, 206
Interstitial dendritic cells, 2
ISCOM, 62
　as adjuvants, 168
　alum, comparison with, 166
　antigen presentation by, 65-67
　antigens in, 61
　biotin-labelled, 62
　cell mediated immune response, 65
　cytotoxic T cells induction by, 67
　EBV envelope protein gp340, 67
　in endosomes, 62-63
　haemagglutinin in, 68
　HIV-1 gp160 in, 67
　immune response, induction of,63
　immune response, mechanism of, 63
　influenza virus antigens in, 64
　interleukin-1 production, 64-65
　in macrophages,62
　MHC Class I reponse, 63, 67
　MHC Class II response, 63, 67
　measles virus antigens in, 67
　melanoma antigens in, 67
　neuraminidase in, 68
　peptides in, 68
　side effects, 63
ISCOM peptides, 68
　immune response to, 68, 69

Keyhole limpet heamocyanin, *see also* KLH, 175
KLH, 175
　carrier protein, 175
Kupffer cells, 6
　localization, 6

Lipids, 157
　TNF, triggering of, 157
Lipopolysaccharide, *see also* LPS, 157
　TNF, triggering of, 157
Liposomal, 204
　cytokine formulations, 204
Liposomal antigens, 15

Liposomal antigens (cont'd)
　targeting of, 15
Liposomal cytokines, 206
　biological activity, 206
　interleukin-2
　therapeutic performance, 206
Liposomal IL-2, 207, 208
　in cancer treatment, 208
　chemoimmunotherapy with, 208
　immunomodulation in mice, 207
　pharmacokinetics, 207
Liposomal S peptide, 44
　immune responses, against, 44-45
Liposomal pre-S_1 peptide, 44
　immune responses, against, 44-45
Liposomal HBV vaccine, 52
　activity, 52
　liposome lipids, 52
　preparation, 52
Liposomal vaccines, 56
　characterization, 56
　efficacy, 56
　against hepatitis B, 56
　preparation, 56
Liposome-based, 51
　vaccines, 51
Liposome-entrapped, 44, 206
　cytokines, 201-203
　HBsAg, 56
　interleukin-2
　pre-S_1 peptide, 44
　S-peptide, 44
Liposomes, 14, 43, 44, 201, 203, 203, 206
　biodegradability, 14
　cytokines in, 201, 203
　extravasation, 206
　half-life, 203
　as immunological adjuvants, 43
　localization, 203
　phagocytosis of, 14
　RES, uptake by, 204
　size, 203
　Stealth™, 203
　toxicity, 14
Live attenuated rabies virus, 132
　virulence, 132

Macrophages, 6, 11, 12, 13, 14, 15, 16, 17
　accessory cells, 16
　alveolar, 14
　antigen degradation by, 11
　antigen presentation, 15
　in antigen processing, 12, 13
　antigen targeting to, 17
　antigen uptake, 6
　cytokine production by, 6
　dendritic cells, effect on, 6
　depletion of, 14

Macrophages (cont'd)
elimination of, 13
functions in vivo, 13
in immune responses, 13
immune responses, role in, 15
localization, 6
peritoneal, 14
as scavengers, 11
and specialized accessory cells, 15
splenic, 14, 18
in tissues, 6
Malaria, 28, 155, 156, 158, 160
anaemia, 158
"anti-disease", 160
cerebral, 156
clinical immunity, 160
clinical "tolerance", 155
complications, 156, 158
exoantigens, 158
human "anti-toxic" immunity, 159
immune response, 28
immunity to, 155
mouse model, 156
P. falciparum, 156
protective immunity, 159
P. yoelii, 156, 158
tolerance, 159
as toxic disease, 155
tumour necrosis factor in, 156
vaccination against, 159
Malaria vaccine, 155
antigenic variation, 155
immune response required, 155
Marker vaccines, 120, 125
deletion mutant, 120
IBRV/FMDV vaccine, 125
production of, 120
Maternal antibodies, 126
interference with vaccination, 126
Memory B cells, 17
generation, 17
Meningitis, 185
causes of, 185
Meningococcal disease, 192
vaccines against, 192
Meningococcal vaccines, 192
polysaccharide antigens, 192, 193
MLV vaccines, 119, 120
rational design, 120
safety of, 119
virulence genes, deletion of, 120
Monoclonal antibodies, 175
non-internal image, 175
Monoclonal antibody, 176
KLH, coupled to, 176
Muramyl dipeptide, 168
as adjuvants, 168
Nanoparticles, 73-76

Nanoparticles (cont'd)
adjuvanticity, 76-78
adjuvanticity, effect of hydrophobicity on, 76
adjuvanticity, effect of size on, 75, 76
as adjuvants, 73
biodegradability, 73
characterization, 75
composition, 73-76
fate in vivo, 79
HIV antigens in, 78, 79
influenza vaccines in, 76
preparation, 74
structure, 73-76
Neuraminidase-galactose oxidase,195 see also
NAGO,
as an adjuvant, 195
adjuvanticity, 195
potency, 195
preparation, 195
Neutralization enzyme immunoassay, 177
New generation vaccines, 140, 186
advantages of, 140
Hib vaccines, 186
immunogenicity, 186
poxvirus-vectored vaccines, 140
vaccination of wildlife, 140

Oligosaccharide vaccines, 188
immunogenicity, 188, 189
in infants, 188
OMNIVAC-PR vaccine, 120, 121
as a vaccine, 120
neutralizing antibodies induced by, 121
OMNIMARK-PR vaccine, 121, 122
antibodies to, 121
efficacy of, 121
engineering of, 121
neutralizing antibodies induced by, 121
protection of newborn pigs, 122
safety of, 121
Ovalbumin, 27
Th-cell peptide, 27

P. berghei, 157
P. falciparum, 157
P. vivax, 157
P. yoelii, 157
Pegylated IL-2, 207
pharmacokinetics, 207
Peptide carriers, 39
clasification, 39
Peptide copolymers, 24
chemical linkage, 24
immunogenicity of, 24
Peptide-fusion proteins, 38
immunogenicity, 38
Peptide vaccines, 168
T and B cell epitopes, 168

222

Peptides, 23, 26, 27, 28, 33, 35, 36, 43, 85,
    see also Synthetic peptides
  B-cell, 43
  carriers for, 33
  charge, 36
  co-linear, 27
  as haptens, 23
  human rhinovirus NRV, 28
  immune response, 26
  as immunogens, 36
  in ISCOM, 68
  in liposomes, 43
  optimum length, 36
  orientation, 35
  surrogate immunogens, 23
  synthesis, 61
  synthetic, 23
  T-cell, 43
  T-helper cell determinants, 23
  in vaccines, 23, 85
Polyethylene glycol, see also PEG, 206
  IL-2 conjugate, 206
Polymerisation, 23
  of B and T cell peptides, 23
Polysaccharide vaccine, 185, 187
  immunogenicity, 187
  in children, 185
  efficacy, 185
  H. influenza, 185
  PRP oligosaccharide-diphtheria toxoid
      mutant (HbOC), 187
Poxvirus, 139
  as a vector, 139
Poxvirus vaccines, 139
  recombinant, 139
Poxvirus vectored vaccines, 140
  in wildlife vaccination, 140
Poxvirus vectors, 167
  effectiveness, 167
Pre-S$_1$ peptide, 44
  antibodies against, 53
  liposomes,entrapment in, 44
Protein carriers,23
  for peptides,23
Pseudorabies, 122
  outbreak of, 122
Pseudorabies marker vaccine, 122
  gene-deleted, 122
Pseudorabies vaccine, 127
  construction of, 127

Quil A, 175
  as an adjuvant, 175

Rabies, 131, 132, 144, 145
  behavioural changes, 132
  control, 144
  oral vaccination, 132

Rabies (cont'd)
  perpetuation, 144
  prevalence, 131, 132
  propagation, 131, 132
  prophylactic measures, 132
  in raccoons, 145
  safety issues, 144
  transmission, 132
  vaccinia virus recombinant, 144
  in wildlife species, 144
Rabies vaccine, 133, 134, 135
  in cats, 134
  in cattle, 134
  in dogs, 134
  experiments in foxes, 133
  in ferrets, 134
  field trials, 134, 135
  formulation, 133
  foxes, vaccination of, 133, 134
  in hamsters, 134
  large-scale vaccination, 135
  in mice, 134
  oral administration, 133
  pathogenicity, 134
  in pigs, 134
  protective ability, 133
  in rabbits, 134
  raccoons, vaccination of, 134
  in sheep, 134
  use of baits, 134
  vaccination of animals, 133
Rabies virus, 132, 139
  inactivated, 132
  live attenuated, 132
  properties, 132
  vaccination with, 139
Recombinant DNA technology, 149
  cloning of proteins, 149
Recombinant protein antigens, 149
  with "built in" adjuvanticity, 149
Recombinant vaccines, 125, 139, 144, 145
  against raccoonpox, 145
  against skunks, 145
  for Aujeszky's diseases, 145
  IBRV/FMDV recombinants, 125
  poxvirus, 139
  preparation, 125
  vaccinia virus, 139
Recombinant virus vectors, 168
  expression of gp340, 168
  immunization with, 168

Semliki Forest Virus, 84, 88, 91, see also SFV
  as an antigen, 84
  avirulent prototype strain, 175
  epitopes, 84
  immunogenicity, 91
  linear epitopes, 88

Semliki Forest Virus (cont'd)
   synthetic peptides from, 91
   T-cell epitopes, 88
   virulent strain, 175
SFV, 176, 177, 178
   B-cell epitope, 176
   mice, immunization of, 178-181
   neutralizing antibodies, 177, 178
   neutralizing MAb, 176
   synthetic peptides, 176
Smallpox, 139
   eradication of, 139
S-peptide, 44
   liposomes, entrapment in, 44
Stealth™ liposomes, 203, 207
Streptococcus mutans, 28
   immune responses, 28
Subunit vaccines, 165
Sylvatic rabies, 131
   eradication, 131
Synthetic peptides, 23, 25, 33, 92
   animal species used for immunization, 34, 35
   B-cell epitopes, 25
   carriers, choice of, 34
   chemical coupling, 25
   colinear synthesis, 25
   conjugation method of, 34
   delivery, 33
   foot-and-mouth disease, 25
   haemagglutinin residues 111-120, 25
   as haptens, 23
   human rhinovirus, 25
   immunization; mode of, 34
   immunization with, 23
   immunogenicity, 33, 92
   as immunogens, 33
   malaria-encoded sequence, 25
   polymerisation, 25
   primary sequence, role of, 34
   T-cell epitopes, 25
Syntex adjuvant formulation, 61

T cells, 7, 11, 15, 47
   activated, 7
   antigen presentation to, 7, 15
   cooperation with B-cells, 15
   cytokine secretion, 7
   cytotoxic, 7
   as effector cells, 11
   helper, 7
   memory, 7
   proliferative response, 47
   responses, 7
   resting, 7
T-cell epitopes, 27, 44, 101
   B-cell epitopes, help for, 44
   location of, 27
   reactive, 27

T-cell epitopes (cont'd)
   synthesis, 101
Tetanus toxin, 28
   immune responses, 28
Th-cell epitopes, 23, 29, 101
   identification of, 29
   in synthetic peptides, 23
Th-cell sites, 27
   "promiscuous", 27
TNF, 158
   antibodies against, 158
   hypoglycaemia, 158
   induction by exoantigens, 158
   protection against, 158
   side effects, 158
Tuberculosis, 144
   in badgers, 144
   brush-tailed possum, 144
   in cattle, 144
   epidemiology, 144
   eradication programs, 144
   in wildlife, 144
Tumour necrosis factor, see also TNF, 157
   induction by antigens, 156
   induction by phospholipid, 157
   in malaria, 156
Tumour necrosis factor alpha, 149
   as an adjuvant, 149

Vaccination, 139, 141, 142, 165
   criteria for intervention, 142
   decision making in, 141
   economics of, 142
   herpes saimiri, 165
   Marek's disease, 165
   objectives, 141
   of wildlife, 139, 141
Vaccination programs, 141
   for wildlife species, 141
Vaccination strategies, 126
   maternal antibodies, role of, 126, 127
   of pregnant sows, 126
Vaccine conjugates, 192
   meningococcal B conjugates, 192
Vaccine markers, 125
   gI, 125
   PRV gIII, 125
Vaccine vectors, 167
   viral, 167
Vaccines, 11, 29, 51, 83, 84, 85, 94, 95, 101, 111,
       117, 119, 120, 121, 127, 139, 140, 155,
       159, 160, 163, 164, 165, 166, 168, 185,
       188, 192
   against Epstein-Barr virus, 163, 164, 166
   against herpesviridae, 117
   against malaria, 155
   "anti-disease", 159
   the anti-disease concept, 155

Vaccines (cont'd)
 anti-human malaria, 160
 artificial, 168
 attenuated MLV, 119
 Bacillus Calmette-Guerin, 111
 in baits, 139
 canine distemper, 143
 in children, 185
 conjugates, 188, 189
 for endangered species, 140
 gene-deleted herpesvirus, 117
 genetically engineered, 83, 127
 gp340 subunit, 166
 IBRV, 127
 liposomal, 51
 live pseudorabies, 127
 meningococcal, 192
 multiple accessory cell concept, 11
 new generation, 101, 139
 oligosaccharides, 188
 OMNIMARK-PR, 121
 OMNIVAC-PR, 120, 121
 oral, 139
 peptide vaccines, 94
 polysaccharide, 185
 recombinant, 95
 subunit, 165
 synthetic, 29, 94
 synthetic peptides, 85
 targeting of, 101
 Th-cell epitopes, role of, 29
 TK⁺ gIII⁻ OMNIMARK, 127
 vaccinia virus, 139
 viral, advantages of, 84
 viral, problems of, 84
 viral, production of, 84
Vaccines carriers, *see also* Carriers, 83
Vaccinia, 166
 recombinants, 166, 167
Vaccinia-rabies recombinant, 132
 construction of, 132
Vaccinia-rabies vaccine, 131
 administration, 131
 efficiency, 131
 field trials of, 131
 innocuity, 131
 live recombinant, 131
Vaccinia recombinants, 167
 as vaccines, 167
Vaccinia virus, 132
 as a live vector, 132
Vaccinology, 139
 history of, 139
Vehicles, 9
 properties of, 9
Viral disease, 122
 diagnostic tests, 122-123
Viral envelope, 165

Viral envelope (cont'd)
 glycoproteins, 165
Viral vaccines, 119
 criteria for, 119
 "marker", 119
 overattenuated, 119
 underattenuated, 119
 vaccination with, 119
Virus membrane, 126
 fusion with cell membrane, 126
Virus vectors, 119, 166, 167
 adenovirus, 166
 bovine-rhinotracheitis virus, 119
 disadvantages, 166
 Epstein-Barr, 119
 herpes simplex virus, 119
 herpesvirus saimiri, 119
 recombinant, 166
 safety, 167
 swine pseudorabies virus, 119
 vaccinia, 166
 varicella-zoster, 119
Virus virulence factors, 119
 deletion of, 119

Wildlife, 141, 142, 144
 disease epidemics, 142
 diseases in, 141, 144
 epidemiology of, 141
 vaccination of, 141, 142
Wildlife vaccination, 142, 143, 144
 approaches, 143, 144
 assessment, 144
 black-footed ferrets, 143
 criteria for intervention, 142
 criteria for success, 143
 decision making, 144
 in domestic cats, 142
 of domestic species, 143
 of endangered species, 143
 European ferrets, 143
 European rabbit, 142
 feasibility, 142
 in Florida panther, 142
 live vaccines, 143
 oral, 143
 programmes, 144
 research and development, 144
 safety issues, 144

Yeast Ty protein, 37
 as fusion proteins, 37